Lecture Notes in Computer Science 9118

Commenced Publication in 1973
Founding and Former Series Editors:
Gerhard Goos, Juris Hartmanis, and Jan van Leeuwen

More information about this series at http://www.springer.com/series/7407

Jeffrey Shallit · Alexander Okhotin (Eds.)

Descriptional Complexity of Formal Systems

17th International Workshop, DCFS 2015
Waterloo, ON, Canada, June 25–27, 2015
Proceedings

 Springer

Editors
Jeffrey Shallit
School of Computer Science
University of Waterloo
Waterloo, ON
Canada

Alexander Okhotin
University of Turku
Department of Mathematics
Turku
Finland

ISSN 0302-9743   ISSN 1611-3349   (electronic)
Lecture Notes in Computer Science
ISBN 978-3-319-19224-6   ISBN 978-3-319-19225-3   (eBook)
DOI 10.1007/978-3-319-19225-3

Library of Congress Control Number: 2015939170

LNCS Sublibrary: SL1 – Theoretical Computer Science and General Issues

Springer Cham Heidelberg New York Dordrecht London

Printed on acid-free paper

Springer International Publishing AG Switzerland is part of Springer Science+Business Media
(www.springer.com)

Preface

The 17th International Workshop on Descriptional Complexity of Formal Systems (DCFS 2015) was held in Waterloo, Ontario, Canada, during June 25–27, 2015. It was organized by the David R. Cheriton School of Computer Science at the University of Waterloo.

The subject of the workshop was descriptional complexity. Roughly speaking, this field is concerned with the size of objects in various mathematical models of computation, such as finite automata, pushdown automata, and Turing machines. Descriptional complexity serves as a theoretical representation of physical realizations, such as the engineering complexity of computer software and hardware. It also models similar complexity phenomena in other areas of computer science, including unconventional computing and bioinformatics.

The DCFS workshop series is a result of merging together two workshop series: *Descriptional Complexity of Automata, Grammars and Related Structures* (DCAGRS) and *Formal Descriptions and Software Reliability* (FDSR). These precursor workshops were DCAGRS 1999 in Magdeburg, Germany, DCAGRS 2000 in London, Ontario, Canada, and DCAGRS 2001 in Vienna, Austria, as well as FSDR 1998 in Paderborn, Germany, FSDR 1999 in Boca Raton, Florida, USA, and FSDR 2000 in San Jose, California, USA. These workshops were merged in DCFS 2002 in London, Ontario, Canada, which is regarded as the 4th DCFS. Since then, DCFS workshops were held in Budapest, Hungary (2003), in London, Ontario, Canada (2004), in Como, Italy (2005), in Las Cruces, New Mexico, USA (2006), in Nový Smokovec, Slovakia (2007), in Charlottetown, Canada (2008), in Magdeburg, Germany (2009), in Saskatoon, Canada (2010), in Limburg, Germany (2011), in Braga, Portugal (2012), in London, Ontario, Canada (2013), and in Turku, Finland (2014).

In 2015, the DCFS workshop was held for the first time in Waterloo, Ontario. There were 29 submissions, with authors from Austria, Canada, Estonia, Germany, Italy, Japan, Korea, Latvia, The Netherlands, Poland, Portugal, The Russian Federation, Slovakia, Switzerland, UK, and USA.

Of the 29 submissions, 23 were accepted for presentation at the conference (79.3 % acceptance rate). These papers were selected from the submissions by the Program Committee on the basis of at least three reviews per submission.

This volume contains extended abstracts of the two invited talks and 23 contributed talks presented at the workshop.

This workshop is a result of the combined efforts of many people, to whom we wish to express our gratitude. In particular, we are indebted to our invited speakers—Rajeev Alur and Thomas Colcombet—and to all the speakers and participants of the workshop. We are grateful to all our Program Committee members and to all reviewers for their work in selecting the workshop program, which was carried out using the EasyChair conference management system. We also thank the staff of the University of Waterloo for administrative assistance.

We gratefully acknowledge the financial support from our sponsor, the Fields Institute for Research in Mathematical Sciences (www.fields.utoronto.ca). Their support was crucial to pay for the invited speakers and to support the travel of junior researchers. This conference was an official event of the International Federation for Information Processing and IFIP Working Group 1.2 (Descriptional Complexity). Finally, we wish to thank the editorial team at Springer, specifically Alfred Hofmann and Anna Kramer, for their efficient production of this volume.

April 2015   Jeffrey Shallit
                                                   Alexander Okhotin

Organization

Program Committee

Rusins Freivalds	University of Latvia, Latvia
Yo-Sub Han	Yonsei University, South Korea
Markus Holzer	Universität Giessen, Germany
Artur Jeż	University of Wrocław, Poland
Galina Jirásková	Slovak Academy of Sciences, Slovakia
Lila Kari	Western University, London, Ontario, Canada
Manfred Kufleitner	University of Stuttgart, Germany
Hing Leung	New Mexico State University, USA
Ian McQuillan	University of Saskatchewan, Canada
Nelma Moreira	Universidade do Porto, Portugal
Alexander Okhotin	University of Turku, Finland
Jean-Éric Pin	LIAFA, CNRS and University Paris 7, France
Daniel Reidenbach	Loughborough University, UK
Kai Salomaa	Queen's University, Canada
Jeffrey Shallit	University of Waterloo, Canada

Additional Reviewers

Barash, Mikhail
Câmpeanu, Cezar
Čevorová, Kristína
Cho, Dajung
Day, Joel
Domaratzki, Mike
Enaganti, Srujan Kumar
Eom, Hae-Sung
Freydenberger, Dominik D.
Gruber, Hermann
Hertrampf, Ulrich
Hirvensalo, Mika
Jirásek, Jozef Štefan
Kapoutsis, Christos
Karamichalis, Rallis

Kopecki, Steffen
Kothari, Robin
Kulkarni, Manasi
Kutrib, Martin
Lohrey, Markus
Maia, Eva
Masopust, Tomas
Meckel, Katja
Ng, Timothy
Palioudakis, Alexandros
Schmid, Markus L.
Šebej, Juraj
Vorel, Vojtěch
Watrous, John
Zetzsche, Georg

Regular Functions

Rajeev Alur

University of Pennsylvania

Abstract. When should a function mapping strings to strings, or strings to numerical costs, or more generally, strings/trees/infinite-strings to a set of output values with a given set of operations, be considered regular? We have proposed a new machine model of *cost register automata,* a class of write-only programs, to define such a notion of regularity. We are developing theoretical foundations for this new class with a focus on algebraic and logical characterization, and algorithms for transformations and analysis questions. We have also designed a declarative language, **DReX,** for string transformations based on these foundations. In this talk, I will give an overview of theoretical results, emerging applications, and open problems for regular functions.

References

1. Alur, R., Cerny, P.: Streaming transducers for algorithmic verification of single- pass list-processing programs. In: Proceedings of the 38th ACM Symposium on Principles of Programming Languages, pp. 599–610 (2011)
2. Alur, R., D'Antoni, L.: Streaming tree transducers. In: Czumaj, A., Mehlhorn, K., Pitts, A., Wattenhofer, R. (eds.) ICALP 2012, Part II. LNCS, vol. 7392, pp. 42–53. Springer, Heidelberg (2012)
3. Alur, R., D'Antoni, L., Deshmukh, J.V., Raghothaman, M., Yuan, Y.: Regular functions and cost register automata. In: 28th Annual ACM/IEEE Symposium on Logic in Computer Science, pp. 13–22 (2013)
4. Alur, R., Freilich, A., Raghothaman, M.: Regular combinators for string transformations. In: 29th Annual ACM/IEEE Symposium on Logic in Computer Science, 9 p (2014)
5. Alur, R., D'Antoni, L., Raghothaman, M.: DReX: a declarative language for efficiently evaluating regular string transformations. In: Proceedings of the 42nd ACM Symposium on Principles of Programming Languages, pp. 125–137 (2015)

Unambiguity in Automata Theory

Thomas Colcombet

CNRS, Université Paris 7 - Paris Diderot
thomas.colcombet@liafa.univ-paris-diderot.fr

Abstract. Determinism of devices is a key aspect throughout all of computer science, simply because of considerations of efficiency of the implementation. One possible way (among others) to relax this notion is to consider unambiguous machines: non-deterministic machines that have at most one accepting run on each input.

In this paper, we will investigate the nature of unambiguity in automata theory, presenting the cases of standard finite words up to infinite trees, as well as data-words and tropical automata. Our goal is to show how this notion of unambiguity is so far not well understood, and how embarrassing open questions remain open.

Contents

Invited Talk

Unambiguity in Automata Theory

Thomas Colcombet[(✉)]

CNRS, Université Paris 7 – Paris Diderot, Paris, France
`thomas.colcombet@liafa.univ-paris-diderot.fr`

Abstract. Determinism of devices is a key aspect throughout all of computer science, simply because of considerations of efficiency of the implementation. One possible way (among others) to relax this notion is to consider unambiguous machines: non-deterministic machines that have at most one accepting run on each input.

In this paper, we will investigate the nature of unambiguity in automata theory, presenting the cases of standard finite words up to infinite trees, as well as data-words and tropical automata. Our goal is to show how this notion of unambiguity is so far not well understood, and how embarrassing open questions remain open.

1 Introduction

In many areas of computer science, the relationship between deterministic and non-deterministic devices is extensively studied. This is in particular the case in complexity theory, and also in automata theory. The notion of unambiguous devices, i.e., non-deterministic devices that have at most one accepting execution for each accepted input, is a natural intermediate class that is potentially more expressive (or succinct) than deterministic devices, while behaviorally easier to handle than general non-deterministic machines.

One specificity of this class is that it is a semantic one: a priori, nobody knows whether a given Turing machine is unambiguous or not. This is undecidable, and even providing a witness of unambiguity is not possible.

Even for weaker complexity classes, such as logspace, the status of the unambiguous machines is not settled. Indeed, unambiguous logspace (UL) is located somewhere between deterministic logspace (L) and non-deterministic logspace NL. Since L and NL are not known to be separated, the separation of UL with respect to either L or NL is also open. This class UL is also interesting, since it is known to contain planar reachability, while the main complete problem for NL is general reachability in a directed-graph. Interestingly, Allender and Reinhardt have shown that in non-uniform complexity classes (i.e., in the presence of advice), logspace and non-deterministic logspace coincide [35].

In the world of automata, the picture is better understood. As a first key difference, unambiguity becomes easily decidable, and furthermore, it is possible to compute and work with witnesses of unambiguity. Nevertheless, many questions related to unambiguity are embarrassingly open and surprisingly complicated.

© Springer International Publishing Switzerland 2015
J. Shallit and A. Okhotin (Eds.): DCFS 2015, LNCS 9118, pp. 3–18, 2015.
DOI: 10.1007/978-3-319-19225-3_1

In fact, the subject of unambiguity related to automata is so wide that it would require a much larger and ambitious presentation in itself. The reader interested in pursuing these subjects further will find a lot of material in [15, 36], and in surveys such as [17] for standard word automata.

Many subjects involving unambiguity cannot even be mentioned in this paper. This includes unambiguous non-finite state machines (such as pushdown automata, see e.g., [17, 31] or less standard forms of automata such as constrained automaton in [6]) or unambiguous regular expressions. Even the theory of codes is in essence a study of unambiguity. There are also some intermediate forms of restricted ambiguity, such as m-ambiguity or polynomial ambiguity (when the number of accepting runs are bounded by m, or by a polynomial in the length of the input) [24], as well as restricted syntactic variants of unambiguous automata ([25] among others). The unambiguous polynomial closure of a family of languages has also been characterized [32]. Other algorithmic questions have also be addressed, such as the inference of automata [12]. Unambiguity is also a very important and a well studied subject in connection with transducers since unambiguous transducers are very close to functional ones (transducers which, instead of a relation, recognize a function) [2, 39, 43]. Unambiguity can also be considered in the analysis of rational subsets of monoids in the absolute [3]. Unambiguity can also be studied for extended notions of words, and in particular infinite words. Over infinite words of length ω, a very important notion of unambiguous automata is the one of prophetic automata [9] (a semantic version of determinism from right to left). Also, on infinite words of unrestricted length, and even for more general classes of automata, compiling temporal logics yields unambiguous automata [13]. None of these topics will be addressed in this paper.

This paper does not intend to make any exhaustive survey of the large body of works related to unambiguity. It rather offers a tour, visiting several arbitrarily chosen topics, involving significantly different situations. This tour starts with standard non-deterministic word automata and their unambiguous subclasses. Several complexity arguments are elegant and worth knowing in this context, such as the use of communication complexity, and the counting principle for deciding universality in polynomial time. We continue with tropical automata. This specific kind of weighted automata computes functions from words to integers. We will see that some elementary problems are undecidable for such automata, and that unambiguous automata are a subclass that appears naturally and has good decidability properties. The description then proceeds with the infinite tree case, in which the story is completely different. There, unambiguity is related to the problem of existence of choice functions. We will finish with a study of register automata, where unambiguity turns out to be a very important subclass. Recent unpublished results obtained with Puppis and Skrcypczak sustain this idea.

In this paper, we will consider the case of finite-state automata (Sect. 2), of tropical automata (Sect. 3), of automata over infinite trees (Sect. 4) and of automata over data words (Sect. 5). In the first three situations, we will see that difficult questions remain open. The last case will report on recent unpublished work in which a constructive understanding of the notion of unambiguity yields new results.

2   Unambiguous Word Automata

In this section, we consider unambiguous automata over finite words. Let us first briefly recall some standard notation.

In this section, we adopt the standard terminology concerning (finite-state) non-deterministic automata. A *non-deterministic automaton* $\mathcal{A}$ reading words over the *alphabet* A has a finite set of *states* Q, a set of *initial states* I, a set of *final states* F, and a *transition relation* $\Delta \subseteq Q \times A \times \Delta$. A *run* ρ over the input word $a_1 \cdots a_n$ of the automaton is a sequence of transitions of the form $(q_0, a_1, q_1)(q_1, a_2, q_2) \cdots (q_{n-1}, a_n, q_n) \in \Delta^*$. It is *accepting* if furthermore $q_0 \in I$ and $q_n \in F$. If an accepting run exists over an input word u, then u is *accepted*. The language *recognized* by the automaton is the set of accepted words. It is denoted $L(\mathcal{A})$. Languages recognized by an automaton are called *regular*. An automaton is *deterministic* if for all states p and all letters $a \in A$ there exists at most one state r such that (p, a, r) is a transition. An automaton is *unambiguous* if for all input words there exists at most one accepting run over it.

One of the first results we learn in an automata course is the inherent exponential blowup of determinization.

Theorem 1 ([27–29, 34]). *Non-deterministic word automata of size n can be transformed into deterministic and complete automata of size at most 2^n for the same language. This bound is tight.*

The witness automaton for the lower bound is very natural: it checks that the n-th letter from the end is a b (over the alphabet a, b):

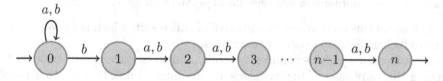

However, though this example is non-deterministic, if we reverse the orientation of its edges (yielding the *mirror automaton*), we obtain a deterministic automaton. This implies that it has at most one accepting run over each input: it is unambiguous. It happens also, as a consequence, that it is very easy to complement it: One adds a new state $\perp$ with self loops labelled a, b, and linked to 1 by an a transition. Then, it is sufficient to complement the set of initial states for complementing the accepted language.

This example shows that the class of unambiguous automata is potentially an interesting compromise between succinctness (these can be exponentially smaller than deterministic automata) and tractability of fundamental problems.

Theorem 2 ([22–24]). *Unambiguous automata can be exponentially more succinct than deterministic automata. Non-deterministic automata can be exponentially more succinct than unambiguous ones.*

Remark 1. Deciding if a non-deterministic word automaton $\mathcal{A}$ is unambiguous is doable in polynomial time. The principle is as follows: consider the product of the automaton $\mathcal{A}$ with itself. Over a given input, an accepting run of this new automaton can be seen as the pair of two accepting runs of $\mathcal{A}$ over this input. It is easy to slightly modify this automaton in such a way that it accepts an input if and only if the input is accepted by two distinct accepting runs of $\mathcal{A}$ (this can be achieved, e.g., by adding one extra bit to each state storing whether the two runs have differed so far). This new automaton has quadratic size in the original one. It accepts an input if and only if the original automaton is ambiguous. Thus, the non-deterministic automaton $\mathcal{A}$ is unambiguous if and only if the language recognized by this new automaton is empty. This can be tested in polynomial time (NL more precisely).

Using variations around these ideas, we can show that unambiguity is decidable for all the classes of automata considered in this paper (tropical automata, infinite tree automata as well as register automata).

There is at least one strong evidence that unambiguous automata are inherently simpler than general non-deterministic automata. This is the complexity of the equivalence, containment, and universality problems. In general, given two non-deterministic automata recognizing the languages K, L respectively, the problem of *equivalence* "$K = L$?", of *containment* "$K \subseteq L$?" and of *universality* "$K = A^*$?" are known to be PSPACE-complete. This is not the case for unambiguous automata, as shown in the following theorem.

Theorem 3. ([40,41]). *The problems of universality and equivalence of unambiguous automata as well as containment of a non-deterministic automaton in an unambiguous automaton are solvable in polynomial time.*

We shall see in this section a complete proof of this result, which is a good excuse for introducing several important techniques.

Of course, knowing this complexity result, and since universality amounts to checking the emptiness of the complement, one might think that another proof of this result could be as follows: complement the unambiguous automaton with a polynomial blowup of states, and then test for emptiness in polynomial time. However, the question of whether unambiguous automata can be complemented with a polynomial blowup in the number of states is an open problem.

Conjecture 1. It is possible to complement unambiguous automata of size n into unambiguous automata of size polynomial in n.

In fact, even whether we can complement an unambiguous automaton into a non-deterministic automaton of polynomial size is open. We lack techniques for addressing this question. In particular, how can we prove a lower bound on the size of an unambiguous automaton for a given language?

Communication Complexity and the Rank Technique [23,24,37]. There is a nice technique for proving lower bounds on the size of an unambiguous automaton for a language, based on communication complexity. Consider a language

$L \subseteq A^*$. Define the *communication relation* $\text{Com}(L) \subseteq A^* \times A^*$ to be the set of ordered pairs (u, v) such that $uv \in L$. A subset of $A^* \times A^*$ is called a *rectangle* if it is of the form $M \times N$ for $M, N \subseteq A^*$. A *non-deterministic decomposition* of $R \subseteq A^* \times A^*$ is a finite union of rectangles, and its *complexity* is the number of rectangles involved in the union. An *unambiguous decomposition* is a non-deterministic decomposition into disjoint rectangles. The *non-deterministic complexity* of R (resp., *unambiguous complexity*) is the minimal complexity nd-comp(R) (resp., unamb-comp(R)) of a non-deterministic decomposition (resp., unambiguous decomposition) of R.

It is easy to show that a language L accepted by a non-deterministic automaton with n states is such that nd-comp($\text{Com}(L)$) $\leq n$. Indeed, define $L_{I,q}$ to be the language recognized by the automaton when the set of final states is set to $\{q\}$, and $L_{q,F}$ to be the language recognized by the automaton when the set of initial states is set to be $\{q\}$. Clearly, for all words u, v, $uv \in L$ if and only if there exists a state q such that $u \in L_{I,q}$ and $v \in L_{q,F}$. This means that $\text{Com}(L) = \cup_{q \in Q} L_{I,q} \times L_{q,F}$. We have found a non-deterministic decomposition for $\text{Com}(L)$ of complexity n.

Pushing further, a language L accepted by an unambiguous automaton of size n is such that unamb-comp($\text{Com}(L)$) $\leq n$. Indeed, consider the non-deterministic decomposition $\text{Com}(L) = \cup_{q \in Q} L_{I,q} \times L_{q,F}$ as in the non-deterministic case, and assume it would be ambiguous. This would mean that there are two distinct states p, q such that $L_{I,p} \cap L_{I,q} \neq \emptyset$ and $L_{p,F} \cap L_{q,F} \neq \emptyset$. Let u be a word in the first intersection, and v be a word in the second. Then the word uv is accepted by two distinct runs: one that reaches state p after reading u, and the other that reaches state q at the same position. This contradicts the unambiguity assumption. Thus, $\cup_{q \in Q} L_{I,q} \times L_{q,F}$ is an unambiguous decomposition of complexity n.

Linear algebra offers an elegant way to bound the unambiguous complexity of a relation from below. Indeed, we can identify a relation $R \subseteq E \times F$ with its *characteristic matrix* : rows are indexed by E and columns by F, and the entry indexed by words x, y is 1 if $(x, y) \in R$ and 0 otherwise.

Lemma 1. rank(R) $\leq$ unamb-comp(R).

Proof. A union of disjoint rectangles can be understood as the sum of the characteristic matrices representing them. Since the rank of a matrix that "contains only one rectangle" is 1 (or 0 if the rectangle is empty), and rank is subadditive, the rank of a matrix is smaller than its unambiguous complexity. □

From this we can derive an upper bound on non-universality witnesses.

Lemma 2 ([37]). *Any shortest witness of non-universality for an unambiguous automaton with n states has length at most n.*

Proof. Let $\mathcal{A}$ be an unambiguous automaton, and let $a_1 a_2 \cdots a_n \notin L(\mathcal{A})$ be the shortest witness of non-universality (if it exists). Consider the matrix N obtained from $\text{Com}(L(\mathcal{A}))$ by restricting the rows to $v_0 = \varepsilon, v_1 = a_1, v_2 = a_1 a_2, \ldots, v_n = a_1 \cdots a_n$ and the columns to $w_0 = a_1 \cdots a_n, w_1 = a_2 \cdots a_n, \ldots, w_n = \varepsilon$.

Since $v_i w_i = u \notin L(\mathcal{A})$ for all $i = 0 \ldots n$, the diagonal of this matrix consists only of 0's. However, for all $0 \leq i < j \leq n$, we have $|v_i w_j| < n$. Hence the "upper right" part of the matrix consists solely of ones. We claim that this matrix has rank at least n. Indeed, that $(1) - N$ (where (1) is the matrix using with 1's on all its entries) is lower triangular with a diagonal of 1. Thus $(1) - N$ has rank $n + 1$. Since (1) has rank 1, we obtain that the rank of N is at least n by subadditivity.

It follows that an unambiguous automaton for $L(\mathcal{A})$, and thus in particular $\mathcal{A}$ has at least n states.   □

Of course, from Lemma 2, one immediately gets a CoNP procedure for deciding whether an unambiguous automaton is universal. However, it is possible to do better, and prove Theorem 3.

Proof. *(of Theorem 3).* For each letter of the alphabet a, consider the matrix $\eta(a) \in \mathbb{N}^{Q \times Q}$ that describes the transition relation of an unambiguous automaton $\mathcal{A}$: the entry p, q of $\eta(a)$ is 1 if there is a transition labelled a in $\mathcal{A}$ from state p to state q, and 0 otherwise. Let us extend η into a morphism from A^* to $\mathbb{N}^{Q \times Q}$ using the standard matrix multiplication: $\eta(\varepsilon) = \mathrm{Id}_Q$, and $\eta(ua) = \eta(u)\eta(a)$. It is easy to prove by induction that the entry p, q of $\eta(u)$ is the number of runs from state p to state q over the word u. Let also I, F be the characteristic vectors of the initial and final states of $\mathcal{A}$ respectively. Thus, ${}^t I \eta(u) F$ is the number of accepting runs of $\mathcal{A}$ over the word u $(\star)$.

Now let us count the number of accepted words up to length n. Define

$$B(n) = \sum_{u \in A^*, \, |u| \leq n} {}^t I \eta(u) F.$$

From $(\star)$, B_n is the number of accepting runs over words up to length n. Since furthermore the automaton $\mathcal{A}$ is unambiguous, B_n is also the number of accepted words up to length n.

Let us show that this quantity can be computed in time polynomial in n. Indeed, define for all $m = 0, \ldots, n$ the matrices:

$$E_m = \sum_{u \in A^*, \, |u| = m} \eta(u), \qquad \text{and} \quad F_m = \sum_{u \in A^*, \, |u| \leq m} \eta(u).$$

These are such that $F_0 = \mathrm{Id}_Q$, $E_1 = \sum_{a \in A} \eta(a)$, and for all $m \geq 1$, $F_m = F_{m-1} + E_m$ and $E_{m+1} = E_m E_1$. These equations can be used to compute F_n in polynomial time (and the numbers in the matrices have a linear number of digits). Thus $B_{n+1} = {}^t I F_{n+1} F$ is computable in polynomial time.

To conclude, an algorithm that decides universality is as follows: compute B_n and check that it equals the number of words of length at most n (i.e., $(|A|^{n+1} - 1)/(|A| - 1)$). This procedure succeeds if an only if all words are accepted up to length n, which in turns holds (by Lemma 2) if and only if $\mathcal{A}$ is universal.   □

What we have seen in this proof is that it is reasonable to conjecture that unambiguous automata are closed under complement with polynomial blowup. Indeed, this is consistent with (1) the complexity of universality, and (2) the size of witnesses of non-universality. In fact, we can also report on another related conjecture:

Conjecture 2. Given two regular languages K, L of empty intersection, there is an unambiguous automaton of polynomial size that recognizing U that separates them, i.e., such that $K \subseteq U$ and $U \cap L = \emptyset$.

In particular, this would imply that all regular languages L can be turned into an unambiguous automaton of size polynomial in the size of a non-deterministic automaton for L and a non-deterministic automaton for $L^{\complement}$.

We will indeed see later that the separation of classes of non-deterministic automata is often related to unambiguous automata (cf. Theorems 5 and 13).

3   Unambiguous Tropical Automata

We pursue our investigation of unambiguity in the world of automata theory with the more exotic context of tropical automata. Tropical automata belong to the wider class of weighted automata as introduced by Schützenberger [38]. We are interested here in min-plus and max-plus automata.

Min-plus and *max-plus automata* are non-deterministic automata that have their transitions labelled with integers (reals would not make a difference) called *weights*. Given an accepting run of such an automaton, its *weight* is the sum of the weights of the transitions seen along the run. The semantic of a min-plus automaton is to *recognize* the function:

$$[\![A]\!]_{\min} : \quad A^* \to \mathbb{Z} \cup \{+\infty\}$$

$$u \mapsto \begin{cases} +\infty & \text{if there are no accepting runs of } A \text{ over } u, \\ \min \{\text{weight}(\rho) \mid \rho \text{ accepting run of } A \text{ over } u\} & \text{otherwise.} \end{cases}$$

Dually, the semantic of a max-plus automaton is to *recognize* the function:

$$[\![A]\!]_{\max} : \quad A^* \to \mathbb{Z} \cup \{-\infty\}$$

$$u \mapsto \begin{cases} -\infty & \text{if there are no accepting runs of } A \text{ over } u, \\ \max \{\text{weight}(\rho) \mid \rho \text{ acceptingrun of } A \text{ over } u\} & \text{otherwise.} \end{cases}$$

These two notions are formally dual in the following sense: Define $-A$ to be the automaton A in which the weights of all transitions are the opposed weight, then $[\![-A]\!]_{\min} = -[\![A]\!]_{\max}$. There are several other ways to define the functions recognized by such automata, in particular using a matrix presentation. These automata appear in many applications. By *tropical automata* we refer indistinctly to either max-plus automata or min-plus automata.

A non-deterministic automaton $\mathcal{A}$ can be viewed as a min-plus automaton with all its transitions labelled with weight 0. In this case, it recognizes the function $[\![\mathcal{A}]\!]_{\min}$ which maps a word to 0 if it belongs to $L(\mathcal{A})$, and to $+\infty$ otherwise. Symmetrically, a non-deterministic automaton can be viewed as max-plus automaton that recognizes the function which maps a word to 0 if it belongs to $L(\mathcal{A})$, and to $-\infty$ otherwise.

Example 1. The following tropical automaton uses its non-determinism for choosing a segment of consecutive a's surrounded by two b's, and computing its length.

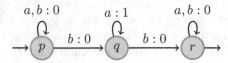

If this automaton is a min-plus automaton, then it maps every word of the form $a^{n_0}ba^{n_1}b\cdots ba_k$ to $\min(n_1,\ldots,n_{k-1})$ if $k \geq 2$, and $+\infty$ otherwise. If this automaton is a max-plus automaton, then it maps every word of the form $a^{n_0}ba^{n_1}b\cdots ba_k$ to $\max(n_1,\ldots,n_{k-1})$ if $k \geq 2$, and $-\infty$ otherwise.

In the world of tropical automata, things are not as nice as for classical finite-state automata, in the sense that undecidability results occur immediately. The central result in this direction is the one of Krob.

Theorem 4 (Krob [21], and [1,14] for simple proofs). *Given a min-plus automaton recognizing a function f, it is undecidable whether $f \leq 0$. Given a max-plus automaton recognizing a function f, it is undecidable whether $f \geq 0$.*

In particular, this means that $f \leq g$ is undecidable for f, g recognized by tropical automata. In fact, this is not completely true: there is one case when this question is decidable, when f is recognized by a max-plus automaton and g by a min-plus automaton, while all other combinations are undecidable by the above theorem.

A tropical automaton is *unambiguous* if the underlying non-deterministic automaton is unambiguous. For instance, in the above example, the automaton is ambiguous. A more careful analysis would show that no unambiguous automata could recognize these functions.

The class of unambiguous tropical automata is interesting since in the definition of $[\![\cdot]\!]_{\min}$ and $[\![\cdot]\!]_{\max}$, the min and the max range over at most one accepting run. Hence, as long as we identify $+\infty$ and $-\infty$, $[\![\cdot]\!]_{\min}$ and $[\![\cdot]\!]_{\max}$ coincide. For this reason, we allow ourselves to simply mention *unambiguous tropical automata* without further mentioning whether these are min-plus or max-plus.

Theorem 5 ([26]). *Functions that are both recognized by min-plus automata and max-plus automata are recognized by unambiguous tropical automata.*

Very informally, if we interpret max-plus automata as a form of complement of min-plus automata, then we can see unambiguous tropical automata as automata that correspond to be both non-deterministic and of non-deterministic complement. We will see a similar phenomenon in the context of register automata in Sect. 5.

A natural question arises: can we decide whether an automaton is equivalent to an unambiguous one? Some first results were obtained in [20]. The best known result is the following:

Theorem 6 ([19]). *There is an algorithm which, given a polynomially ambiguous[1] tropical automaton decides whether there is an unambiguous tropical automaton recognizing the same function.*

Quite naturally, the most important question in this context is to lift the polynomial ambiguity assumption.

Question 1. Can we decide, given a tropical automaton whether it is equivalent to an unambiguous one?

4   Unambiguous Infinite Tree Automata

Another situation where the notion of unambiguity is worth noticing is the context of infinite trees. To keep the presentation light, we expect the reader to know the notion of non-deterministic automaton over infinite trees. An introduction can, for instance, be found in [42].

Let us start by recalling some definitions. An *infinite tree* labelled by the alphabet A is a map from $\{0,1\}^*$ to A. The elements of $\{0,1\}^*$ are *nodes*. The node ε is the *root* of the infinite tree. Given a node u, $u0$ is its *left child* and $u1$ its *right child*. The transitive closure of the child relation is the *descendant relation*. A *branch* is a maximal set of nodes totally ordered under the descendant relation. In this section, *languages* are sets of infinite trees. An *infinite tree automaton* has a finite set of states Q, a set of *initial states* $I \subseteq Q$, and a set of *transitions* $\Delta \subseteq Q \times A \times Q \times Q$. A *run* of an automaton over an infinite tree t is an infinite tree ρ labelled by Q such that $(\rho(u), t(u), \rho(u0), \rho(u1)) \in \Delta$ for all nodes u. The run is *accepting* if $\rho(\varepsilon) \in I$, and for all branches B the set of states assumed on infinitely many nodes by ρ belongs to a given set $M \subseteq 2^Q$, called the *Muller acceptance condition*[2]. If there is an accepting run of the automaton over some input infinite tree, then the infinite tree is *accepted*. The set of infinite trees accepted is the language *recognized* by the automaton.

The central result concerning infinite tree automata is without any question Rabin's theorem stating that infinite tree automata have effectively the same expressive power as monadic second-order logic over infinite trees, and that, as a consequence, this monadic second-order logic is decidable over infinite trees. This logical aspect is certainly far beyond the topic of this paper, but the main lemma in the proof is very relevant:

Lemma 3 (Rabin main Lemma [33]). *Infinite tree automata are effectively closed under complement.*

---

[1] An automaton is *polynomially ambiguous* if the number of accepting runs over an input is bounded by a polynomial in the length of the input.

[2] Other choices are possible, but these distinctions do not make any difference here.

Once more, for this class of automata, the unambiguity notion is natural. An *unambiguous infinite tree automaton* is an infinite tree automaton such that for all input infinite trees, there is at most one accepting run. To start with, there are languages which are recognized by infinite tree automata, but by no unambiguous infinite tree automata, and there are languages that are recognized by unambiguous infinite tree automata, and by no deterministic automata[3]. However, the status of unambiguous automata is very different here than in simpler contexts. In particular, it is not clear whether all regular languages can be recognized by unambiguous automata.

A first answer has been given by Niwiński and Walukiewicz:

Theorem 7 ([30]). *Consider the language "there is a node labelled by the letter a". If this language is recognized by an unambiguous infinite tree automaton, then there exists a regular choice function.*

Informally, a *choice function* is a language that implements the notion of "choice", i.e., given a non-empty set, it selects a unique element in it. One way to formalize this is as follows: consider the alphabet a, b, a^c (a stands for the set in which choice has to be performed, and a^c is the chosen node). A language C of a, b, a^c-labelled infinite trees is a *choice function* if:

– All infinite trees in C contain exactly one occurrence of the letter a^c,
– For all a, b-labelled infinite tree t containing at least one occurrence of the letter a, there exists one and only one a-labelled node x such that $t[x \leftarrow a^c]$ is accepted, where $t[x \leftarrow a^c]$ is the infinite tree t in which the label of the node x is changed into a^c.

A *regular choice function* is a choice function which is recognized by an infinite tree automaton. The existence of a regular choice function has been first studied in [16], where the non-existence of such function is established. However, there is a known unrecoverable hole in the proof. The result was established by Carayol and Löding using much simpler automata-theoretic arguments.

Theorem 8 ([7]). *There does not exist any regular choice function over infinite trees.*

These results were finally published together.

Theorem 9 ([8]). *The language of infinite trees "there is a node labelled by the letter a" is regular, but intrinsically ambiguous; i.e., there exists no unambiguous automaton for this language.*

As it is the case for tropical automata, deciding if a language can be recognized unambiguously is an open problem.

---

[3] In the context of trees, two forms of determinism for automata are possible: *top-down determinism*, i.e., from root to leaves, and *bottom-up determinism*, i.e., from leaves to root. The former (considered here) is known to be strictly weaker than general automata, even over finite trees. The later does not make real sense over infinite trees, since there may be no leaves.

Question 2. Given a infinite tree automaton, can we decide whether its recognized language can be recognized by an unambiguous automaton?

However, if we come back to simpler classes of models, namely finite words, infinite words of length ω or finite trees, it is very easy to have a regular choice function, and also to transform any automaton into an equivalent one that is unambiguous. Nevertheless, there is still an unclear situation. Call *tamed* (or scattered, or thin) an infinite tree that has countably many branches (the definition of an infinite tree needs to be slightly generalized for that, and has to allow leaves). This class is very important, and such infinite trees are significantly simpler than general ones (in particular automata are simpler). To some extend, tamed infinite trees can be understood as the joint extension of infinite words and finite trees.

Question 3. Can we separate unambiguous automata from general automata over tamed trees? Does there exist an automaton, unambiguous over tamed trees that recognizes the language "there is a node labelled by a"? Does there exist a regular choice function over tamed trees?

Let us conclude with another, intriguing, relation linking unambiguity over infinite trees and the existence of regular choice functions over tamed trees.

Theorem 10 ([4]). *Under the assumption that there are no regular choice functions over tamed trees, there is an algorithm which decides whether a regular language of infinite trees is bi-unambiguous[4].*

5   Unambiguous Register Automata

In this last section, we concentrate our attention to data languages. Once more the questions raised are of a slightly different nature. More positively, this is an instance of a situation where new results can be obtained thanks to a careful analysis of the nature of unambiguity. This section will mainly be a report on recent unpublished results obtained in collaboration with Gabriele Puppis and Michał Skrzypczak, in particular establishing conjectures raised in [10].

Originally, register automata were introduced by Kaminski and Francez [18] and were the subject of much attention. There are various ways to introduce this model, including the very interesting "atom approach" [5]. We adopt here a more model-theoretic presentation.

Let us fix ourselves an infinite set of *data values* $\mathbb{D}$. We are only allowed to compare such values using equalities, and as a consequence, the exact set $\mathbb{D}$ does not really matter. Depending on the context, data values can be the identifiers in a database, simply numbers, the agent in a concurrent system, and so on... *Data words* are words over $\mathbb{D}$, i.e., elements of $\mathbb{D}^*$. It is also often convenient to consider slightly richer data words which are elements of $(A \times \mathbb{D})^*$.

---

[4] A language of infinite trees is *bi-unambiguous* if it is accepted by a unambiguous infinite tree automaton as well as its complement.

This distinction has essentially no impact in what follows. Sets of data words are named *data languages*.

A (non-deterministic) *register automaton* has *states*, *initial states*, *final states* and *transitions* as a non-deterministic finite automaton, and furthermore:

- there is a finite set of *registers* $r, s, \ldots$, with values ranging in $\mathbb{D}$, and;
- transitions are equipped of *guards* that are boolean combinations of properties of the form
  - $r = s'$ for r, s registers, signifying that the value of the register r before the transition should be the same as the value of the register s after the transition,
  - $r = d$ for r a register, signifying that the value assumed by the register r before the transition is equal to the data read during the transition,
  - $d = s'$ is defined similarly.

This description should give a fairly good intuition of what is going on. A *run* of a register automaton is a sequence of *configurations* consisting of a state and a valuation of the registers, that respect the transitions and the guards. A run is *accepting* if it starts in an initial state and ends in a final state. It should be clear what an unambiguous register automaton is: an *unambiguous register automaton* is a register automaton such that on every input there is at most one accepting run.

Let us proceed with an example.

Example 2. Consider the following register automaton. We use two registers, r, s, and all transitions are assumed to preserve the values of these counters. Thus we only write on the transitions whether r and s should be equal or not-equal to the read data value.

Note first that this register automaton is non-deterministic: at the very beginning, we do not know the values of the registers (these have to be non-deterministically guessed in some sense). But even without this problem, it is hard to know when in state 2, whether the run should stay in state 2, or proceed with state 3.

In fact, at the same time that we describe the behavior of this automaton, we will see that it is unambiguous. Let us recall that in this register automaton, the values of the registers do not change along the run (we do not know these values a *priori*). While processing an input, this automaton has to take the as first transition the one from state 1 to state 2. Since this transition is guarded by $r = d$, it enforces the value of register r to be the first data value occurring in the input data word. Nox note that all transitions with both extremities among states $2, 3, 4$ have $r \neq d$ in their guard, and this enforces that the data value read to be different from the value of r. Note furthermore that the only transition that exits state 4 (and go to state 5) enforces $r = d$ in its guard. Hence, necessarily, the transition from 4 to 5 has to be taken the first time the value of r is seen again in the data word. This means that this position is unambiguously determined. This means also that the moment the transition from 3 to 4 is used is also unambiguously determined (just one step before). Since furthermore the transition from 3 to 4 has guard $s = d$, the value of register s also has to be unambiguously determined. Overall, this means that, if there is an accepting run over some data word, then the values of r and s are uniquely determined: the value of r is the first data value in the input data word, and the value of s is the data value that occurs just before the second occurrence of the first data value. Once these values fixed, it is easy to see that this automaton is unambiguous, and the language it accepts can be described as follows:

$$\bigcup_{\substack{r, s \in \mathbb{D} \\ r \neq s}} r(\mathbb{D} \setminus \{r\})^* s(\mathbb{D} \setminus \{r, s\})^* sr\mathbb{D}^*.$$

It should also be clear that such a language cannot be determinized. Indeed, while reading an input word from left to right, it is not possible to know what the value of s should be as long as the second occurrence of the first data value is not met. Hence, a deterministic device should memorize all possible data values seen up to that moment. A similar argument prevents to determinize it from right to left.

When working with register automata, the undecidability is again close. The essential results are as follows.

Theorem 11 ([18]). *The languages recognized by register automata are effectively closed under union and intersection, and emptiness is decidable. The universality problem for register automata is undecidable.*

In [10], some conjectures were raised concerning the class of unambiguous register automata. These conjectures are now all established[5]. Let us briefly present these results.

Theorem 12 ([11]). *Unambiguous register automata are effectively closed under complement, and hence universality, containment and equivalence are decidable.*

However, in fact, using the same techniques, we obtain a separation result.

Theorem 13 ([11]). *Given two languages of data words K, L recognized by register automata of empty intersection, there exists a language of data words U recognized by an unambiguous register automaton that separates K and L, i.e., $K \subseteq U$ and $U \cap L = \emptyset$.*

In particular, a language of data words is recognized by an unambiguous register automaton if and only if both itself and its complement are recognized by register automata.

---

[5] Strictly speaking, Conjecture 6 is wrong, but has a corrected version.

6 Conclusion

In this paper, we have tried to present the notion of unambiguity following a rather non-standard path, in particular considering models that are usually not studied together. Along this presentation, we have seen several difficult open questions concerning unambiguous devices. These questions are natural, and show that unambiguity is still quite poorly understood. We have also seen several results that show that unambiguity arises sometimes from characterization reasons: (1) unambiguous tropical automata correspond to functions that are recognized by both min-plus and max-plus automata, and (2) unambiguous register automata correspond to languages that are both recognized as well as their complement by non-deterministic register automata. We believe that this characterization is more than a mere coincidence, and corresponds to the intrinsic nature of unambiguity.

Acknowledgment. I am really grateful to Jean-Éric Pin, Gabriele Puppis and Michał Skrypczak for their precious help and their discussions on the topic.

References

1. Almagor, S., Boker, U., Kupferman, O.: What's decidable about weighted automata? In: Bultan, T., Hsiung, P.-A. (eds.) ATVA 2011. LNCS, vol. 6996, pp. 482–491. Springer, Heidelberg (2011)
2. Berstel, J.: Transductions and Context-Free Languages. Leitfäden der Angewandten Mathematik und Mechanik [Guides to Applied Mathematics and Mechanics], vol. 38. B.G. Teubner, Stuttgart (1979)
3. Berstel, J., Sakarovitch, J.: Recent results in the theory of rational sets. In: Gruska, J., Rovan, B., Wiedermann, J. (eds.) Mathematical Foundations of Computer Science (Bratislava 1986). LNCS, vol. 233, pp. 15–28. Springer, Berlin (1986)
4. Bilkowski, M., Skrzypczak, M.: Unambiguity and uniformization problems on infinite trees. In: CSL, volume of LIPIcs, pp. 81–100 (2013)
5. Bojańczyk, M., Lasota, S.: Fraenkel-mostowski sets with non-homogeneous atoms. In: Finkel, A., Leroux, J., Potapov, I. (eds.) RP 2012. LNCS, vol. 7550, pp. 1–5. Springer, Heidelberg (2012)
6. Cadilhac, M., Finkel, A., McKenzie, P.: Unambiguous constrained automata. Int. J. Found. Comput. Sci. **24**(7), 1099–1116 (2013)
7. Carayol, A., Löding, C.: MSO on the infinite binary tree: choice and order. In: Duparc, J., Henzinger, T.A. (eds.) CSL 2007. LNCS, vol. 4646, pp. 161–176. Springer, Heidelberg (2007)
8. Carayol, A., Löding, C., Niwiński, D., Walukiewicz, I.: Choice functions and well-orderings over the infinite binary tree. Cent. Eur. J. Math. **8**(4), 662–682 (2010)
9. Carton, O., Michel, M.: Unambiguous Büchi automata. Theor. Comput. Sci. **297**(1–3), 37–81 (2003)
10. Colombet, T.: Forms of determinism for automata. In: Dürr, C., Wilke, T. (eds.) STACS 2012: 29th International Symposium on Theoretical Aspects of Computer Science, Volume 14 of LIPIcs, pp. 1–23. Schloss Dagstuhl - Leibniz-Zentrum für Informatik (2012)

11. Colcombet, T., Puppis, G., Skrypczak, M.: Unambiguous register automata. Unpublished
12. Coste, F., Fredouille, D.C.: Unambiguous automata inference by means of state-merging methods. In: Lavrač, N., Gamberger, D., Todorovski, L., Blockeel, H. (eds.) ECML 2003. LNCS (LNAI), vol. 2837, pp. 60–71. Springer, Heidelberg (2003)
13. Cristau, J.: Automata and temporal logic over arbitrary linear time (2011). CoRR, abs/1101.1731
14. Droste, M., Kuske, D.: Weighted automata. To appear in Handbook AutoMathA (2013)
15. Goldstine, J., Kappes, M., Kintala, C.M.R., Leug, H., Malcher, A., Wotschke, D.: Descriptional complexity of machines with limited resources. J. Univ. Comput. Sci. 8, 193–234 (2002)
16. Gurevich, Y., Shelah, S.: Rabin's uniformization problem. J. Symb. Log. 48(4), 1105–1119 (1983)
17. Holzer, M., Kutrib, M.: Descriptional complexity of (un)ambiguous finite state machines and pushdown automata. In: Kučera, A., Potapov, I. (eds.) RP 2010. LNCS, vol. 6227, pp. 1–23. Springer, Heidelberg (2010)
18. Kaminski, M., Francez, N.: Finite-memory automata. Theor. Comput. Sci. 134(2), 329–363 (1994)
19. Kirsten, D., Lombardy, S.: Deciding unambiguity and sequentiality of polynomially ambiguous min-plus automata. In: 26th International Symposium on Theoretical Aspects of Computer Science, STACS 2009, 26–28 February 2009, Freiburg, Germany, Proceedings, pp. 589–600 (2009)
20. Klimann, I., Lombardy, S., Mairesse, J., Prieur, C.: Deciding unambiguity and sequentiality from a finitely ambiguous max-plus automaton. Theor. Comput. Sci. 327(3), 349–373 (2004)
21. Krob, D.: The equality problem for rational series with multiplicities in the tropical semiring is undecidable. Int. J. Algebra Comput. 4(3), 405–425 (1994)
22. Leiss, E.L.: Succint representation of regular languages by boolean automata. Theor. Comput. Sci. 13, 323–330 (1981)
23. Leung, H.: Separating exponentially ambiguous finite automata from polynomially ambiguous finite automata. SIAM J. Comput. 27(4), 1073–1082 (1998)
24. Leung, H.: Descriptional complexity of NFA of different ambiguity. Int. J. Found. Comput. Sci. 16(5), 975–984 (2005)
25. Leung, H.: Structurally unambiguous finite automata. In: Ibarra, O.H., Yen, H.-C. (eds.) CIAA 2006. LNCS, vol. 4094, pp. 198–207. Springer, Heidelberg (2006)
26. Lombardy, S., Mairesse, J.: Series which are both max-plus and min-plus are unambiguous. RAIRO - Theor. Inf. Appl. 40(1), 1–14 (2006)
27. Lupanov, O.B.: A comparison of two types of finite sources. Problemy Kybernetiki 9, 321–326 (1963)
28. Meyer, A.R., Fischer, M.J.: Economy of description by automata, grammars, and formal systems. In: Symposium on Switching and Automata Theory, pp. 188–191. IEEE (1971)
29. Moore, F.R.: On the bounds for state-set size in the proofs of equivalence between deterministic, nondeterministic, and two-way finite automata. IEEE Trans. Comput. 20(10), 1211–1214 (1971)
30. Niwiński, D., Walukiewicz, I.: Ambiguity problem for automata on infinite trees (1996). unpublished
31. Okhotin, A., Salomaa, K.: Descriptional complexity of unambiguous input-driven pushdown automata. Theor. Comput. Sci. 566, 1–11 (2015)

32. Pin, J.É., Weil, P.: Polynomial closure and unambiguous product. Theory Comput. Syst. **30**(4), 383–422 (1997)
33. Rabin, M.O.: Decidability of second-order theories and automata on infinite trees. Trans. Amer. Math. Soc. **141**, 1–35 (1969)
34. Rabin, M.O., Scott, D.: Finite automata and their decision problems. IBM J. Res. Dev. **3**, 114–125 (1959)
35. Reinhardt, K., Allender, E.: Making nondeterminism unambiguous. SIAM J. Comput. **29**(4), 1118–1131 (2000)
36. Sakarovitch, J.: Elements of Automata Theory. Cambridge University Press, Cambridge (2009)
37. Schmidt, E.M.: Succinctness of descriptions of context-free, regular and finite languages. Ph.D. thesis, Cornell University (1977)
38. Schützenberger, M.-P.: On the definition of a family of automata. Inf. Control **4**, 245–270 (1961)
39. Schützenberger, M.-P.: Sur les relations fonctionnelles. In: Brakhage, H. (ed.) GI-Fachtagung 1975. LNCS, vol. 33, pp. 209–213. Springer, Heidelberg (1975)
40. Seidl, H.: Deciding equivalence of finite tree automata. SIAM J. Comput. **19**(3), 424–437 (1990)
41. Stearns, R.E., Hunt III, H.B.: On the equivalence and containment problems for unambiguous regular expressions, grammars, and automata. In: 22nd Annual Symposium on Foundations of Computer Science, Nashville, Tennessee, USA, 28–30 October 1981, pp. 74–81 (1981)
42. Thomas, W.: Languages, automata and logic. In: Rozenberg, G., Salomaa, A. (eds.) Handbook of Formal Languages, vol. 3, pp. 389–455. Springer, New York (1997). Chap. 7
43. Weber, A.: Decomposing a k-valued transducer into k unambiguous ones. ITA **30**(5), 379–413 (1996)

Contributed Papers

Partial Derivative Automaton for Regular Expressions with Shuffle

Sabine Broda[✉], António Machiavelo, Nelma Moreira, and Rogério Reis

CMUP and DM, Faculdade de Ciências da Universidade do Porto,
Rua Do Campo Alegre, 4169-007 Porto, Portugal
{sbb,nam,rvr}@dcc.fc.up.pt, ajmachia@fc.up.pt

Abstract. We generalize the partial derivative automaton to regular expressions with shuffle and study its size in the worst and in the average case. The number of states of the partial derivative automata is in the worst case at most 2^m, where m is the number of letters in the expression, while asymptotically and on average it is no more than $(\frac{4}{3})^m$.

1 Introduction

The class of regular languages is closed under shuffle (or interleaving operation), and extended regular expressions with shuffle can be much more succinct than the equivalent ones with disjunction, concatenation, and star operators. For the shuffle operation, Mayer and Stockmeyer [14] studied the computational complexity of membership and inequivalence problems. Inequivalence is exponential-time-complete, and membership is NP-complete for some classes of regular languages. In particular, they showed that for regular expressions (REs) with shuffle, of size n, an equivalent nondeterministic finite automaton (NFA) needs at most 2^n states, and presented a family of REs with shuffle, of size $\mathcal{O}(n)$, for which the corresponding NFAs have at least 2^n states. Gelade [10], and Gruber and Holzer [11,12] showed that there exists a double exponential trade-off in the translation from REs with shuffle to stantard REs. Gelade also gave a tight double exponential upper bound for the translation of REs with shuffle to DFAs. Recently, conversions of shuffle expressions to finite automata were presented by Estrade et al. [7] and Kumar and Verma [13]. In the latter paper the authors give an algorithm for the construction of an ε-free NFA based on the classic Glushkov construction, and the authors claim that the size of the resulting automaton is at most 2^{m+1}, where m is the number of letters that occur in the RE with shuffle.

In this paper we present a conversion of REs with shuffle to ε-free NFAs, by generalizing the partial derivative construction for standard REs [1,15]. For standard REs, the partial derivative automaton ($\mathcal{A}_{pd}$) is a quotient of the Glushkov automaton ($\mathcal{A}_{pos}$), and Broda et al. [2,3] showed that, asymptotically and on average, the size of $\mathcal{A}_{pd}$ is half the size of $\mathcal{A}_{pos}$. In the case of REs with shuffle

Authors partially funded by the European Regional Development Fund through the programme COMPETE and by the Portuguese Government through the FCT under projects UID/MAT/00144/2013 and FCOMP-01-0124-FEDER-020486.

© Springer International Publishing Switzerland 2015
J. Shallit and A. Okhotin (Eds.): DCFS 2015, LNCS 9118, pp. 21–32, 2015.
DOI: 10.1007/978-3-319-19225-3_2

we show that the number of states of the partial derivative automaton is, in the worst-case, 2^m (with m as before) and an upper bound for the average size is, asymptotically, $(\frac{4}{3})^m$.

This paper is organized as follows. In the next section we review the shuffle operation and regular expressions with shuffle. In Sect. 3 we consider equation systems, for languages and expressions, associated with nondeterministic finite automata and define a solution for a system of equations for a shuffle expression. An alternative and equivalent construction, denoted by $\mathcal{A}_{pd}$, is given in Sect. 4 using the notion of partial derivative. An upper bound for the average number of states of $\mathcal{A}_{pd}$ using the framework of analytic combinatorics is given in Sect. 5. We conclude in Sect. 6 with some considerations about how to improve the presented upper bound and related future work.

2   Regular Expressions with Shuffle

Given an alphabet Σ, the shuffle of two words in $\Sigma^\star$ is a finite set of words defined inductively as follows, for $x, y \in \Sigma^\star$ and $a, b \in \Sigma$

$$x \ш \varepsilon = \varepsilon \ш x = \{x\}$$
$$ax \ш by = \{\, az \mid z \in x \ш by \,\} \cup \{\, bz \mid z \in ax \ш y \,\}.$$

This definition is extended to sets of words, i.e., languages, in the natural way:

$$L_1 \ш L_2 = \{\, x \ш y \mid x \in L_1, y \in L_2 \,\}.$$

It is well known that if two languages $L_1, L_2 \subseteq \Sigma^\star$ are regular then $L_1 \ш L_2$ is regular. One can extend regular expressions to include the $\ш$ operator. Given an alphabet Σ, we let $\mathsf{T}_\ш$ denote the set containing $\emptyset$ plus all terms finitely generated from $\Sigma \cup \{\varepsilon\}$ and operators $+, \cdot, \ш, {}^\star$, that is, the expressions τ generated by the grammar

$$\tau \rightarrow \emptyset \mid \alpha \tag{1}$$
$$\alpha \rightarrow \varepsilon \mid a \mid \alpha + \alpha \mid \alpha \cdot \alpha \mid \alpha \ш \alpha \mid \alpha^\star \quad (a \in \Sigma). \tag{2}$$

As usual, the (regular) language $\mathcal{L}(\tau)$ represented by an expression $\tau \in \mathsf{T}_\ш$ is inductively defined as follows: $\mathcal{L}(\emptyset) = \emptyset$, $\mathcal{L}(\varepsilon) = \{\varepsilon\}$, $\mathcal{L}(a) = \{a\}$ for $a \in \Sigma$, $\mathcal{L}(\alpha^\star) = \mathcal{L}(\alpha)^\star$, $\mathcal{L}(\alpha + \beta) = \mathcal{L}(\alpha) \cup \mathcal{L}(\beta)$, $\mathcal{L}(\alpha\beta) = \mathcal{L}(\alpha)\mathcal{L}(\beta)$, and $\mathcal{L}(\alpha \ш \beta) = \mathcal{L}(\alpha) \ш \mathcal{L}(\beta)$. We say that two expressions $\tau_1, \tau_2 \in \mathsf{T}_\ш$ are equivalent, and write $\tau_1 = \tau_2$, if $\mathcal{L}(\tau_1) = \mathcal{L}(\tau_2)$.

Example 1. Consider $\alpha_n = a_1 \ш \cdots \ш a_n$, where $n \geq 1$, $a_i \neq a_j$ for $1 \leq i \neq j \leq n$. Then,

$$\mathcal{L}(\alpha_n) = \{\, a_{i_1} \cdots a_{i_n} \mid i_1, \ldots, i_n \text{ is a permutation of } 1, \ldots, n\}.$$

We recall that standard regular expressions constitute a Kleene algebra and the shuffle operator ɰ is commutative, associative, and distributes over $+$. It also follows that for all $a, b \in \Sigma$ and $\tau_1, \tau_2 \in \mathsf{T}_{ɰ}$,

$$a\tau_1 ɰ b\tau_2 = a(\tau_1 ɰ b\tau_2) + b(a\tau_1 ɰ \tau_2).$$

Given a language L, we define $\varepsilon(\tau) = \varepsilon(\mathcal{L}(\tau))$, where, $\varepsilon(L) = \varepsilon$ if $\varepsilon \in L$ and $\varepsilon(L) = \emptyset$ otherwise. A recursive definition of $\varepsilon : \mathsf{T}_{ɰ} \longrightarrow \{\emptyset, \varepsilon\}$ is given by the following: $\varepsilon(a) = \varepsilon(\emptyset) = \emptyset$, $\varepsilon(\varepsilon) = \varepsilon(\alpha^*) = \varepsilon$, $\varepsilon(\alpha + \beta) = \varepsilon(\alpha) + \varepsilon(\beta)$, $\varepsilon(\alpha\beta) = \varepsilon(\alpha)\varepsilon(\beta)$, and $\varepsilon(\alpha ɰ \beta) = \varepsilon(\alpha)\varepsilon(\beta)$.

3   Automata and Systems of Equations

We first recall the definition of an NFA as a tuple $\mathcal{A} = \langle S, \Sigma, S_0, \delta, F \rangle$, where S is a finite set of states, Σ is a finite alphabet, $S_0 \subseteq S$ a set of initial states, $\delta : S \times \Sigma \longrightarrow \mathcal{P}(S)$ the transition function, and $F \subseteq S$ a set of final states. The extension of δ to sets of states and words is defined by $\delta(X, \varepsilon) = X$ and $\delta(X, ax) = \delta(\cup_{s \in X}\delta(s, a), x)$. A word $x \in \Sigma^*$ is accepted by $\mathcal{A}$ if and only if $\delta(S_0, x) \cap F \neq \emptyset$. The *language of* $\mathcal{A}$ is the set of words accepted by $\mathcal{A}$ and denoted by $\mathcal{L}(\mathcal{A})$. The *right language of a state* s, denoted by $\mathcal{L}_s$, is the language accepted by $\mathcal{A}$ if we take $S_0 = \{s\}$. The class of languages accepted by all the NFAs is precisely the set of regular languages.

It is well known that, for each n-state NFA $\mathcal{A}$ over $\Sigma = \{a_1, \dots, a_k\}$, where $S = [1, n]$, having right languages $\mathcal{L}_1, \dots, \mathcal{L}_n$, it is possible to associate a system of linear language equations

$$\mathcal{L}_i = a_1\mathcal{L}_{1i} \cup \dots \cup a_k\mathcal{L}_{ik} \cup \varepsilon(\mathcal{L}_i), \quad i \in [1, n]$$

where each $\mathcal{L}_{ij}$ is a (possibly empty) union of elements in $\{\mathcal{L}_1, \dots, \mathcal{L}_n\}$, and $\mathcal{L}(\mathcal{A}) = \bigcup_{i \in S_0} \mathcal{L}_i$.

In the same way, it is possible to associate with each regular expression a system of equations on expressions. Here, we extend this notion to regular expressions with shuffle.

Definition 2. *Consider* $\Sigma = \{a_1, \dots, a_k\}$ *and* $\alpha_0 \in \mathsf{T}_{ɰ}$. *A support of* α_0 *is a set* $\{\alpha_1, \dots, \alpha_n\}$ *that satisfies a system of equations*

$$\alpha_i = a_1\alpha_{1i} + \dots + a_k\alpha_{ki} + \varepsilon(\alpha_i), \quad i \in [0, n] \tag{3}$$

for some $\alpha_{1i}, \dots, \alpha_{ki}$, *each one a (possibly empty) sum of elements in* $\{\alpha_1, \dots, \alpha_n\}$. *In this case* $\{\alpha_0, \alpha_1, \dots, \alpha_n\}$ *is called a* prebase *of* α_0.

It is clear from what was just said above, that the existence of a support of α implies the existence of an NFA that accepts the language determined by α.

Note that the system of Eq. (3) can be written in matrix form $\mathsf{A}_\alpha = \mathsf{C} \cdot \mathsf{M}_\alpha + \mathsf{E}_\alpha$, where M_α is the $k \times (n+1)$ matrix with entries α_{ij}, and A_α, C and E_α denote respectively the following three matrices,

$$\mathsf{A}_\alpha = [\alpha_0 \cdots \alpha_n], \quad \mathsf{C} = [a_1 \cdots a_k], \quad \text{and} \quad \mathsf{E}_\alpha = [\varepsilon(\alpha_0) \cdots \varepsilon(\alpha_n)],$$

where, $C \cdot M_\alpha$ denotes the matrix obtained from C and M_α applying the standard rules of matrix multiplication, but replacing the multiplication by concatenation. This notation will be used below.

A support for an expression $\alpha \in T_{\text{⧢}}$ can be computed using the function $\pi : T_{\text{⧢}} \longrightarrow \mathcal{P}(T_{\text{⧢}})$ recursively given by the following.

Definition 3. *Given $\tau \in T_{\text{⧢}}$, the set $\pi(\tau)$ is inductively defined by,*

$$\pi(\emptyset) = \pi(\varepsilon) = \emptyset \qquad \pi(\alpha + \beta) = \pi(\alpha) \cup \pi(\beta)$$
$$\pi(a) = \{\varepsilon\} \quad (a \in \Sigma) \qquad \pi(\alpha\beta) = \pi(\alpha)\beta \cup \pi(\beta)$$
$$\pi(\alpha^*) = \pi(\alpha)\alpha^* \qquad \pi(\alpha \text{⧢} \beta) = \pi(\alpha) \text{⧢} \pi(\beta)$$
$$\cup \, \pi(\alpha) \text{⧢} \{\beta\} \cup \{\alpha\} \text{⧢} \pi(\beta),$$

where, given $S, T \subseteq T_{\text{⧢}}$ and $\beta \in T_{\text{⧢}} \setminus \{\emptyset, \varepsilon\}$, $S\beta = \{\, \alpha\beta \mid \alpha \in S \,\}$ and $S \text{⧢} T = \{\, \alpha \text{⧢} \beta \mid \alpha \in S, \beta \in T \,\}$, $S\varepsilon = \{\varepsilon\} \text{⧢} S = S \text{⧢} \{\varepsilon\} = S$, and $S\emptyset = \emptyset S = \emptyset$.

The following lemma follows directly from the definitions and will be used in the proof of Proposition 5.

Lemma 4. *If $\alpha, \beta \in T_{\text{⧢}}$, then $\varepsilon(\beta) \cdot \mathcal{L}(\alpha) \subseteq \mathcal{L}(\alpha \text{⧢} \beta)$.*

Proposition 5. *If $\alpha \in T_{\text{⧢}}$, then the set $\pi(\alpha)$ is a support of α.*

Proof. We proceed by induction on the structure of α. Excluding the case where α is $\alpha_0 \text{⧢} \beta_0$, the proof can be found in [6,15]. We now describe how to obtain a system of equations corresponding to an expression $\alpha_0 \text{⧢} \beta_0$ from systems for α_0 and β_0. Suppose that $\pi(\alpha_0) = \{\alpha_1, \ldots, \alpha_n\}$ is a support of α_0 and $\pi(\beta_0) = \{\beta_1, \ldots, \beta_m\}$ is a support of β_0. For α_0 and β_0 consider $C, A_{\alpha_0}, M_{\alpha_0}, E_{\alpha_0}$ and $A_{\beta_0}, M_{\beta_0}, E_{\beta_0}$ as above. We wish to show that

$$\pi(\alpha_0 \text{⧢} \beta_0) = \{\alpha_1 \text{⧢} \beta_1, \ldots, \alpha_1 \text{⧢} \beta_m, \ldots, \alpha_n \text{⧢} \beta_1, \ldots, \alpha_n \text{⧢} \beta_m\}$$
$$\cup \, \{\alpha_1 \text{⧢} \beta_0, \ldots, \alpha_n \text{⧢} \beta_0\} \cup \{\alpha_0 \text{⧢} \beta_1, \ldots, \alpha_0 \text{⧢} \beta_m\}$$

is a support of $\alpha_0 \text{⧢} \beta_0$. Let $A_{\alpha_0 \text{⧢} \beta_0}$ be the $(n+1)(m+1)$-entry row-matrix whose entires are

$$\begin{bmatrix} \alpha_0 \text{⧢} \beta_0 \; \alpha_1 \text{⧢} \beta_1 \; \cdots \; \alpha_n \text{⧢} \beta_m \; \alpha_1 \text{⧢} \beta_0 \; \cdots \; \alpha_n \text{⧢} \beta_0 \; \alpha_0 \text{⧢} \beta_1 \; \cdots \; \alpha_0 \text{⧢} \beta_m \end{bmatrix}.$$

Then, $E_{\alpha_0 \text{⧢} \beta_0}$ is defined as usual, i.e. containing the values of $\varepsilon(\alpha)$ for all entries α in $A_{\alpha_0 \text{⧢} \beta_0}$.

Finally, let $M_{\alpha_0 \text{⧢} \beta_0}$ be the $k \times (n+1)(m+1)$ matrix whose entries $\gamma_{l,(i,j)}$, for $l \in [1,k]$ and $(i,j) \in [0,n] \times [0,m]$, are defined by

$$\gamma_{l,(i,j)} = \alpha_{li} \text{⧢} \beta_j + \alpha_i \text{⧢} \beta_{lj}.$$

Note that, since by the induction hypothesis each α_{li} is a sum of elements in $\pi(\alpha)$ and each β_{lj} is a sum of elements in $\pi(\beta)$, after applying distributivity of ⧢ over $+$ each element of $M_{\alpha_0 \text{⧢} \beta_0}$ is in fact a sum of elements in $\pi(\alpha_0 \text{⧢} \beta_0)$. We will show that $A_{\alpha_0 \text{⧢} \beta_0} = C \cdot M_{\alpha_0 \text{⧢} \beta_0} + E_{\alpha_0 \text{⧢} \beta_0}$. For this, consider $\alpha_i \text{⧢} \beta_j$

for some $(i,j) \in [0,n] \times [0,m]$. We have $\alpha_i = a_1\alpha_{1i} + \cdots + a_k\alpha_{ki} + \varepsilon(\alpha_i)$ and $\beta_j = a_1\beta_{1j} + \cdots + a_k\beta_{kj} + \varepsilon(\beta_j)$. Consequently, using properties of ɰ, namely distributivity over $+$, as well as Lemma 4,

$$
\begin{aligned}
\alpha_i \amalg \beta_j &= (a_1\alpha_{1i} + \cdots + a_k\alpha_{ki} + \varepsilon(\alpha_i)) \amalg (a_1\beta_{1j} + \cdots + a_k\beta_{kj} + \varepsilon(\beta_j)) \\
&= a_1\left(\alpha_{1i} \amalg \beta_j + \alpha_i \amalg \beta_{1j} + \varepsilon(\beta_j)\alpha_{1i} + \varepsilon(\alpha_i)\beta_{1j}\right) + \cdots \\
&\quad + a_k\left(\alpha_{ki} \amalg \beta_j + \alpha_i \amalg \beta_{kj} + \varepsilon(\beta_j)\alpha_{ki} + \varepsilon(\alpha_i)\beta_{kj}\right) + \varepsilon(\alpha_i \amalg \beta_j) \\
&= a_1\left(\alpha_{1i} \amalg \beta_j + \alpha_i \amalg \beta_{1j}\right) + \cdots \\
&\quad + a_k\left(\alpha_{ki} \amalg \beta_j + \alpha_i \amalg \beta_{kj}\right) + \varepsilon(\alpha_i \amalg \beta_j) \\
&= a_1\gamma_{1,(i,j)} + \cdots + a_k\gamma_{k,(i,j)} + \varepsilon(\alpha_i \amalg \beta_j).
\end{aligned}
$$

$\square$

It is clear from its definition that $\pi(\alpha)$ is finite. In the following proposition, an upper bound for the size of $\pi(\alpha)$ is given. Example 7 is a witness that this upper bound is tight.

Proposition 6. *Given* $\alpha \in T_\amalg$, *one has* $|\pi(\alpha)| \le 2^{|\alpha|_\Sigma} - 1$, *where* $|\alpha|_\Sigma$ *denotes the number of alphabet symbols in* α.

Proof. The proof proceeds by induction on the structure of α. It is clear that the result holds for $\alpha = \emptyset$, $\alpha = \varepsilon$ and for $\alpha = a \in \Sigma$. Now, suppose the claim is true for α and β. There are four induction cases to consider. We will make use of the fact that, for $m, n \ge 0$ one has $2^m + 2^n - 2 \le 2^{m+n} - 1$. For $\alpha^\star$, one has $|\pi(\alpha^\star)| = |\pi(\alpha)\alpha^\star| = |\pi(\alpha)| \le 2^{|\alpha|_\Sigma} - 1 = 2^{|\alpha^\star|_\Sigma} - 1$. For $\alpha + \beta$, one has $|\pi(\alpha+\beta)| = |\pi(\alpha) \cup \pi(\beta)| \le 2^{|\alpha|_\Sigma} - 1 + 2^{|\beta|_\Sigma} - 1 \le 2^{|\alpha|_\Sigma + |\beta|_\Sigma} - 1 = 2^{|\alpha+\beta|_\Sigma} - 1$. For $\alpha\beta$, one has $|\pi(\alpha\beta)| = |\pi(\alpha)\beta \cup \pi(\beta)| \le 2^{|\alpha|_\Sigma} - 1 + 2^{|\beta|_\Sigma} - 1 \le 2^{|\alpha\beta|_\Sigma} - 1$. Finally, for $\alpha \amalg \beta$, one has $|\pi(\alpha \amalg \beta)| = |\pi(\alpha) \amalg \pi(\beta) \cup \pi(\alpha) \amalg \{\beta\} \cup \{\alpha\} \amalg \pi(\beta)| \le (2^{|\alpha|_\Sigma} - 1)(2^{|\beta|_\Sigma} - 1) + 2^{|\alpha|_\Sigma} - 1 + 2^{|\beta|_\Sigma} - 1 = 2^{|\alpha|_\Sigma + |\beta|_\Sigma} - 1 = 2^{|\alpha\amalg\beta|_\Sigma} - 1$. $\square$

Example 7. Considering $\alpha_n = a_1 \amalg \cdots \amalg a_n$, where $n \ge 1$, $a_i \ne a_j$ for $1 \le i \ne j \le n$ again, one has

$$
|\pi(\alpha_n)| = |\{ \amalg_{i \in I} a_i \mid I \subsetneq \{1, \ldots, n\} \}| = 2^n - 1,
$$

where by convention $\amalg_{i \in \emptyset} a_i = \varepsilon$.

The proof of Proposition 5 gives a way to construct a system of equations for an expression $\tau \in T_\amalg$, corresponding to an NFA that accepts the language represented by τ. This is done by recursively computing $\pi(\tau)$ and the matrices A_τ and E_τ, obtaining the whole NFA in the final step.

In the next section we will show how to build the same NFA in a more efficient way using the notion of partial derivative.

4   Partial Derivatives

Recall that the *left quotient* of a language L w.r.t. a symbol $a \in \Sigma$ is

$$a^{-1}L = \{ x \mid ax \in L \}.$$

The left quotient of L w.r.t. a word $x \in \Sigma^\star$ is then inductively defined by $\varepsilon^{-1}L = L$ and $(xa)^{-1}L = a^{-1}(x^{-1}L)$. Note that for $L_1, L_2 \subseteq \Sigma^\star$ and $a, b \in \Sigma$ the shuffle operation satisfies $a^{-1}(L_1 \amalg L_2) = (a^{-1}L_1) \amalg L_2 \cup L_1 \amalg (a^{-1}L_2)$.

Definition 8. *The set of partial derivatives of a term $\tau \in \mathsf{T}_\amalg$ w.r.t. a letter $a \in \Sigma$, denoted by $\partial_a(\tau)$, is inductively defined by*

$$\partial_a(\emptyset) = \partial_a(\varepsilon) = \emptyset \qquad\qquad \partial_a(\alpha^*) = \partial_a(\alpha)\alpha^*$$
$$\partial_a(\alpha + \beta) = \partial_a(\alpha) \cup \partial_a(\beta)$$
$$\partial_a(b) = \begin{cases} \{\varepsilon\} & \textit{if } b = a \\ \emptyset & \textit{otherwise} \end{cases} \qquad \partial_a(\alpha\beta) = \partial_a(\alpha)\beta \cup \varepsilon(\alpha)\partial_a(\beta)$$
$$\partial_a(\alpha \amalg \beta) = \partial_a(\alpha) \amalg \{\beta\} \cup \{\alpha\} \amalg \partial_a(\beta).$$

The set of partial derivatives of $\tau \in \mathsf{T}_\amalg$ w.r.t. a word $x \in \Sigma^\star$ is inductively defined by $\partial_\varepsilon(\tau) = \{\tau\}$ and $\partial_{xa}(\tau) = \partial_a(\partial_x(\tau))$, where, given a set $S \subseteq \mathsf{T}_\amalg$, $\partial_a(S) = \bigcup_{\tau \in S} \partial_a(\tau)$.

We let $\partial(\tau)$ denote the set of all partial derivatives of an expression τ, i.e. $\partial(\tau) = \bigcup_{x \in \Sigma^\star} \partial_x(\tau)$, and by $\partial^+(\tau)$ the set of partial derivatives excluding the trivial derivative by ε, i.e. $\partial^+(\tau) = \bigcup_{x \in \Sigma^+} \partial_x(\tau)$. Given a set $S \subseteq \mathsf{T}_\amalg$, we define $\mathcal{L}(S) = \bigcup_{\tau \in S} \mathcal{L}(\tau)$. The following result has a straightforward proof.

Proposition 9. *Given $x \in \Sigma^\star$ and $\tau \in \mathsf{T}_\amalg$, one has $\mathcal{L}(\partial_x(\tau)) = x^{-1}\mathcal{L}(\tau)$.*

The following properties of $\partial^+(\tau)$ will be used in the proof of Proposition 11.

Lemma 10. *For $\tau \in \mathsf{T}_\amalg$, the following hold.*

1. *If $\partial^+(\tau) \neq \emptyset$, then there is $\alpha_0 \in \partial^+(\tau)$ with $\varepsilon(\alpha_0) = \varepsilon$.*
2. *If $\partial^+(\tau) = \emptyset$ and $\tau \neq \emptyset$, then $\mathcal{L}(\tau) = \{\varepsilon\}$ and $\varepsilon(\tau) = \varepsilon$.*

Proof. 1. From the grammar rule (2) it follows that $\emptyset$ cannot appear as a subexpression of a larger term. Suppose that there is some $\gamma \in \partial^+(\tau)$. We conclude, from Definition 8 and from the previous remark, that there is some word $x \in \Sigma^+$ such that $x \in \mathcal{L}(\gamma)$. This is equivalent to $\varepsilon \in \mathcal{L}(\partial_x(\gamma))$, which means that there is some $\alpha_0 \in \partial_x(\gamma) \subseteq \partial^+(\tau)$ such that $\varepsilon(\alpha_0) = \varepsilon$.

2. $\partial^+(\tau) = \emptyset$ implies that $\partial_x(\tau) = \emptyset$ for all $x \in \Sigma^+$. Thus, $\mathcal{L}(\partial_x(\tau)) = \{ y \mid xy \in \mathcal{L}(\tau) \} = \emptyset$, and consequently there is no word $z \in \Sigma^+$ in $\mathcal{L}(\tau)$. On the other hand, since $\emptyset$ does not appear in τ, it follows that $\mathcal{L}(\tau) \neq \emptyset$. Thus, $\mathcal{L}(\tau) = \{\varepsilon\}$. $\square$

Proposition 11. *∂^+ satisfies the following:*

$$\partial^+(\emptyset) = \partial^+(\varepsilon) = \emptyset \qquad\qquad \partial^+(\alpha + \beta) = \partial^+(\alpha) \cup \partial^+(\beta)$$
$$\partial^+(a) = \{\varepsilon\} \quad (a \in \Sigma) \qquad\qquad \partial^+(\alpha\beta) = \partial^+(\alpha)\beta \cup \partial^+(\beta)$$
$$\partial^+(\alpha^*) = \partial^+(\alpha)\alpha^* \qquad\qquad \partial^+(\alpha \amalg \beta) = \partial^+(\alpha) \amalg \partial^+(\beta)$$
$$\cup\ \partial^+(\alpha) \amalg \{\beta\} \cup \{\alpha\} \amalg \partial^+(\beta).$$

Proof. The proof proceeds by induction on the structure of α. It is clear that $\partial^+(\emptyset) = \emptyset$, $\partial^+(\varepsilon) = \emptyset$ and, for $a \in \Sigma$, $\partial^+(a) = \{\varepsilon\}$.

In the remaining cases, to prove that an inclusion $\partial^+(\gamma) \subseteq E$ holds for some expression E, we show by induction on the length of x that for every $x \in \Sigma^+$ one has $\partial_x(\gamma) \subseteq E$. We will therefore just indicate the corresponding computations for $\partial_a(\gamma)$ and $\partial_{xa}(\gamma)$, for $a \in \Sigma$. We also make use of the fact that, for any expression γ and letter $a \in \Sigma$, the set $\partial^+(\gamma)$ is closed for taking derivatives w.r.t. a, i.e., $\partial_a(\partial^+(\gamma)) \subseteq \partial^+(\gamma)$.

Now, suppose the claim is true for α and β. There are four induction cases to consider.

- For $\alpha + \beta$, we have $\partial_a(\alpha + \beta) = \partial_a(\alpha) + \partial_a(\beta) \subseteq \partial^+(\alpha) \cup \partial^+(\beta)$, as well as $\partial_{xa}(\alpha + \beta) = \partial_a(\partial_x(\alpha + \beta)) \subseteq \partial_a(\partial^+(\alpha) \cup \partial^+(\beta)) \subseteq \partial_a(\partial^+(\alpha)) \cup \partial_a(\partial^+(\beta)) \subseteq \partial^+(\alpha) \cup \partial^+(\beta)$. Similarly, one proves that $\partial_x(\alpha) \in \partial^+(\alpha + \beta)$ and $\partial_x(\beta) \in \partial^+(\alpha + \beta)$, for all $x \in \Sigma^+$.

- For $\alpha^\star$, we have $\partial_a(\alpha^\star) = \partial_a(\alpha)\alpha^\star \subseteq \partial^+(\alpha)\alpha^\star$, as well as

$$\partial_{xa}(\alpha^\star) = \partial_a(\partial_x(\alpha^\star)) \subseteq \partial_a(\partial^+(\alpha)\alpha^\star) \subseteq \partial_a(\partial^+(\alpha))\alpha^\star \cup \partial_a(\alpha^\star)$$
$$\subseteq \partial^+(\alpha)\alpha^\star \cup \partial_a(\alpha)\alpha^\star \subseteq \partial^+(\alpha)\alpha^\star.$$

Furthermore, $\partial_a(\alpha)\alpha^\star = \partial_a(\alpha^\star) \subseteq \partial^+(\alpha^\star)$ and $\partial_{xa}(\alpha)\alpha^\star = \partial_a(\partial_x(\alpha))\alpha^\star \subseteq \partial_a(\partial_x(\alpha)\alpha^\star) \subseteq \partial_a(\partial^+(\alpha^\star)) \subseteq \partial^+(\alpha^\star)$.

- For $\alpha\beta$, we have $\partial_a(\alpha\beta) = \partial_a(\alpha)\beta \cup \varepsilon(\alpha)\partial_a(\beta) \subseteq \partial^+(\alpha)\beta \cup \partial^+(\beta)$ and

$$\partial_{xa}(\alpha\beta) = \partial_a(\partial_x(\alpha\beta)) \subseteq \partial_a(\partial^+(\alpha)\beta \cup \partial^+(\beta)) = \partial_a(\partial^+(\alpha)\beta) \cup \partial_a(\partial^+(\beta))$$
$$\subseteq \partial_a(\partial^+(\alpha))\beta \cup \partial_a(\beta) \cup \partial_a(\partial^+(\beta)) \subseteq \partial^+(\alpha)\beta \cup \partial^+(\beta).$$

Also, $\partial_a(\alpha)\beta \subseteq \partial_a(\alpha\beta) \subseteq \partial^+(\alpha\beta)$ and

$$\partial_{xa}(\alpha)\beta = \partial_a(\partial_x(\alpha))\beta \subseteq \partial_a(\partial_x(\alpha)\beta) \subseteq \partial_a(\partial^+(\alpha\beta)) \subseteq \partial^+(\alpha\beta).$$

Finally, if $\varepsilon(\alpha) = \varepsilon$, then $\partial_a(\beta) \subseteq \partial_a(\alpha\beta)$ and $\partial_{xa}(\beta) = \partial_a(\partial_x(\beta)) \subseteq \partial_a(\partial_x(\alpha\beta)) = \partial_{xa}(\alpha\beta)$. We conclude that $\partial_x(\beta) \subseteq \partial_x(\alpha\beta)$ for all $x \in \Sigma^+$, and therefore $\partial^+(\beta) \subseteq \partial^+(\alpha\beta)$. Otherwise, $\varepsilon(\alpha) = \emptyset$, and it follows from Lemma 10 that $\partial^+(\alpha) \neq \emptyset$, and that there is some $\alpha_0 \in \partial^+(\alpha)$ with $\varepsilon(\alpha_0) = \emptyset$. As above, this implies that $\partial_x(\beta) \subseteq \partial_x(\alpha_0\beta)$ for all $x \in \Sigma^+$. On the other hand, have already shown that $\partial^+(\alpha)\beta \subseteq \partial^+(\alpha\beta)$. In particular, $\alpha_0\beta \in \partial^+(\alpha\beta)$. From these two facts, we conclude that $\partial_x(\beta) \subseteq \partial_x(\alpha_0\beta) \subseteq \partial_x(\partial^+(\alpha\beta)) \subseteq \partial^+(\alpha\beta)$, which finishes the proof for the case of concatenation.

- For $\alpha \shuffle \beta$, we have

$$\partial_a(\alpha \shuffle \beta) = \partial_a(\alpha) \shuffle \{\beta\} \cup \{\alpha\} \shuffle \partial_a(\beta)$$
$$\subseteq \partial^+(\alpha) \shuffle \partial^+(\beta) \cup \partial^+(\alpha) \shuffle \{\beta\} \cup \{\alpha\} \shuffle \partial^+(\beta)$$

and

$$\partial_{xa}(\alpha \shuffle \beta) \subseteq \partial_a(\partial^+(\alpha) \shuffle \partial^+(\beta) \cup \partial^+(\alpha) \shuffle \{\beta\} \cup \{\alpha\} \shuffle \partial^+(\beta))$$
$$= \partial_a(\partial^+(\alpha) \shuffle \partial^+(\beta)) \cup \partial_a(\partial^+(\alpha) \shuffle \{\beta\}) \cup \partial_a(\{\alpha\} \shuffle \partial^+(\beta))$$
$$= \partial_a(\partial^+(\alpha)) \shuffle \partial^+(\beta) \cup \partial^+(\alpha) \shuffle \partial_a(\partial^+(\beta)) \cup \partial_a(\partial^+(\alpha)) \shuffle \{\beta\}$$
$$\cup \partial^+(\alpha) \shuffle \partial_a(\beta) \cup \partial_a(\alpha) \shuffle \partial^+(\beta) \cup \{\alpha\} \shuffle \partial_a(\partial^+(\beta))$$
$$\subseteq \partial^+(\alpha) \shuffle \partial^+(\beta) \cup \partial^+(\alpha) \shuffle \{\beta\} \cup \{\alpha\} \shuffle \partial^+(\beta).$$

Now we prove that for all $x \in \Sigma^+$, one has $\partial_x(\alpha) \shuffle \{\beta\} \subseteq \partial_x(\alpha \shuffle \beta)$, which implies $\partial^+(\alpha) \shuffle \{\beta\} \subseteq \partial^+(\alpha \shuffle \beta)$. In fact, we have $\partial_a(\alpha) \shuffle \{\beta\} \subseteq \partial_a$ $(\alpha \shuffle \beta)$ and

$$\partial_{xa}(\alpha) \shuffle \{\beta\} \subseteq \partial_a(\partial_x(\alpha)) \shuffle \{\beta\}$$
$$\subseteq \partial_a(\partial_x(\alpha) \shuffle \{\beta\}) \subseteq \partial_a(\partial_x(\alpha \shuffle \beta)) = \partial_{xa}(\alpha \shuffle \beta).$$

Showing that $\{\alpha\} \shuffle \partial_x(\beta) \subseteq \partial_x(\alpha \shuffle \beta)$ is analogous. Finally, for $x, y \in \Sigma^+$ we have $\partial_x(\alpha) \shuffle \partial_y(\beta) \subseteq \partial_y(\partial_x(\alpha) \shuffle \{\beta\}) \subseteq \partial_y(\partial_x(\alpha \shuffle \beta)) = \partial_{xy}(\alpha \shuffle \beta) \subseteq \partial^+(\alpha \shuffle \beta).$ □

Corollary 12. *Given $\alpha \in T_\shuffle$, one has $\partial^+(\alpha) = \pi(\alpha)$.*

We conclude that $\partial(\alpha)$ corresponds to the set $\{\alpha\} \cup \pi(\alpha)$, as is the case for standard regular expressions. It is well known that the set of partial derivatives of a regular expression gives rise to an equivalent NFA, called the Antimirov automaton or partial derivative automaton, that accepts the language determined by that expression. This remains valid in our extension of the partial derivatives to regular expressions with shuffle.

Definition 13. *Given $\tau \in T_\shuffle$, we define the partial derivative automaton associated with τ by*
$$\mathcal{A}_{pd}(\tau) = \langle \partial(\tau), \Sigma, \{\tau\}, \delta_\tau, F_\tau \rangle,$$
where $F_\tau = \{ \gamma \in \partial(\tau) \mid \varepsilon(\gamma) = \varepsilon \}$ and $\delta_\tau(\gamma, a) = \partial_a(\gamma)$.

It is easy to see that the following holds.

Proposition 14. *For every state $\gamma \in \partial(\tau)$, the right language $\mathcal{L}_\gamma$ of γ in $\mathcal{A}(\tau)$ is equal to $\mathcal{L}(\gamma)$, the language represented by γ. In particular, the language accepted by $\mathcal{A}_{pd}(\tau)$ is exactly $\mathcal{L}(\tau)$.*

Note that for the REs α_n considered in Examples 1 and 7, $\mathcal{A}_{pd}(\alpha_n)$ has 2^n states which is exactly the bound presented by Mayer and Stockmeyer [14].

5   Average State Complexity of the Partial Derivative Automaton

In this section, we estimate the asymptotic average size of the number of states in partial derivative automata. This is done by the use of the standard methods of analytic combinatorics as expounded by Flajolet and Sedgewick [9], which apply to generating functions $A(z) = \sum_n a_n z^n$ associated with combinatorial classes. Given some measure of the objects of a class $\mathcal{A}$, the coefficient a_n represents the sum of the values of this measure for all objects of size n. We will use the notation $[z^n]A(z)$ for a_n. For an introduction of this approach applied to formal languages, we refer to Broda *et al.* [4]. In order to apply this method, it is necessary to have an unambiguous description of the objects of the combinatorial class, as is the case for the specification of $\mathsf{T}_{\sqcup\!\sqcup}$-expressions without $\emptyset$ in (2). For the length or size of a $\mathsf{T}_{\sqcup\!\sqcup}$-expression α we will consider the number of symbols in α, not counting parentheses. Taking $k = |\Sigma|$, we compute from (2) the generating functions $R_k(z)$ and $L_k(z)$, for the number of $\mathsf{T}_{\sqcup\!\sqcup}$-expressions without $\emptyset$ and the number of alphabet symbols in $\mathsf{T}_{\sqcup\!\sqcup}$-expressions without $\emptyset$, respectively. Note that excluding one object, $\emptyset$, of size 1 has no influence on the asymptotic study.

According to the specification in (2) the generating function $R_k(z)$ for the number of $\mathsf{T}_{\sqcup\!\sqcup}$-expressions without $\emptyset$ satisfies

$$R_k(z) = z + kz + 3zR_k(z)^2 + zR_k(z),$$

thus,

$$R_k(z) = \frac{(1-z) - \sqrt{\Delta_k(z)}}{6z}, \text{ where } \Delta_k(z) = 1 - 2z - (11 + 12k)z^2.$$

The radius of convergence of $R_k(z)$ is $\rho_k = \frac{-1+2\sqrt{3+3k}}{11+12k}$. Now, note that the number of letters $l(\alpha)$ in an expression α satisfies: $l(\varepsilon) = 0$, in $l(a) = 1$, for $a \in \Sigma$, $l(\alpha + \beta) = l(\alpha) + l(\beta)$, etc. From this, we conclude that the generating function $L_k(z)$ satisfies

$$L_k(z) = kz + 3zL_k(z)R_k(z) + zL_k(z),$$

thus,

$$L_k(z) = \frac{(-kz)}{6zR_k(z) + z - 1} = \frac{kz}{\sqrt{\Delta_k(z)}}.$$

Now, let $P_k(z)$ denote the generating function for the size of $\pi(\alpha)$ for $\mathsf{T}_{\sqcup\!\sqcup}$-expressions without $\emptyset$. From Definition 3 it follows that, given an expression α, an upper bound, $p(\alpha)$, for the number of elements[1] in the set $\pi(\alpha)$ satisfies:

$$
\begin{aligned}
p(\varepsilon) &= 0 & p(\alpha + \beta) &= p(\alpha) + p(\beta) \\
p(a) &= 1, \text{ for } a \in \Sigma & p(\alpha\beta) &= p(\alpha) + p(\beta) \\
p(\alpha^\star) &= p(\alpha) & p(\alpha \sqcup\!\sqcup \beta) &= p(\alpha)p(\beta) + p(\alpha) + p(\beta).
\end{aligned}
$$

From this, we conclude, using the symbolic method [9], that the generating function $P_k(z)$ satisfies

$$P_k(z) = kz + 6zP_k(z)R_k(z) + zP_k(z) + zP_k(z)^2,$$

---

[1] This upper bound corresponds to the case where all unions in $\pi(\alpha)$ are disjoint.

thus

$$P_k(z) = Q_k(z) + S_k(z),$$

where

$$Q_k(z) = \frac{\sqrt{\Delta_k(z)}}{2z}, \qquad S_k(z) = -\frac{\sqrt{\Delta'_k(z)}}{2z},$$

and $\Delta'_k(z) = 1 - 2z - (11 + 16k)z^2$. The radii of convergence of $Q_k(z)$ and $S_k(z)$ are respectively ρ_k (defined above) and $\rho'_k = \frac{-1 + 2\sqrt{3 + 4k}}{11 + 16k}$.

5.1 Asymptotic Analysis

A generating function f can be seen as a complex analytic function, and the study of its behaviour around its dominant singularity ρ (in case there is only one, as it happens with the functions considered here) gives us access to the asymptotic form of its coefficients. In particular, if $f(z)$ is analytic in some appropriate neighbourhood of ρ, then one has the following [4,9,16]:

1. if $f(z) = a - b\sqrt{1 - z/\rho} + o\left(\sqrt{1 - z/\rho}\right)$, with $a, b \in \mathbb{R}$, $b \neq 0$, then

$$[z^n]f(z) \sim \frac{b}{2\sqrt{\pi}}\rho^{-n}n^{-3/2};$$

2. if $f(z) = \frac{a}{\sqrt{1 - z/\rho}} + o\left(\frac{1}{\sqrt{1 - z/\rho}}\right)$, with $a \in \mathbb{R}$, and $a \neq 0$, then

$$[z^n]f(z) \sim \frac{a}{\sqrt{\pi}}\rho^{-n}n^{-1/2}.$$

Hence, by 1. one has for the number of $T_\sqcup$-expressions of size n,

$$[z^n]R_k(z) = \frac{(3 + 3k)^{\frac{1}{4}}}{6\sqrt{\pi}}\rho_k^{-n - \frac{1}{2}}(n + 1)^{-\frac{3}{2}} \tag{4}$$

and by 2. for the number of alphabet symbols in all expression of size n,

$$[z^n]L_k(z) = \frac{k}{2\sqrt{\pi}(3 + 3k)^{\frac{1}{4}}}\rho_k^{-n + \frac{1}{2}}n^{-\frac{1}{2}}. \tag{5}$$

Consequently, the average number of letters in an expression of size n, which we denote by avL, is asymptotically given by

$$avL = \frac{[z^n]L_k(z)}{[z^n]R_k(z)} = \frac{3k\rho_k}{\sqrt{3 + 3k}}\frac{(n + 1)^{\frac{3}{2}}}{n^{\frac{1}{2}}}.$$

Finally, by 1., one has for the size of expressions of size n,

$$[z^n]P_k(z) = [z^n]Q_k(z) + [z^n]S_k(z)$$
$$= \frac{-(3 + 3k)^{\frac{1}{4}}\rho_k^{-n - \frac{1}{2}} + (3 + 4k)^{\frac{1}{4}}(\rho'_k)^{-n - \frac{1}{2}}}{2\sqrt{\pi}}(n + 1)^{-\frac{3}{2}},$$

and the average size of $\pi(\alpha)$ for an expression α of size n, denoted by avP, is asymptotically given by

$$avP = \frac{[z^n]P_k(z)}{[z^n]R_k(z)}.$$

Taking into account Proposition 6, we want to compare the values of $\log_2 avP$ and avL. In fact, one has

$$\lim_{n,k\to\infty} \frac{\log_2 avP}{avL} = \log_2 \frac{4}{3} \sim 0.415.$$

This means that,

$$\lim_{n,k\to\infty} avP^{1/avL} = \frac{4}{3}.$$

Therefore, one has the following significant improvement, when compared with the worst case, for the average case upper bound.

Proposition 15. *For large values of k and n an upper bound for the average number of states of $\mathcal{A}_{pd}$ is $(\frac{4}{3} + o(1))^{|\alpha|_\Sigma}$.*

6   Conclusion and Future Work

We implemented the construction of the $\mathcal{A}_{pd}$ for REs with shuffle in the FAdo system [8] and performed some experimental tests for small values of n and k. Those experiments over statistically significant samples of uniform random generated REs suggest that the upper bound obtained in the last section falls far short of its true value. This is not surprising as in the construction of $\pi(\alpha) \cup \{\alpha\}$ repeated elements can occur.

In previous work [2], we identified classes of standard REs that capture a significant reduction on the size of $\pi(\alpha)$. In the case of REs with shuffle, those classes enforce only a marginal reduction in the number of states, but a drastic increase in the complexity of the associated generating function. Thus the expected gains don't seem to justify its quite difficult asymptotic study.

Sulzmann and Thiemann [17] extended the notion of Brzozowski derivative for several variants of the shuffle operator. It will be interesting to carry out a descriptional complexity study of those constructions and to see if it is interesting to extend the notion of partial derivative to those shuffle variants.

An extension of the partial derivative construction for extended REs with intersection and negation was recently presented by Caron *et al.* [5]. It will be also interesting to study the average complexity of this construction.

References

1. Antimirov, V.M.: Partial derivatives of regular expressions and finite automaton constructions. Theoret. Comput. Sci. **155**(2), 291–319 (1996)
2. Broda, S., Machiavelo, A., Moreira, N., Reis, R.: On the average state complexity of partial derivative automata. Int. J. Found. Comput. Sci. **22**(7), 1593–1606 (2011)

3. Broda, S., Machiavelo, A., Moreira, N., Reis, R.: On the average size of Glushkov and partial derivative automata. Int. J. Found. Comput. Sci. **23**(5), 969–984 (2012)
4. Broda, S., Machiavelo, A., Moreira, N., Reis, R.: A Hitchhiker's guide to descriptional complexity through analytic combinatorics. Theor. Comput. Sci. **528**, 85–100 (2014)
5. Caron, P., Champarnaud, J.-M., Mignot, L.: Partial derivatives of an extended regular expression. In: Dediu, A.-H., Inenaga, S., Martín-Vide, C. (eds.) LATA 2011. LNCS, vol. 6638, pp. 179–191. Springer, Heidelberg (2011)
6. Champarnaud, J.M., Ziadi, D.: From Mirkin's prebases to Antimirov's word partial derivatives. Fundam. Inform. **45**(3), 195–205 (2001)
7. Estrade, B.D., Perkins, A.L., Harris, J.M.: Explicitly parallel regular expressions. In: Ni, J., Dongarra, J. (eds.) 1st IMSCCS, pp. 402–409. IEEE Computer Society (2006)
8. FAdo, P.: FAdo: tools for formal languages manipulation. http://fado.dcc.fc.up.pt/. Accessed October 01 2014
9. Flajolet, P., Sedgewick, R.: Analytic Combinatorics. CUP (2008)
10. Gelade, W.: Succinctness of regular expressions with interleaving, intersection and counting. Theor. Comput. Sci. **411**(31–33), 2987–2998 (2010)
11. Gruber, H.: On the descriptional and algorithmic complexity of regular languages. Ph.D. thesis, Justus Liebig University Giessen (2010)
12. Gruber, H., Holzer, M.: Finite automata, digraph connectivity, and regular expression size. In: Aceto, L., Damgård, I., Goldberg, L.A., Halldórsson, M.M., Ingólfsdóttir, A., Walukiewicz, I. (eds.) ICALP 2008, Part II. LNCS, vol. 5126, pp. 39–50. Springer, Heidelberg (2008)
13. Kumar, A., Verma, A.K.: A novel algorithm for the conversion of parallel regular expressions to non-deterministic finite automata. Appl. Math. Inf. Sci. **8**, 95–105 (2014)
14. Mayer, A.J., Stockmeyer, L.J.: Word problems-this time with interleaving. Inf. Comput. **115**(2), 293–311 (1994)
15. Mirkin, B.G.: An algorithm for constructing a base in a language of regular expressions. Eng. Cybern. **5**, 51–57 (1966)
16. Nicaud, C.: On the average size of Glushkov's automata. In: Dediu, A.H., Ionescu, A.M., Martín-Vide, C. (eds.) LATA 2009. LNCS, vol. 5457, pp. 626–637. Springer, Heidelberg (2009)
17. Sulzmann, M., Thiemann, P.: Derivatives for regular shuffle expressions. In: Dediu, A.-H., Formenti, E., Martín-Vide, C., Truthe, B. (eds.) LATA 2015. LNCS, vol. 8977, pp. 275–286. Springer, Heidelberg (2015)

Upper Bound on Syntactic Complexity of Suffix-Free Languages

Janusz Brzozowski[1] and Marek Szykuła[2]([⊠])

[1] David R. Cheriton School of Computer Science, University of Waterloo, Waterloo, ON N2L 3G1, Canada
brzozo@uwaterloo.ca
[2] Institute of Computer Science, University of Wrocław, Joliot-Curie 15, PL-50-383 Wrocław, Poland
msz@cs.uni.wroc.pl

Abstract. We solve an open problem concerning syntactic complexity: We prove that the cardinality of the syntactic semigroup of a suffix-free language with n left quotients (that is, with state complexity n) is at most $(n-1)^{n-2} + n - 2$ for $n \geqslant 7$. Since this bound is known to be reachable, this settles the problem. We also reduce the alphabet of the witness languages reaching this bound to five letters instead of $n + 2$, and show that it cannot be any smaller. Finally, we prove that the transition semigroup of a minimal deterministic automaton accepting such a witness language is unique for each $n \geqslant 7$.

Keywords: Regular language · Suffix-free · Syntactic complexity · Transition semigroup · Upper bound

1 Preliminaries

1.1 Introduction

The *syntactic complexity* [7] $\sigma(L)$ of a regular language L is the size of its syntactic semigroup [10]. This semigroup is isomorphic to the transition semigroup of the quotient automaton $\mathcal{D}$ (a minimal deterministic finite automaton) accepting the language. The number n of states of $\mathcal{D}$ is the *state complexity* of the language [12], and it is the same as the *quotient complexity* [3] (number of left quotients) of the language. The *syntactic complexity of a class* of regular languages is the maximal syntactic complexity of languages in that class expressed as a function of the quotient complexity n.

If $w = uxv$ for some $u, v, x \in \Sigma^*$, then u is a *prefix* of w, v is a *suffix* of w and x is a *factor* of w. A suffix of w is also a factor of w. A language L is *prefix-free* (respectively, *suffix-free, factor-free*) if $w, u \in L$ and u is a prefix

---

This work was supported by the Natural Sciences and Engineering Research Council of Canada grant No. OGP000087, and by Polish NCN grant DEC-2013/09/N/ST6/01194.

J. Shallit and A. Okhotin (Eds.): DCFS 2015, LNCS 9118, pp. 33–45, 2015.
DOI: 10.1007/978-3-319-19225-3_3

(respectively, *suffix, factor*) of w, implies that $u = w$. A language is *bifix-free* if it is both prefix- and suffix-free. These languages play an important role in coding theory, have applications in such areas as cryptography, data compression, and information transmission, and have been studied extensively; see [2] for example. In particular, suffix-free languages (with the exception of $\{\varepsilon\}$, where ε is the empty word) are suffix codes. Moreover, suffix-free languages are special cases of suffix-convex languages, where a language is *suffix-convex* if it satisfies the condition that, if a word w and its suffix u are in the language, then so is every suffix of w that has u as a suffix [1,11]. We are interested only in regular suffix-free languages.

The syntactic complexity of prefix-free languages was proved to be n^{n-2} in [4]. The syntactic complexities of suffix-, bifix-, and factor-free languages were also studied in [4], and the following lower bounds were established $(n-1)^{n-2} + n - 2$, $(n-1)^{n-3} + (n-2)^{n-3} + (n-3)2^{n-3}$, and $(n-1)^{n-3} + (n-3)2^{n-3} + 1$, respectively. It was conjectured that these bounds are also upper bounds; we prove the conjecture for suffix-free languages in this paper.

A full version of the paper is available in [5].

1.2   Languages, Automata and Transformations

Let Σ be a finite, non-empty alphabet and let $L \subseteq \Sigma^*$ be a language. The *left quotient* or simply *quotient* of a language L by a word $w \in \Sigma^*$ is denoted by $L.w$ and defined by $L.w = \{x \mid wx \in L\}$. A language is regular if and only if it has a finite number of quotients. We denote the set of quotients by $K = \{K_0, \ldots, K_{n-1}\}$, where $K_0 = L = L.\varepsilon$ by convention. Each quotient K_q can be represented also as $L.w_q$, where $w_q \in \Sigma^*$ is such that $L.w_q = K_q$.

A *deterministic finite automaton (DFA)* is a quintuple $\mathcal{D} = (Q, \Sigma, \delta, q_0, F)$, where Q is a finite non-empty set of *states*, Σ is a finite non-empty *alphabet*, $\delta \colon Q \times \Sigma \to Q$ is the *transition function*, $q_0 \in Q$ is the *initial* state, and $F \subseteq Q$ is the set of *final* states. We extend δ to a function $\delta \colon Q \times \Sigma^* \to Q$ as usual.

The *quotient DFA* of a regular language L with n quotients is defined by $\mathcal{D} = (K, \Sigma, \delta_{\mathcal{D}}, K_0, F_{\mathcal{D}})$, where $\delta_{\mathcal{D}}(K_q, w) = K_p$ if and only if $K_q.w = K_p$, and $F_{\mathcal{D}} = \{K_q \mid \varepsilon \in K_q\}$. To simplify the notation, without loss of generality we use the set $Q = \{0, \ldots, n-1\}$ of subscripts of quotients as the set of states of $\mathcal{D}$; then $\mathcal{D}$ is denoted by $\mathcal{D} = (Q, \Sigma, \delta, 0, F)$, where $\delta(q, w) = p$ if $\delta_{\mathcal{D}}(K_q, w) = K_p$, and F is the set of subscripts of quotients in $F_{\mathcal{D}}$. The quotient corresponding to $q \in Q$ (known also as the *right language* of q) is $K_q = \{w \mid \delta_{\mathcal{D}}(K_q, w) \in F_{\mathcal{D}}\}$. The quotient $K_0 = L$ is the *initial* quotient. A quotient is *final* if it contains ε. A state q is *empty* if its quotient K_q is empty. The quotient DFA of L is isomorphic to each complete minimal DFA of L. The number of states in the quotient DFA of L (the quotient complexity of L) is therefore equal to the state complexity of L.

In any DFA, each letter $a \in \Sigma$ induces a transformation of the set Q of n states. Let $\mathcal{T}_Q$ be the set of all n^n transformations of Q; then $\mathcal{T}_Q$ is a monoid under composition. The *image* of $q \in Q$ under transformation t is denoted by qt. If s, t are transformations of Q, their composition is denoted $s \circ t$ and defined by $q(s \circ t) = (qs)t$; the $\circ$ is usually omitted. The *in-degree* of a state q in a transformation t is the cardinality of the set $\{p \mid pt = q\}$.

The *identity* transformation $\mathbf{1}$ maps each element to itself. For $k \geqslant 2$, a transformation (permutation) t of a set $P = \{q_0, q_1, \ldots, q_{k-1}\} \subseteq Q$ is a *k-cycle* if $q_0 t = q_1, q_1 t = q_2, \ldots, q_{k-2} t = q_{k-1}, q_{k-1} t = q_0$. A k-cycle is denoted by $(q_0, q_1, \ldots, q_{k-1})$. If a transformation t of Q is a k-cycle of some $P \subseteq Q$, then t *has a k-cycle*. A transformation *has a cycle* if it has a k-cycle for some $k \geqslant 2$. A 2-cycle (q_0, q_1) is called a *transposition*. A transformation is *unitary* if it changes only one state p to a state $q \neq p$; it is denoted by $(p \to q)$. A transformation mapping a subset P of Q to a single state and acting as the identity on $Q \setminus P$ is denoted by $(P \to q)$.

The binary relation ω_t on $Q \times Q$ is defined as follows: For any $p, q \in Q$, $p\,\omega_t\,q$ if and only if $pt^k = qt^\ell$ for some $k, \ell \geqslant 0$. This is an equivalence relation, and each equivalence class is called an *orbit* [8] of t. For any $q \in Q$, the orbit of t containing q is denoted by $\omega_t(q)$. An orbit contains either exactly once cycle and no fixed points or exactly one fixed point and no cycles. The set of all orbits of t is a partition of Q.

If $w \in \Sigma^*$ induces a transformation t, we denote this by $w \colon t$.

The *transition semigroup* of a DFA $\mathcal{D} = (Q, \Sigma, \delta, 0, F)$ is the semigroup of transformations of Q generated by the transformations induced by the letters of Σ. Since the transition semigroup of a minimal DFA of a language L is isomorphic to the syntactic semigroup of L [10], syntactic complexity is equal to the cardinality of the transition semigroup.

1.3 Suffix-Free Languages

For any transformation t, consider the sequence $(0, 0t, 0t^2, \ldots)$; we call it the 0-*path* of t. Since Q is finite, there exist i, j such that $0, 0t, \ldots, 0t^i, 0t^{i+1}, \ldots, 0t^{j-1}$ are distinct but $0t^j = 0t^i$. The integer $j - i$ is the *period* of t and if $j - i = 1$, t is *initially aperiodic*. Let $Q = \{0, \ldots, n-1\}$, let $\mathcal{D}_n = (Q, \Sigma, \delta, 0, F)$ be a minimal DFA accepting a language L, and let T_n be its transition semigroup. The following is known [4,9]:

Lemma 1. *If L is a suffix-free language, then*

1. *There exists $w \in \Sigma^*$ such that $L.w = \emptyset$; hence $\mathcal{D}_n$ has an empty state, which is state $n-1$ by convention.*
2. *For $w, x \in \Sigma^+$, if $L.w \neq \emptyset$, then $L.w \neq L.xw$.*
3. *If $L.w \neq \emptyset$, then $L.w = L$ implies $w = \varepsilon$.*
4. *For any $t \in T_n$, the 0-path of t in $\mathcal{D}_n$ is aperiodic and ends in $n-1$.*

An (unordered) pair $\{p, q\}$ of distinct states in $Q \setminus \{0, n-1\}$ is *colliding* (or p *collides* with q) in T_n if there is a transformation $t \in T_n$ such that $0t = p$ and $rt = q$ for some $r \in Q \setminus \{0, n-1\}$. A pair of states is *focused* by a transformation u of Q if u maps both states of the pair to a single state $r \notin \{0, n-1\}$. We then say that $\{p, q\}$ is *focused to state r*. If L is a suffix-free language, then from Lemma 1 (2) it follows that if $\{p, q\}$ is colliding in T_n, there is no transformation

$t' \in T_n$ that focuses $\{p, q\}$. So colliding states can be mapped to a single state by a transformation in T_n only if that state is the empty state $n - 1$.

Remark 1. If $n = 1$, the only suffix-free language is the empty language $\emptyset$ and $\sigma(\emptyset) = 1$. If $n \geq 2$ and $\Sigma = \{a\}$, the language $L = a^{n-2}$ is the only suffix-free language of quotient complexity n, and its syntactic complexity is $\sigma(L) = n - 1$.

Assume now that $|\Sigma| \geq 2$. If $n = 2$, the language $L = \varepsilon$ is the only suffix-free language, and $\sigma(L) = 1$. If $n = 3$, the tight upper bound on syntactic complexity of suffix-free languages is 3, and $L = ab^*$ over $\Sigma = \{a, b\}$ meets this bound [4].

If $n = 4$ and $n = 5$, the tight upper bounds are 13, and 73 [4]. In [4] it was shown that there is a suffix-free witness DFA with n states and an alphabet of size $n + 2$ that meets the bound $(n - 1)^{n-2} + n - 2$ for $n \geq 4$. For $n = 4$ and $n = 5$, these bounds are 11 and 67, and so are smaller than the bounds above. For $n \geq 6$, $(n - 1)^{n-2} + n - 2$ is the largest known lower bound. ∎

2 Lower Bound for Suffix-Free Languages

The lower bound of $(n-1)^{n-2} + n - 2$ on the complexity of suffix-free languages was established in [4] using a witness DFA with an alphabet with $n + 2$ letters. Our first contribution is to simplify the witness of [4] by using an alphabet with only five letters, as stated in Definition 1. The transitions induced by inputs a, b, c, and e are the same as in [4].

Definition 1 (Witness). *For $n \geq 4$ define the DFA $\mathcal{W}_n = (Q, \Sigma_{\mathcal{W}}, \delta_{\mathcal{W}}, 0, \{1\})$, where $Q = \{0, \ldots, n-1\}$, $\Sigma_{\mathcal{W}} = \{a, b, c, d, e\}$, and $\delta_{\mathcal{W}}$ is defined by the transformations $a\colon (0 \to n-1)(1, \ldots, n-2)$, $b\colon (0 \to n-1)(1, 2)$, $c\colon (0 \to n-1)(n-2 \to 1)$, $d\colon (\{0, 1\} \to n-1)$, and $e\colon (Q \setminus \{0\} \to n-1)(0 \to 1)$. For $n = 4$, a and b coincide, and we can use $\Sigma_{\mathcal{W}} = \{b, c, d, e\}$. Let S_n be the transition semigroup of $\mathcal{W}_n$.*

The structure of $\mathcal{W}_n$ is illustrated in Fig. 1 for $n = 5$. We claim that no pair of states from Q is colliding in S_n. If $0t = p \notin \{0, n-1\}$, then t is not the identity but must be induced by a word of the form ew for some $w \in \Sigma^*$. Such a word maps every $r \notin \{0, n-1\}$ to $n-1$; so $q = rt = n-1$, and p and q do not collide.

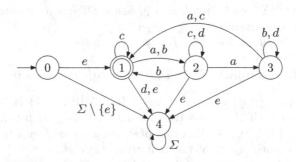

Fig. 1. Witness DFA $\mathcal{W}_5$.

Proposition 1. *For $n \geqslant 4$ the DFA of Definition 1 is minimal, suffix-free, and its transition semigroup S_n has cardinality $(n-1)^{n-2} + n - 2$. In particular, S_n contains (a) all $(n-1)^{n-2}$ transformations that send 0 and $n-1$ to $n-1$ and map $Q \setminus \{0, n-1\}$ to $Q \setminus \{0\}$, and (b) all $n-2$ transformations that send 0 to a state in $Q \setminus \{0, n-1\}$ and map all the other states to $n-1$.*

3 Upper Bound for Suffix-Free Languages

Our second result shows that the lower bound $(n-1)^{n-2} + n - 2$ on the syntactic complexity of suffix-free languages is also an upper bound. Our approach is as follows: We consider a minimal DFA $\mathcal{D}_n = (Q, \Sigma, \delta, 0, F)$, where $Q = \{0, \ldots, n-1\}$, of an arbitrary suffix-free language with n quotients and let T_n be the transition semigroup of $\mathcal{D}_n$. We also deal with the witness DFA $\mathcal{W}_n = (Q, \Sigma_\mathcal{W}, \delta_\mathcal{W}, 0, \{1\})$ of Definition 1 that has the same state set as $\mathcal{D}_n$ and whose transition semigroup is S_n. We shall show that there is an injective mapping $\varphi : T_n \to S_n$, and this will prove that $|T_n| \leqslant |S_n|$.

The image of our mapping φ of a transition t in T_n depends on the properties of t. We separate these properties into 12 mutually disjoint cases that cover all the possibilities. The cases are structured as follows: We begin with an arbitrary transformation $t \in T_n$. Case 1 consists of transformations t that are also in S_n, and the remainder, R_1, of the cases has $t \notin S_n$. Having reached Case i, we define Case $(i+1)$ as all the transformations that do not fit in Cases 1 to i and satisfy a property P_{i+1}. The remainder R_{i+1} consists of all the transformations that do not fit in Cases 1 to i, and do not satisfy P_{i+1}. Because of this structure it is evident that the cases are mutually disjoint. In view of Case 12, they exhaust all the possibilities. The proof for each case is similar: we prove that $s = \varphi(t)$ differs from all the images s defined in previous cases and also from all the other images defined in the present case.

A note about terminology may be helpful to the reader. The semigroups T_n and S_n share the set Q. When we say that a pair of states from Q is *colliding* we mean that it is colliding in T_n; there is no room for confusion because no pair of states is colliding in S_n. Since we are dealing with suffix-free languages, a transformation that focuses a colliding pair cannot belong to T_n.

In Cases 2–11 of the proof p always denotes $0t$.

Theorem 1 (Tight Bound). *For $n \geqslant 6$ the syntactic complexity of the class of suffix-free languages with n quotients is $(n-1)^{n-2} + n - 2$.*

Proof. The case $n = 6$ has been proved in [4]; hence assume that $n \geqslant 7$. In [4] and in Proposition 1 it was shown that $(n-1)^{n-2} + n - 2$ is a lower bound for $n \geqslant 7$; hence it remains to prove that it is also an upper bound, and we do this here. We have the following cases:

Case 1: $t \in S_n$. Let $\varphi(t) = t$; obviously φ is injective.

Case 2: $t \notin S_n$, and t has a cycle. By Lemma 1 (4) we have the chain $0 \overset{t}{\to} p \overset{t}{\to} pt \overset{t}{\to} \cdots \overset{t}{\to} pt^k \overset{t}{\to} n - 1$, where $k \geqslant 0$. Observe that pairs $\{pt^i, pt^j\}$ for

$0 \leqslant i < j \leqslant k$ are colliding, since transformation t^{i+1} maps 0 to pt^i and pt^{j-i-1} to pt^j. Also, p collides with any state from a cycle of t and any fixed point of t other than $n - 1$.

Let r be minimal among the states that appear in cycles of t, that is, $r = \min\{q \in Q \mid q \text{ is in a cycle of } t\}$. Let s be the transformation illustrated in Fig. 2 and defined by

$$0s = n - 1, \quad ps = r, \quad (pt^i)s = pt^{i-1} \text{ for } 1 \leqslant i \leqslant k,$$
$$qs = qt \text{ for the other states } q \in Q.$$

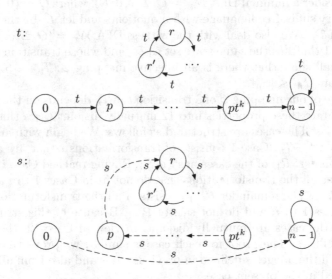

Fig. 2. Case 2 in the proof of Theorem 1.

By Proposition 1, $\varphi(t) = s$ is in S_n, since it maps 0 to $n - 1$, fixes $n - 1$, and does not map any states to 0. Note that the sets of cyclic states in both t and s are the same. Let r' be the state from the cycle of t such that $r't = r$; then transformation s has the following properties:

(a) Since p collides with any state in a cycle of t, $\{p, r'\}$ is a colliding pair focused by s to state r in the cycle. Moreover, if q' is a state in a cycle of s, and $\{q, q'\}$ is colliding and focused by s to a state in a cycle, then that state must be r (the minimal state in the cycles of s), q must be p, and q' must be r'. This follows from the definition of s. Since s differs from t only in the mapping of states pt^i and 0, any colliding pair focused by s contains pt^i for some i, $0 \leqslant i \leqslant k$. Only p is mapped to r, which is in a cycle of t, and r' is the only state in that cycle that is mapped to r.

(b) For each i with $1 \leqslant i < k$, there is precisely one state q colliding with pt^{i-1} and mapped by s to pt^i, and that state is $q = pt^{i+1}$. Clearly $q = pt^{i+1}$

satisfies this condition. Suppose that $q \neq pt^{i+1}$. Since pt^{i+1} is the only state mapped to pt^i by s and not by t, it follows that $qt = qs = pt^i$. So q and pt^{i-1} are focused to pt^i by t; since they collide, this is a contradiction.

(c) Every focused colliding pair consists of states from the orbit of p. This follows from the fact that all the states except 0 that are mapped by s differently than by t belong to the orbit of p.

(d) s has a cycle.

From (a), $s \notin T_n$ and so s is different from the transformations of Case 1.

Given a transformation s from this case we will construct a unique t that results in s when the definition of s given above is applied. This will show that our mapping φ has an inverse, and so is injective. From (a) there is the unique colliding pair focused to a state in a cycle. Moreover, one of the states in the pair, say p, is not in this cycle and another one, say r', is in this cycle. It follows that $0t = p$. Since there is no state $q \neq 0$ such that $qt = p$, the only state mapped to p by s is pt. From (b) for $i = 1, \ldots, k - 1$ state pt^{i+1} is uniquely determined. Finally, for $i = k$ there is no state colliding with pt^{k-1} and mapped to pt^k; so $pt^{k+1} = n - 1$. Since the other transitions in s are defined exactly as in t, this procedure defines the inverse function φ^{-1} for the transformations of this case.

Case 3: $t \notin S_n$, t has no cycles, but $pt \neq n - 1$. Let s be the transformation defined by

$$0s = n - 1, \quad ps = p, \quad (pt^i)s = pt^{i-1} \text{ for } 1 \leqslant i \leqslant k,$$
$$qs = qt \text{ for the other states } q \in Q.$$

Observe that s has the following properties:

(a) $\{p, pt\}$ is the only colliding pair focused by s to a fixed point. Moreover the fixed point is contained in the pair, and has in-degree 2. This follows from the definition of s, since any colliding pair focused by s contains pt^i, for some i with $0 \leqslant i \leqslant k$, and only pt is mapped to p, which is a fixed point. Also, no state except 0 is mapped to p by t since this would violate suffix-freeness; so only p and pt are mapped by s to p, and p has in-degree 2.

(b) For each i with $1 \leqslant i < k$, there is precisely one state q colliding with pt^{i-1} and mapped to pt^i, and that state is $q = pt^{i+1}$. This follows exactly like Property (b) from Case 2.

(c) Every colliding pair focused by s consists of states from the orbit of p. This follows exactly like Property (c) from Case 2.

(d) s does not have a cycle, but has a fixed point $f \neq n - 1$ with in-degree $\geqslant 2$, which is p.

From (a), $s \notin T_n$ and so s is different from the transformations of Case 1. Here s does not have a cycle in contrast with the transformations of Case 2.

As before, s uniquely defines the transformation t from which it is obtained: From (a) there is the unique colliding pair $\{p, pt\}$ focused to the fixed point p. Thus $0t = p$. Then, as in Case 2, for $i = 1, \ldots, k - 1$ state pt^{i+1} is uniquely defined, and $pt^k = n - 1$. Since the other transitions in s are defined exactly as in t, this procedure yields the inverse function φ^{-1} for this case.

Case 4: t does not fit in any of the previous cases, but there is a fixed point $r \in Q \setminus \{0, n - 1\}$ with in-degree $\geqslant 2$. Let s be the transformation defined by

$$0s = n - 1, \quad ps = r,$$
$$qs = qt \text{ for the other states } q \in Q.$$

Observe that s has the following properties:

(a) $\{p, r\}$ is the only colliding pair focused by s to a fixed point, where the fixed point is contained in the pair. Also, the fixed point has in-degree at least 3. Since s differs from t only by the mapping of states 0 and p, it follows that all focused colliding pairs contain p. Since p is mapped to r, the second state in the pair must be the fixed point r. Since r has in-degree at least 2 in t, and s additionally maps p to r, r has in-degree at least 3.

(b) s does not have a cycle, but has a fixed point other than $n - 1$ with in-degree $\geqslant 3$, which is r.

From (a) we have $s \notin T_n$, and so s is different from the transformations of Case 1. Here s does not have a cycle in contrast with the transformations of Case 2. Also from (a) we know that the fixed point in the distinguished colliding pair has in-degree $\geqslant 3$, whereas in Case 3 it has in-degree 2. From (a) we see that the colliding pair $\{p, r\}$ in which r is a fixed point and p is not is uniquely defined. Hence $0t = p$ and $pt = n - 1$, and t is again uniquely defined from s.

Case 5: t does not fit in any of the previous cases, but there is a state r with in-degree $\geqslant 1$ that is not a fixed point and satisfies $rt \neq n - 1$.

Since there are no fixed points in s with in-degree $\geqslant 2$ other than $n - 1$, and there are no cycles, it follows that r belongs to the orbit of $n - 1$. Hence we can choose r such that $rt \neq n - 1$ and $rt^2 = n - 1$.

Let s be the transformation defined by

$$0s = n - 1, \quad ps = rt,$$
$$qs = qt \text{ for the other states } q \in Q.$$

Observe that s has the following properties:

(a) All focused colliding pairs contain p, and the second state from such a pair has in-degree $\geqslant 1$.

This follows since s differs from t only in the mapping of 0 and p.

(b) The smallest i with $ps^i = n - 1$ is 2.

(c) s has neither a cycle nor a fixed point with in-degree $\geqslant 2$ other than $n - 1$.

Note that p and r collide. Since $\{p, r\}$ is focused to rt, we have $s \notin T_n$ and so s is different from the transformations of Case 1. Here s does not have a cycle in contrast with the transformations of Case 2. Also s does not have a fixed point other than $n - 1$, and so is different from the transformations of Cases 3 and 4.

From (a) all focused colliding pairs contain p. If there are two or more such pairs, p is the only state in their intersection. If there is only one such pair, then it must be $\{p, r\}$, and p is uniquely determined, since it has in-degree 0 and r has in-degree $\geqslant 1$. Hence $0t = p$ and $pt = n - 1$, and again t is uniquely defined from s.

Case 6: t does not fit in any of the previous cases, but there is a state $r \in Q \setminus \{0, n - 1\}$ with in-degree $\geqslant 2$. Clearly $r \neq p$, since the in-degree of p is 1. Also $rt = n - 1$, as otherwise t would fit in Case 5. Let $R = \{r' \in Q \mid r't = r\}$; then $|R| \geqslant 2$. We consider the following two sub-cases. If $p < r$, let q_1 be the smallest state in R and let q_2 be the second smallest state; so $q_1 < q_2$. If $p > r$, let q_1 be the second smallest state in R, and let q_2 be the smallest state; so $q_2 < q_1$. Let s be the transformation defined by

$$0s = n - 1, \quad ps = q_1, \quad rs = q_1, \quad q_1 s = q_2, \quad q_2 s = n - 1,$$
$$qs = qt \text{ for the other states } q \in Q.$$

Observe that s has the following properties:

(a) There is only one focused colliding pair, namely $\{p, r\}$, mapped to q_1. Clearly p and r collide. Note that no state can be mapped by t to q_1 or q_2, since this would satisfy Case 5. Because q_1 is the only state mapped by s to q_2, it does not belong to a focused colliding pair. Also 0 and q_2 are mapped to $n - 1$. Since the other states are mapped exactly as in t, it follows that s does not focus any other colliding pairs.
(b) The smallest i with $ps^i = n - 1$ is 3.
(c) s has neither a cycle nor a fixed point $\neq n - 1$ with in-degree $\geqslant 2$. This follows since t does not have a cycle, and the states $0, p, r, q_1, q_2$ that are mapped differently by s are in the orbit of $n - 1$.

Since s focuses the colliding pair $\{p, r\}$, s is different from the transformations of Case 1. Also s has neither a cycle nor a fixed point $\neq n - 1$ and so is different from the transformations of Cases 2, 3 and 4. In Case 5, transformation s^2 maps a colliding pair to $n - 1$, and here s^2 maps the unique colliding pair to $q_2 \neq n - 1$. Thus, s is different from the transformations of Case 5.

From (a) we have the unique colliding pair $\{p, r\}$ focused to q_1. Then $q_1 < q_1 s = q_2$ means that $p < r$, and so p is distinguished from r. Similarly, $q_1 > q_2$ means that $p > r$. Thus $0t = p$, $pt = n - 1$, $q_1 t = r$, $q_2 t = r$, and $rt = n - 1$, and t is again uniquely defined from s.

Case 7: t does not fit in any of the previous cases, but there are two states $q_1, q_2 \in Q \setminus \{0, n - 1\}$ that are not fixed points and satisfy $q_1 t \neq n - 1$ and $q_2 t \neq n - 1$. Since this is not Case 5 we may assume that $q_1 t^2 = n - 1$ and $q_2 t^2 = n - 1$. Let $r_1 = q_1 t$ and $r_2 = q_2 t$; clearly $p \neq r_1$ and $p \neq r_2$. The in-degree of both q_1 and q_2 is 0; otherwise t would fit in Case 5. We consider the following two sub-cases. If $p < r_1$ then **(i)** let s be the transformation defined by

$$0s = n-1, \quad ps = q_1, \quad r_1 s = q_1, \quad q_1 s = n - 1,$$
$$qs = qt \text{ for the other states } q \in Q.$$

If $p > r_1$ then **(ii)** let s be the transformation defined by

$$0s = n - 1, \quad ps = q_1, \quad r_1 s = q_1, \quad q_1 s = q_2,$$
$$qs = qt \text{ for the other states } q \in Q.$$

Case 8: t does not fit in any of the previous cases, but it has two fixed points r_1 and r_2 in $Q \setminus \{0, n-1\}$ with in-degree 1; assume that $r_1 < r_2$.

Let s be the transformation defined by

$$0s = n - 1, \quad ps = r_2, \quad r_1 s = r_2, \quad r_2 s = r_1,$$
$$qs = qt \text{ for the other states } q \in Q.$$

Case 9: t does not fit in any of the previous cases, but there is a state $q \in Q \setminus \{0, n-1\}$ that is not a fixed point and satisfies $qt \neq n-1$, $p < qt$, and there is a fixed point $f \neq n-1$.

Let $r = qt$; then $rt = n - 1$ because otherwise this would fit in Case 5. Here q is the only state from $Q \setminus \{0\}$ that is not a fixed point and is not mapped to $n - 1$, as otherwise t would fit in Case 7. Similarly, f is the only fixed point $\neq n - 1$, as otherwise t would fit in either Case 4 or Case 8.

Let s be the transformation defined by

$$0s = n - 1, \quad ps = r, \quad rs = q, \quad qs = p, \quad fs = r,$$
$$qs = qt \text{ for the other states } q \in Q.$$

Case 10: t does not fit in any of the previous cases, but there is a state $q \in Q \setminus \{0, n-1\}$ that is not a fixed point and satisfies $qt \neq n-1$, and a fixed point $f \in Q \setminus \{0, n-1\}$. Let $r = qt$; then $rt = n - 1$ since this is not Case 5. Now, in contrast to Case 9, we have $p > r$. Let s be the transformation defined by

$$0s = n - 1, \quad ps = q, \quad rs = q, \quad qs = n - 1,$$
$$qs = qt \text{ for the other states } q \in Q.$$

Case 11: t does not fit in any of the previous cases, but there is a state $q \in Q \setminus \{0, n-1\}$ that is not a fixed point and satisfies $qt \neq n-1$.

As shown in Case 9, q is the only state from $Q \setminus \{0\}$ that is not mapped to $n-1$, and also t has no fixed points other than $n-1$, as otherwise it would fit in one of the previous cases. Hence, all states from $Q \setminus \{0, q\}$ are mapped to $n-1$. Let $r = qt$. Here we use the assumption that $n \geqslant 7$. So in $Q \setminus \{0, p, q, r, n-1\}$ we have at least 2 states, say r_1 and r_2, that are mapped to $n-1$.

Sub-case (i): $p < r$. Let s be the transformation defined by

$$0s = n - 1, \quad ps = q, \quad rs = q, \quad qs = n - 1,$$
$$qs = qt \text{ for the other states } q \in Q.$$

Sub-case (ii): $p > r$. Let s be the transformation defined by

$$0s = n - 1, \quad ps = q, \quad rs = q, \quad qs = n - 1, \quad r_1 s = r_2, \quad r_2 s = r_1,$$
$$qs = qt \text{ for the other states } q \in Q.$$

Case 12: t does not fit in any of the previous cases.

Here t must contain exactly one fixed point $f \in Q \setminus \{n-1\}$, and every state from $Q \setminus \{0, f\}$ is mapped to $n-1$. If all states from $Q \setminus \{0\}$ would be mapped to $n-1$, then by Proposition 1, t would be in S_n and so would fit in Case 1.

Because $n \geqslant 7$, in $Q \setminus \{0, p, f, n-1\}$ we have at least 2 states, say r_1 and r_2, that are mapped to $n-1$. Let s be the transformation defined by

$$0s = n-1, \quad ps = f, \quad r_1 s = r_2, \quad r_2 s = r_1,$$
$$qs = qt \text{ for the other states } q \in Q.$$

$\square$

4   Uniqueness of Maximal Witness

Our third contribution is a proof that the transition semigroup of a DFA $\mathcal{D}_n = (Q, \Sigma, \delta, 0, F)$ of a suffix-free language with syntactic complexity $(n-1)^{n-2}+n-2$ is unique.

Lemma 2. *If $n \geqslant 4$ and $\mathcal{D}_n$ has no colliding pairs, then $|T_n| \leqslant (n-1)^{n-2}+n-2$ and T_n is a subsemigroup of S_n.*

Lemma 3. *If $n \geqslant 7$ and $\mathcal{D}_n$ has at least one colliding pair, then $|T_n| < (n-1)^{n-2} + n - 2$.*

Proof. Let φ be the injective function from the proof of Theorem 1 and assume that there is a colliding pair $\{p, r\}$. Let r_1, r_2 and r_3 be three distinct states from $Q \setminus \{0, p, r, n-1\}$; there are at least 3 such states since $n \geqslant 7$. Let s be the following transformation:

$$0s = n-1, \quad ps = r, \quad rs = r, \quad r_1 s = r_2, \quad r_2 s = r_3, \quad r_3 s = r_1,$$
$$qs = qt \text{ for the other states } q \in Q.$$

We can show that s is not defined in any case in the proof of Theorem 1. Note that s focuses the colliding pair $\{p, r\}$, and so it cannot be present in T_n; hence it is not defined in Case 1. We can follow the proof of injectivity of the transformations in Case 12 of Theorem 1, and show that s is different from all the transformations of Cases 2–11. For a distinction from the transformations of Case 12, observe that they each have a 2-cycle, and here s has a 3-cycle.

Thus $s \notin \varphi(T_n)$, but $s \in S_n$, and so $\varphi(T_n) \subsetneq S_n$. Since φ is injective, it follows that $|T_n| < |S_n| = (n-1)^{n-2} + n - 2$. $\square$

Corollary 1. *For $n \geqslant 7$, the maximal transition semigroups of DFAs of suffix-free languages are unique.*

Finally, we show that Σ cannot have fewer than five letters.

Theorem 2. *If $n \geqslant 7$, $\mathcal{D}_n = (Q, \Sigma, \delta, 0, F)$ is a minimal DFA of a suffix-free language, and $|\Sigma| < 5$, then $|T_n| < (n-1)^{n-2} + n - 2$.*

Proof. DFA $\mathcal{D}_n$ has the initial state 0, and an empty state, say $n-1$. Let M be the set of the remaining $n-2$ "middle" states. From Lemma 1 no transformation can map any state in Q to 0, and every transformation fixes $n-1$.

Suppose the upper bound $(n-1)^{n-2}+n-2$ is reached by T_n. From Proposition 1 and Corollary 1 all transformations of M must be possible, and it is well known that three generators are necessary to achieve this. Let the letters a, b, and c correspond to these three generators, t_a, t_b and t_c. If $0t_a \neq n-1$, then t_a must be a transformation of type (b) from Proposition 1, and so $qt_a = n-1$ for any $q \in M$. So t_a cannot be a generator of a transformation of M. Hence we must have $0t_a = n-1$, and also $0t_b = 0t_c = n-1$.

So far, the states in M are not reachable from 0; hence there must be a letter, say e, such that $0t_e = p$ is in M. This must be a transformation of type (b) from Proposition 1, and all the states of M must be mapped to $n-1$ by t_e.

Finally, to reach the upper bound we must be able to map any proper subset of M to $n-1$. The letter e will not do, since it maps *all* states of M to $n-1$. Hence we require a fifth letter, say d. $\qquad\square$

5   Conclusions

We have shown that the upper bound on the syntactic complexity of suffix-free languages is $(n-1)^{n-2}+n-2$. Since it was known that this is also a lower bound, our result settles the problem. Moreover, we have proved that an alphabet of at least five letters is necessary to reach the upper bound, and that the maximal transition semigroups are unique.

In our proof we exhibited an injective function from the transition semigroup of a minimal DFA of an arbitrary suffix-free semigroup to the transition semigroup of the witness DFA attaining the upper bound for suffix-free languages. This approach is generally applicable for other subclasses of regular languages. For example, in [6] we have used this method to establish the upper bound for left and two-sided ideals.

References

1. Ang, T., Brzozowski, J.: Languages convex with respect to binary relations, and their closure properties. Acta Cybernet. **19**(2), 445–464 (2009)
2. Berstel, J., Perrin, D., Reutenauer, C.: Codes and Automata. Cambridge University Press, UK (2009)
3. Brzozowski, J.: Quotient complexity of regular languages. J. Autom. Lang. Comb. **15**(1/2), 71–89 (2010)
4. Brzozowski, J., Li, B., Ye, Y.: Syntactic complexity of prefix-, suffix-, bifix-, and factor-free regular langauges. Theoret. Comput. Sci. **449**, 37–53 (2012)
5. Brzozowski, J., Szykuła, M.: Upper bound for syntactic complexity of suffix-free languages (2014). http://arxiv.org/abs/1412.2281
6. Brzozowski, J., Szykuła, M.: Upper bounds on syntactic complexity of left and two-sided ideals. In: Shur, A.M., Volkov, M.V. (eds.) DLT 2014. LNCS, vol. 8633, pp. 13–24. Springer, Heidelberg (2014)

7. Brzozowski, J., Ye, Y.: Syntactic complexity of ideal and closed languages. In: Mauri, G., Leporati, A. (eds.) DLT 2011. LNCS, vol. 6795, pp. 117–128. Springer, Heidelberg (2011)
8. Ganyushkin, O., Mazorchuk, V.: Classical Finite Transformation Semigroups: An Introduction. Springer, Heidelberg (2009)
9. Han, Y.S., Salomaa, K.: State complexity of basic operations on suffix-free regular languages. Theoret. Comput. Sci. **410**(27–29), 2537–2548 (2009)
10. Pin, J.E.: Syntactic semigroups. In: Rozenberg, G., Salomaa, A. (eds.) Handbook of Formal Languages. Word, Language, Grammar, vol. 1, pp. 679–746. Springer, New York (1997)
11. Thierrin, G.: Convex languages. In: Nivat, M. (ed.) Automata, Languages and Programming, pp. 481–492. North-Holland, Amsterdam (1973)
12. Yu, S.: State complexity of regular languages. J. Autom. Lang. Comb. **6**, 221–234 (2001)

Nondeterministic Tree Width
of Regular Languages

Cezar Câmpeanu[1][✉] and Kai Salomaa[2]

[1] Department of Computer Science and Information Technology, University of Prince Edward Island, Charlottetown, PE C1A 4P3, Canada
ccampeanu@upei.ca
[2] School of Computing, Queen's University, Kingston, ON K7L 2N8, Canada
ksalomaa@cs.queensu.ca

Abstract. The tree width of a nondeterministic finite automaton (NFA) counts the maximum number of computations the automaton may have on a given input. Here we consider the tree width of a regular language, which, roughly speaking, measures the amount of nondeterminism that a state-minimal NFA for the language needs. We prove that an infinite tree width is obtained from finite tree width, for most operations on regular languages.

Keywords: Regular languages · Nondeterministic finite automata · Measures of nondeterminism

1 Introduction

Various ways to quantify the amount of nondeterminism of a finite automaton have been considered in the literature. The degree of ambiguity of a nondeterministic finite automaton (NFA) counts the number of accepting computations on a given input [18,24], and the tree width of an NFA counts the number of all computations (accepting or non-accepting) [22]. The tree width measure of an NFA A is also known in the literature as 'leaf size' [14], or 'computations(A)' [2]. Additional nondeterminism measures can be based on the amount of nondeterminism on a single best (or worst) computation on a given input [10,23]. The tree width measure of an NFA can also be related to the ambiguity of regular expressions and languages [3].

Here we focus on the tree width measure. Instead of measuring the amount of nondeterminism in individual NFAs, we want to associate a nondeterministic (tree) width measure with a regular language. The nondeterministic width of a language L is defined as the least tree width of any state-minimal NFA recognizing L.

If a DFA recognizing L is minimal also as an NFA, then the nondeterministic width of L is one. We give examples of languages that have infinite width. Next, we consider the associated decision problems. Since minimization of even

© Springer International Publishing Switzerland 2015
J. Shallit and A. Okhotin (Eds.): DCFS 2015, LNCS 9118, pp. 46–57, 2015.
DOI: 10.1007/978-3-319-19225-3_4

constant tree width NFAs is known to be intractable [2,19], it can be expected
that the same holds for deciding the tree width of a regular language.

We show that deciding the nondeterministic width of a language recognized
by an NFA is PSPACE-complete, and the problem is in co-NP, if the input NFA
has constant tree width. For unary NFAs, we show the problem to be NP-hard.

Finally, we consider the nondeterministic tree width of operations on regular languages. The operational state complexity is a much studied topic in the
literature [9,27]. If $\circ$ is a regularity-preserving language operation, and L_1 and
L_2 are regular languages, we consider the question what can be said about the
nondeterministic tree width of $L_1 \circ L_2$ as a function of the nondeterministic
tree width of L_1 and L_2, respectively. The results here are almost exclusively
negative. For most operations $\circ$, the nondeterministic width of $L_1 \circ L_2$ is not
bounded by any function of the widths of the individual languages.

2   Tree Width of a Regular Language

The tree width (or leaf size) measure for nondeterministic finite automata (NFA)
was considered in [14,22]. Roughly speaking, the tree width counts the number
of branches in the computation tree of the NFA on a given input.

Let $A = (Q, \Sigma, \delta, q_0, F)$ be an NFA, and $w \in \Sigma^*$. A partial computation of A
on w is either a computation that consumes the entire string w or a computation
that encounters an undefined transition after reading some prefix of w. The *tree
width of A on w*, $\mathrm{tw}_A(w)$, is the number of partial computations of A on w. Note
that the number of partial computations of A on w equals the number of leaves
of the computation tree of A on w, as defined in [22].

The tree width of the NFA A is defined as

$$\mathrm{tw}(A) = \sup\{\mathrm{tw}_A(w) \mid w \in \Sigma^*\}.$$

We say that A has *finite tree width*, if $\mathrm{tw}(A)$ is finite.

A transition of an NFA A from state q to state p on input b is said to
be *nondeterministic* if $|\delta(q, b)| > 1$; otherwise the transition is deterministic. An
undefined transition is deterministic. When we say that part of a particular computation is deterministic, this means that it uses only deterministic transitions.
The following characterization of finite tree width NFAs is known.

Proposition 2.1 ([22]). *An NFA A has finite tree width if and only if no cycle
of A contains a nondeterministic transition.*

We will later use the following technical lemma. The lemma is based on the
simple idea that the computations of an NFA, on repeated copies of the same
string, must be "essentially unary".

Lemma 2.1. *Let A be a finite tree width NFA with n states, input alphabet Σ,
and let $w \in \Sigma^+$.*

*Consider a (nondeterministic) computation C of A on w^k, where $k \geq n$.
Then, after reading the prefix w^n, C must be entirely deterministic.*

Corollary 2.1. *Let A be a finite tree width NFA with n states, input alphabet Σ, and let $u, w \in \Sigma^+$. Consider a (nondeterministic) computation C of A on uw^k, where $k \geq n$. Then, after reading the prefix uw^n, the computation C must be entirely deterministic.*

We say that an NFA A has *optimal tree width* if $L(A)$ is not recognized by any NFA B, where size$(B) \leq$ size(A), tw$(B) \leq$ tw(A), and at least one of the inequalities is strict.

The *nondeterministic state complexity* of a regular language L, nsc(L), is the size of a minimal NFA recognizing L. Analogously, we define the tree width k nondeterministic state complexity of L, as

$$\text{nsc}_{\text{tw} \leq k}(L) = \inf\{\text{size}(A) \mid A \text{ is an NFA recognizing } L, \text{ and tw}(A) \leq k\}. \quad (1)$$

In relation (1), nsc$_{\text{tw} \leq k}(L)$ is the smallest number of states needed by an NFA of tree width at most k to recognize L. Extending the above notation, we could write nsc(L) as nsc$_{\text{tw} \leq \infty}(L)$.

Definition 2.1. *Let L be a regular language. The tree width of the language L is the minimum tree width of minimal NFAs for L, and it is given by the formula:*

$$\text{tw}(L) = \inf\{\text{tw}(A) \mid L(A) = L, A \text{ is a minimal NFA}\}. \quad (2)$$

Thus, if tw$(L) = k$, then nsc$_{\text{tw} \leq k}(L) = $ nsc(L) and, for all $\ell < k$, nsc$_{\text{tw} \leq \ell}(L) >$ nsc(L). If tw(L) is infinite, all the state minimal NFAs for L have unbounded tree width. Note that tw$(L) = \inf\{k \mid \text{nsc}_{\text{tw} \leq k}(L) = \text{nsc}(L)\}$.

The language L having tree width k means intuitively that a state minimal NFA recognizing L needs to use limited nondeterminism of tree width k, but any additional nondeterminism would not allow any further reduction of the number of states. In particular, if L has tree width 1, this means that the minimal incomplete DFA for L is also minimal as an NFA.

The following lemmas determine the tree width of some languages. These results will be used in later sections.

Lemma 2.2. *For $w \in \Sigma^*$, tw$(\Sigma^* w) = 1$.*

Lemma 2.3. *Let $\Sigma = \{a, b\}$, $k \geq 2$, and define $L_k = \Sigma^* \cdot b \cdot \Sigma^{k-1}$. Then the tree width of L_k is infinite.*

Proof. The language L_k is recognized by an NFA A that guesses the k-th symbol from the end, verifies that it is a b, and it is followed by exactly $k - 1$ symbols. The NFA A has $k+1$ states (and infinite tree width), and it is presented in Fig. 1.

To prove the lemma, it is sufficient to show that L_k cannot be recognized by any finite tree width NFA of size less than $k + 2$.

It is easy to see that an arbitrary NFA B recognizing L_k must reach the end of strings with a suffix, respectively a^k, $a^{k-1}b$, $a^{k-2}ba$, ..., and ba^{k-1} in distinct states, which means that B needs at least $k + 1$ states. However, since

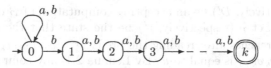

Fig. 1. The automaton A for the language L_k

B is nondeterministic (and the finite degree of nondeterminism can even be larger than the number of states), there is the possibility that after reading, for example, a suffix $a^{k-2}b^2$, the state may depend on the continuation of the input – thus, it does not follow directly that B needs further states. In order to establish that strictly more than $k+1$ states are needed, we consider computations that are forced to be deterministic by repeating the same substring sufficiently many times.

For the sake of contradiction, assume that L_k is recognized by an NFA B, with $k+1$ states, and finite tree width.

We define the following $k+2$ strings of length k:

$$v_0 = a^k, \; v_1 = a^{k-1}b, \; v_2 = a^{k-2}ba, \ldots, v_k = ba^{k-1}, v_{k+1} = a^{k-2}bb.$$

Write $w = v_0 \cdot v_1 \cdots \cdot v_{k+1}$, and choose strings $z_0, z_1, \ldots, z_k$ by setting

$$z_0 = w^{k+1}v_0 \cdot ba^{k-1}, \; z_1 = w^{k+1}v_0v_1 \cdot a^{k-1}, \; z_2 = w^{k+1}v_0v_1v_2a^{k-2},$$
$$\ldots, z_k = w^{k+1}v_0v_1 \cdots v_k \cdot \varepsilon.$$

The prefix w^{k+1} is included, because we want to use Lemma 2.1 to argue that any computation after reading a prefix $w^{k+1}v_0v_1 \cdots v_i$ must be deterministic.

The strings z_i, $0 \le i \le k$, are in L_k. Let C_i be an arbitrary but fixed accepting computation of B on z_i, $0 \le i \le k$, and let q_i be the state that the computation C_i reaches after the prefix $w^{k+1}v_0 \cdots v_i$, $0 \le i \le k$.

Claim 1. The states $q_0, q_1, \ldots, q_k$ must be all distinct.

Proof of the claim. If $q_0 = q_j$, for $1 \le j \le k$, this gives an accepting computation for string $w^{k+1}v_0a^{k-j}$, which is not in the language L_k. For $1 \le i, j \le k$, we have

$$w^{k+1}v_0 \cdots v_i \cdot a^{k-j} \in L_k \text{ iff } i = j, \tag{3}$$

and this implies that also the states $q_1, \ldots, q_k$ must be all distinct. This concludes the proof of the claim.

Since B has $k+1$ states, the state set must be exactly $\{q_0, q_1, \ldots, q_k\}$. Now, from the proof of Claim 1, it follows that any computation of B that processes a prefix $w^{k+1}v_0v_1 \cdots v_i$ of the input, must be in exactly the state q_i, $0 \le i \le k$. (Since these are all the states, as in the proof of the claim, we see that otherwise B accepts illegal strings.)

Next, consider the following two strings of L_k:

$$z_{k+1} = w^{k+1}v_0v_1 \cdots v_kv_{k+1} \cdot a^{k-1} \text{ and } z'_{k+1} = w^{k+1}v_0v_1 \cdots v_kv_{k+1} \cdot a^{k-2},$$

and let D (respectively, D') be an accepting computation of B on z_{k+1} (respectively, on z'_{k+1}). Let p (respectively, p') be the state that the computation D (respectively, p') reaches after reading the prefix $w^{k+1}v_0v_1 \cdots v_kv_{k+1}$.

Now $v_0v_1 \cdots v_kv_{k+1}$ is equal to w. By Lemma 2.1, any computation of B on an input in w^* after the prefix w^{k+1} is deterministic (B has entered a cycle and must continue the same cycle on the "next" w).

Furthermore, above we have observed that any state B can reach after reading the prefix $w^{k+1}v_0v_1 \cdots v_k$ must be q_k. Since p and p' are then both obtained in a deterministic computation from q_k by reading v_{k+1}, it must be the case $p = p'$ (and $p \in \{q_0, q_1, \ldots, q_k\}$).

If $p = q_0$, then B accepts $w^{k+1} \cdot v_0 \cdot a^{k-1}$, which is not in L_k. Finally, assuming that $p = q_r$, $1 \leq r \leq k$, we get that $w^{k+1}v_0v_1 \cdots v_r \cdot a^s \in L_k$ both for $s = k - 1$ and $s = k - 2$, which contradicts (3).   ∎

Corollary 2.2. *Let Σ be a finite alphabet with $\#\Sigma = k$, $k \geq 2$, and define the language $L_{\overline{a,k}} = \Sigma^*(\Sigma - \{a\})\Sigma^{k-1}$. Then the tree width of $L_{\overline{a,k}}$ is infinite.*

It is known that the minimal DFA for L_k has 2^k states, and the size blow-up of determinizing a finite tree width NFA is polynomial [22]. At first sight, these observations could seem to imply that a finite tree width NFA for L_k needs more than $k + 1$ states. However, this argument does not work, because the polynomial giving the size blow-up of determinization depends on the tree width of the given NFA, and for this reason, in the proof of Lemma 2.3, we used an ad-hoc combinatorial argument.

We conjecture that any finite tree width NFA for the language L_k, in fact, needs 2^k states. Lower bound proofs for the sizes of (finite tree width) NFAs typically need to use ad hoc arguments [22], and in order to keep the proof simple, we proved only what is need to establish that L_k has infinite tree width.

2.1   Unary Languages

Recall that a unary DFA consists always of a tail (sequence of states), followed by a cycle of states. The cycle may be empty.

A unary NFA in *Chrobak normal form* [6,7,15] similarly consists of a tail followed by a nondeterministic transition to one or more cycles. Every unary regular language can be recognized by a Chrobak normal form NFA; however, it may not be a state-minimal NFA for that language. When considering finite tree width NFAs, a state-minimal NFA can always be found in Chrobak normal form. Note that the tree width of a Chrobak normal form NFA is simply the number of cycles it has, and this result will be used in the proof of Theorem 3.2.

Proposition 2.2 ([21]). *Let A be a unary n-state NFA with tree width $k \in \mathbb{N}$. Then the language $L(A)$ can be recognized by a Chrobak normal form NFA, with at most n states, and tree width k.*

3 Deciding the Tree Width of a Regular Language

Immediately, by inspecting the transition graph of an NFA A, we can determine whether the tree width of A is finite, and the tree width of a finite tree width NFA can be computed in polynomial time [22, 23]. By modifying the known PSPACE-hardness result of the DFA intersection-emptiness problem [17], we see that finding the tree width of the language recognized by an NFA is PSPACE-complete. As has been done before, we convert the intersection-emptiness problem to the union-universe problem [16].

Theorem 3.1. *Given a finite tree width NFA A, and $\ell \in \mathbb{N}$, the problem of determining whether the tree width of $L(A)$ equals ℓ is PSPACE-complete (with respect to log-space reductions). In particular the question of deciding whether the tree width of $L(A)$ equals one is PSPACE-complete.*

Proof. *(Sketch.)* Given A and ℓ, we can nondeterministically (NPSPACE = PSPACE) choose A' such that $\text{size}(A') \leq \text{size}(A)$, and $\text{tw}(A') < \ell'$, then in PSPACE check that $L(A) = L(A')$.

For the hardness claim, it is sufficient to show that deciding $\text{tw}(L(A)) = 1$ is PSPACE-hard. To this end, we can use, with minor modifications, the construction by Kozen [17] that establishes the PSPACE-hardness of the intersection emptiness problem for DFAs.

Given a deterministic Turing machine M with polynomial space bound $p(n)$, and input $x \in \Sigma^*$, $|x| = n$, where the accepting computations of M on x are assumed to have even length, the proof of Lemma 3.2.3 of [17] constructs $2p(n) - 3$ DFAs F_i (of polynomial size) over an alphabet Ω such that, if M accepts x, the intersection of the languages $L(F_i)$ consists of the encoding of the unique accepting computation of M on x; otherwise the intersection of the languages $L(F_i)$ is empty. Here, Ω is the alphabet used to encode computations of M (i.e., sequences of configurations separated by a marker $\#$).

Simply by interchanging the accepting and non-accepting states of the DFAs F_i, and adding a new initial state q_0, we then construct an NFA D recognizing the union of the complements of the languages $L(F_i)$. If F_i' is the DFA accepting the complement of $L(F_i)$, the NFA D has, for each $b \in \Omega$, a transition from q_0 into F_i' that simulates the transition of F_i' on input b from the initial state. Thus, the tree width of D is $2p(n) - 3$.

Now, $L(D) = \Omega^*$ if M does not accept x, and if M accepts x,

$$L(D) = \Omega^* - \{w_{\text{acc-comp}}\},$$

where $w_{\text{acc-comp}}$ is the encoding of the accepting computation of M on x. In the case where the accepting computation exists, it is easy to see that the minimal DFA for $\Omega^* - \{w_{\text{acc-comp}}\}$ must be larger than the NFA D. The DFA needs to check for mismatches at all positions in each pair of consecutive configurations in $\{w_{\text{acc-comp}}\}$, which means that the number of states of D needs to be roughly $C_\#(w)^2$, where $C_\#(w)$ is the number of configurations occurring in the computation of M on x. The NFA D has separate components that check for mismatches at each individual position (one component verifies a position in odd-numbered computation steps, and another a position in even-numbered computation steps) and the size of D is in $O(p(n)^2)$, where $p(n)$ is the length of the longest configuration in the computation [17]. Note that, without loss of generality, we can modify M so that after the original computation accepts, the modified machine enters a new phase that successively writes on the tape all possible strings over Ω, with length at most $p(n)$. This guarantees that $C_\#(w)$ is at least $|\Omega|^{p(n)}$.

It follows that $L(D)$ has tree width one if and only if M does not accept x. ∎

Note that in the proof of Theorem 3.1, while the constructed NFA D has finite tree width, the value of the tree width is not bounded, that is, the tree width is part of the input instance. Naturally, one may ask the same question for NFAs with fixed tree width k.

Lemma 3.1. Let $k \in \mathbb{N}$ be a constant. The problem of deciding for a given NFA A with tree width k, and for a given $m \le k$ whether $\mathrm{tw}(L(A)) \ge m$ is in coNP.

Proof. In nondeterministic polynomial time, we can guess an NFA A', such that $\mathrm{tw}(A') \le m$, and $\mathrm{size}(A') \le \mathrm{size}(A)$, and one of the inequalities is strict.

From the existence of deterministic decompositions [22, 23], it follows that we can verify in polynomial time that the tree width of A' is at most m, i.e., the algorithm can verify the correctness of the guess. Determinizing an NFA with at most constant k tree width causes only a polynomial size blow-up [22] (where the polynomial depends on k). Now, in deterministic polynomial time, the algorithm can determinize A and A', and test the equality of the resulting DFAs [27]. ∎

It is known that finding the minimal NFA equivalent to a given DFA is PSPACE-hard [16], and even the minimization of constant tree width NFAs is NP-hard [2, 19]. Consequently, it seems likely that determining the tree width of a language specified by a fixed tree width NFA is intractable. However, the hardness of minimization does not directly imply the hardness of computing the tree width of the corresponding regular language, and proving the intractability of deciding the tree width of a language given by a constant tree width NFA (or by a DFA) remains open.

To conclude this section, we consider the decision problem of determining the tree width of a language recognized by a unary NFA.

Theorem 3.2. Given a unary NFA A, it is NP-hard to decide whether or not $\mathrm{tw}(L(A)) \ne 1$.

Proof. The result follows from the construction used in the proof of Theorem 1 of [12]; also see [26]. Given a 3SAT formula F, the proof constructs, in polynomial time, a regular language $L_F \subseteq \{a\}^*$ such that if F is unsatisfiable, then $L_F = \{a\}^*$, and if F is satisfiable, then a minimal NFA for L_F is not a DFA, i.e., must have tree width greater than one. ∎

4   Tree Width of Operations

In operational state complexity we want to determine, for a regularity-preserving language operation $\circ$, the worst-case size of the minimal DFA/NFA for the language $L_1 \circ L_2$ as a function of the sizes of given automata for L_1 and L_2 [27]. The nondeterministic state complexity of unrestricted NFAs was studied by Holzer and Kutrib [13] and independently by Ellul [8]; the study of operational state complexity finite tree width NFAs has been initiated by Palioudakis et al. [20].

Here, instead of state complexity, we consider the effect of language operations on the tree width of a language. The general question is to estimate, for a regularity preserving operation $\circ$, the value of $\mathrm{tw}(L_1 \circ L_2)$ in terms of $\mathrm{tw}(L_1)$ and $\mathrm{tw}(L_2)$.

4.1   Union

For union, we have a strong negative result: the value $\mathrm{tw}(R_1 \cup R_2)$ cannot be bounded by any function on $\mathrm{tw}(R_1)$ and $\mathrm{tw}(R_2)$.

Theorem 4.1. *There exist regular languages R_i with tree width $m_i \in \mathbb{N}$, $i = 1, 2$, such that the tree width of $R_1 \cup R_2$ is infinite.*

Proof. Let $\Sigma = \{a, b\}$, $k \geq 2$, and for $w \in \Sigma^{k-1}$, define

$$L_w = \Sigma^* \cdot b \cdot w.$$

By Lemma 2.2, $\mathrm{tw}(L_w) = 1$ for all $w \in \Sigma^{k-1}$. On the other hand, if we denote

$$L_k = \bigcup_{w \in \Sigma^{k-1}} L_w = \Sigma^* \cdot b \cdot \Sigma^{k-1},$$

we know, by Lemma 2.3, that $\mathrm{tw}(L_k)$ is infinite.

∎

As a corollary of the proof, the negative result can be stated in a slightly stronger form.

Corollary 4.1. *There exists a regular language R_1 with finite tree width, and a regular language R_2 with tree width one such that the tree width of $R_1 \cup R_2$ is infinite.*

Note that the proof of Theorem 4.1 establishes the existence of the languages R_1 and R_2, but does not explicitly tell how the languages R_1 and R_2 are chosen as finite unions of the languages L_w, $w \in \Sigma^*$. With small values like $k = 2$, it should be possible to find an explicit definition of R_1 and R_2, but the verification would still require a fair amount of computation, because there is no efficient algorithm to compute the tree width of a regular language. In the example given by Fig. 2, we give concrete languages of tree width one such that their union has infinite tree width.

Example 4.1. Choose $w_1 = aaa$, $w_2 = aba$. By Lemma 2.2, the languages $L_i = L((a + b)^*bw_i)$, $i = 1, 2$, have tree width 1. We note that

$$L_1 \cup L_2 = L((a + b)^*ba(a + b)a).$$

In Fig. 2 is a minimal NFA for the language $L_1 \cup L_2$. It seems clear that the minimal NFA for $L_1 \cup L_2$ is unique, although to formally verify this would require some effort. This then implies that the tree width of $L_1 \cup L_2$ is infinite.

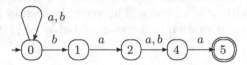

Fig. 2. A minimal NFA for the union of $L((a + b)^*baba)$ and $L((a + b)^*baaa)$

To conclude the subsection, we consider the union of unary languages. Since by Proposition 2.2, we know that a minimal finite tree width unary NFA can always be found in Chrobak normal form, it might be tempting to think that, for unary languages L_1 and L_2, the tree width of $L_1 \cup L_2$ is bounded by $tw(L_1) + tw(L_2)$.

Indeed, if A_i is a minimal Chrobak normal form NFA for L_i, having tree width k_i, $i = 1, 2$, then based on L_1 and L_2, we can easily construct a Chrobak normal form NFA of tree width $k_1 + k_2$, where possibly some of the cycles can be further combined. However, the construction does not yield an upper bound for the tree width of the union of L_1 and L_2, because it is possible that the minimal (Chrobak normal form) NFA for $L_1 \cup L_2$ needs to be differently constructed. Below we give such an example:

Example 4.2. Let $L_1 = (a^{30})^*(a^2 + a^3 + a^4 + a^5)$ and

$$L_2 = (a^{30})^*(\varepsilon + a^6 + a^8 + a^9 + a^{10} + a^{12} + a^{14} + a^{15} + a^{16} + a^{18}$$
$$+ a^{20} + a^{21} + a^{22} + a^{24} + a^{26} + a^{27} + a^{28}).$$

The minimal NFA both for L_1 and L_2 consists of a single deterministic cycle of length 30, hence $tw(L_i) = 1$, $i = 1, 2$. On the other hand,

$$L_1 \cup L_2 = L((a^2)^* + (a^3)^* + (a^5)^*),$$

and it is clear that the tree width of $L_1 \cup L_2$ is 3.

Generalizing the construction of Example 4.2, for any $k \in \mathbb{N}$, we can construct unary regular languages L_1 and L_2 having tree width one, such that $tw(L_1 \cup L_2) = k$. Let $p_1, p_2, \ldots, p_k$ be the first k primes. The language

$$P_k = L((a^{p_1})^* + (a^{p_2})^* + \cdots + (a^{p_k})^*)$$

has tree width k. Now, it is easy to write the language P_k as a union of languages L_1 and L_2, such that the minimal NFA for L_i consists of a single cycle of length $\prod_{i=1}^{k} p_i$; hence $tw(L_i) = 1$, $i = 1, 2$.

4.2   Concatenation and Reversal

The tree width of the concatenation $L_1 \cdot L_2$ is not bounded by any function on the tree widths of L_1 and L_2. In fact, already the concatenation of tree width one languages may have infinite tree width.

Theorem 4.2. *Let $\Sigma = \{a, b\}$. There exist languages $L_1, L_2 \subseteq \Sigma^*$, $\mathrm{tw}(L_i) = 1$, $i = 1, 2$, such that $\mathrm{tw}(L_1 \cdot L_2)$ is infinite.*

Proof. Let $k \geq 2$ and choose $L_1 = \Sigma^*$ and $L_2 = b\Sigma^{k-1}$. The minimal DFA for L_1 and L_2 is minimal also as an NFA, hence $\mathrm{tw}(L_1) = \mathrm{tw}(L_2) = 1$. By Lemma 2.3, the tree width of $L_1 \cdot L_2$ is infinite.   ∎

Again, using Lemma 2.3, we see that the reversal of a tree width one regular language may have infinite tree width.

Theorem 4.3. *There exists a regular language L with $\mathrm{tw}(L) = 1$, such that the tree width of L^R is infinite.*

Proof. Let $\Sigma = \{a, b\}$ and $k \geq 2$. Choose $L = (a+b)^{k-1}b(a+b)^*$. The minimal DFA for L is also a minimal NFA, hence the tree width of L is one. By Lemma 2.3, $\mathrm{tw}(L^R)$ is infinite.   ∎

4.3   Complementation and Intersection

The complement of a tree width one language L may have infinite tree width and, furthermore, we can choose L to be unary.

Lemma 4.1. *Let $\Sigma = \{a\}$, and $L = \{\varepsilon, a, a^2, a^4\}$. Then $\mathrm{tw}(L) = 1$, and $\mathrm{tw}(\overline{L})$ is infinite.*

Proof. The minimal NFA for a finite unary language always consists of a single chain of states. On the other hand,

$$\overline{L} = (a^2)^* \cdot (a^3)^+$$

and $\overline{L}$ has an NFA with infinite tree width consisting of a cycle of length 2, followed by a cycle of length 3, and a total of 5 states. On the other hand, any finite tree width NFA for $\overline{L}$ needs 6 states [21]. This means that the tree width of $\overline{L}$ is infinite.   ∎

Finally we consider the intersection operation. We consider the language $L_1 = L((a^*)) \setminus \{\varepsilon, a, a^2\}$ and $L_2 = L((a^*)) \setminus \{\varepsilon, a^2, a^4\}$. Now, $L_1 \cap L_2 = L((a^*)) \setminus \{\varepsilon, a, a^2, a^4\}$, and according to Lemma 4.1, $\mathrm{tw}(L_1 \cap L_2)$ is infinite. On the other hand, both L_1 and L_2 have a minimal NFA that is a DFA (Fig. 3).

Therefore, we just proved that:

Theorem 4.4. *There exist two regular languages L_1, L_2 with $\mathrm{tw}(L_i) = 1$, $i = 1, 2$, such that the tree width of $L_1 \cap L_2$ is infinite.*

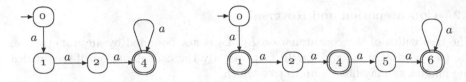

Fig. 3. A minimal NFA for L_1, left, and L_2, right.

5   Conclusion

In this paper we have analyzed the tree width of regular languages, and proved that computing it is computationally hard, even in the case of unary languages. We proved that for common operations, like boolean and regular ones, the tree width of the resulting languages is not bounded by any function which depends on the tree width of the input languages.

There are still many related areas to be considered, and we list here few of them:

1. Given the negative results in Sect. 4, can we find an "interesting" operation $\circ$ and a bounded function f, such that $tw(L_1 \circ L_2) \leq f(tw(L_1), tw(L_2))$?
2. There are other measures of nondeterminism considered in [22,23], such as branching, trace, or the degree of ambiguity. Find the connections between these nondeterminism measures for a given regular language R. Because we have negative results concerning the tree width of language operations, is the same valid for the other non-determinism measures?
3. Relate the ambiguity of regular expressions with the measures of nondeterminism on NFAs.

References

1. Birget, J.C.: Intersection and union of regular languages and state complexity. Inf. Process. Lett. **43**, 185–190 (1992)
2. Björklund, H., Martens, W.: The tractability frontier for NFA minimization. J. Comput. Syst. Sci. **78**, 198–210 (2012)
3. Brüggemann-Klein, A., Wood, D.: One-unambiguous regular languages. Inf. Comput. **142–2**, 182–206 (1998)
4. Câmpeanu, C.: Simplyfying nondeterministic finite cover automata. Electron. Proc. Theoret. Comput. Sci. **151–AFL**, 162–173 (2014)
5. Câmpeanu, C.: Non-deterministic finite cover automata. Sci. Ann. Comput. Sci. **29**, 3–28 (2015)
6. Chrobak, M.: Finite automata and unary languages. Theoret. Comput. Sci. **47**, 149–158 (1986)
7. Eilenberg, S., Schützenberger, M.P.: Rational sets in commutative monoids. J. Algebra **13**(2), 173–191 (1969)
8. Ellul, K.: Descriptional Complexity Measures of Regular Languages. Master's thesis, University of Waterloo (2004)

9. Gao, Y., Moreira, N., Reis, R., Yu, S.: A review on state complexity of individual operations. Faculdade de Ciencias, Universidade do Porto, Technical report DCC-2011-8. www.dcc.fc.up.pt/dcc/Pubs/TReports/TR11/dcc-2011-08.pdf To appear in Computer Science Review
10. Goldstine, J., Kintala, C.M.R., Wotschke, D.: On measuring nondeterminism in regular languages. Inf. Comput. **86**, 179–194 (1990)
11. Gruber, H., Holzer, M.: Finding lower bounds for nondeterministic state complexity is hard. In: Ibarra, O.H., Dang, Z. (eds.) DLT 2006. LNCS, vol. 4036, pp. 363–374. Springer, Heidelberg (2006). http://dx.doi.org/10.1007/11779148_33
12. Gruber, H., Holzer, M.: Computational complexity of NFA minimization for finite and unary languages. In: Proceedings of LATA, pp. 261–272 (2007)
13. Holzer, M., Kutrib, M.: Nondeterministic descriptional complexity of regular languages. Int. J. Found. Comput. Sci. **14**, 1087–1102 (2003)
14. Hromkovič, J., Seibert, S., Karhumäki, J., Klauck, H., Schnitger, G.: Communication complexity method for measuring nondeterminism in finite automata. Inf. Comput. **172**, 202–217 (2002)
15. Jiang, T., McDowell, E., Ravikumar, B.: The structure and complexity of minimal NFAs over a unary alphabet. Int. J. Found. Comput. Sci. **2**, 163–182 (1991)
16. Jiang, T., Ravikumar, B.: Minimal NFA problems are hard. SIAM J. Comput. **22**, 1117–1141 (1993)
17. Kozen, D.: Lower bounds for natural proof systems. In: Proceedings of the 18th Annual Symposium on Foundations of Computer Science, FOCS, pp. 254–266 (1977)
18. Leung, H.: Descriptional complexity of NFA of different ambiguity. Int. J. Found. Comput. Sci. **16**, 975–984 (2005)
19. Malcher, A.: Minimizing finite automata is computationally hard. Theoret. Comput. Sci. **327**, 375–390 (2004)
20. Palioudakis, A., Salomaa, K., Akl, S.G.: State complexity and limited nondeterminism. In: Kutrib, M., Moreira, N., Reis, R. (eds.) DCFS 2012. LNCS, vol. 7386, pp. 252–265. Springer, Heidelberg (2012)
21. Palioudakis, A., Salomaa, K., Akl, S.G.: Unary NFAs with limited nondeterminism. In: Geffert, V., Preneel, B., Rovan, B., Štuller, J., Tjoa, A.M. (eds.) SOFSEM 2014. LNCS, vol. 8327, pp. 443–454. Springer, Heidelberg (2014)
22. Palioudakis, A., Salomaa, K., Akl, S.G.: State complexity of finite tree width NFAs. J. Automata Lang. Comb. **17**(2–4), 245–264 (2012)
23. Palioudakis, A.: State complexity of nondeterministic finite automata with limited nondeterminism. Ph.D. thesis, Queen's University (2014)
24. Ravikumar, B., Ibarra, O.H.: Relating the degree of ambiguity of finite automata to the succinctness of their representation. SIAM J. Comput. **18**, 1263–1282 (1989)
25. Shallit, J.: A Second Course in Formal Languages and Automata Theory. Cambridge University Press, Cambridge (2009)
26. Stockmeyer, L.J., Meyer, A.R.: Word problems requiring exponential time. In: Proceedings of the 5th Symposium on Theory of Computing, pp. 1–9 (1973)
27. Yu, S.: Regular languages. In: Rozenberg, G., Salomaa, A. (eds.) Handbook of Formal Languages, vol. I, pp. 41–110. Springer, Heidelberg (1997)

Integer Complexity: Experimental and Analytical Results II

Juris Čerņenoks[1], Jānis Iraids[1], Mārtiņš Opmanis[2],
Rihards Opmanis[2], and Kārlis Podnieks[1(✉)]

[1] University of Latvia, Raiņa bulvāris 19, Riga LV-1586, Latvia
`karlis.podnieks@lu.lv`
[2] Institute of Mathematics and Computer Science, University of Latvia,
Raiņa bulvāris 29, Riga LV-1459, Latvia

Abstract. We consider representing natural numbers by expressions using only 1's, addition, multiplication and parentheses. Let $\|n\|$ denote the minimum number of 1's in the expressions representing n. The logarithmic complexity $\|n\|_{\log}$ is defined to be $\|n\|/\log_3 n$. The values of $\|n\|_{\log}$ are located in the segment $[3, 4.755]$, but almost nothing is known with certainty about the structure of this "spectrum" (are the values dense somewhere in the segment?, etc.). We establish a connection between this problem and another difficult problem: the seemingly "almost random" behaviour of digits in the base-3 representation of the numbers 2^n.

We also consider representing natural numbers by expressions that include subtraction.

Keywords: Integer complexity · Logarithmic complexity · Spectrum · Powers of two · Ternary representations

1 Introduction

The field explored in this paper is represented in "The On-Line Encyclopedia of Integer Sequences" (OEIS) as the sequences A005245 [15] and A091333 [19]. The topic seems to be gaining popularity — see [2–4,6,7,9,16,18]. The paper continues our previous work [13].

First, in Sect. 2 we consider representing natural numbers by arithmetical expressions using 1's, addition, multiplication and parentheses. Let us call this "representing numbers in the basis $\{1, +, \cdot\}$".

Definition 1. *Let $\|n\|$ denote the **minimum** number of 1's in the expressions representing n in basis $\{1, +, \cdot\}$. We call it the **integer complexity** of n. The logarithmic complexity $\|n\|_{\log}$ is defined to be $\frac{\|n\|}{\log_3 n}$.*

Integer complexity of n corresponds to the sequence A005245 [15] in the OEIS. For quick reference, here are the optimal expressions for some numbers:

$$\|1\| = 1$$
$$\|3\| = 3; 3 = 1 + 1 + 1$$
$$\|6\| = 5; 6 = (1 + 1 + 1) \cdot (1 + 1)$$
$$\|10\| = 7; 10 = 1 + (1 + 1 + 1) \cdot (1 + 1 + 1) = (1 + 1 + 1 + 1 + 1) \cdot (1 + 1)$$

© Springer International Publishing Switzerland 2015
J. Shallit and A. Okhotin (Eds.): DCFS 2015, LNCS 9118, pp. 58–69, 2015.
DOI: 10.1007/978-3-319-19225-3_5

In the following expressions we have used 2, 3, and 5 as a shorthand for $1+1$, $1+1+1$ and $1+1+1+1+1$, respectively, and exponentiation – as a shorthand for repeated multiplication.

$$\|107\| = 16; 107 = 1 + 2 \cdot (1 + 2^2 \cdot (1 + 3 \cdot 2^2))$$
$$\|321\| = 18; 321 = 1 + 5 \cdot 2^6$$
$$\|1439\| = 26; 1439 = 1 + 2 \cdot (1 + 2 \cdot (1 + 1 + 3 \cdot (2^4 + 1) \cdot (3 \cdot 2 + 1)))$$

It is well known that all the values of $\|n\|_{\log}$ are located in the segment $[3, 4.755]$, but almost nothing is known with certainty about the structure of this "spectrum" (are the values dense somewhere in the segment?, etc.). We establish a connection between this problem and another difficult problem: the seemingly "almost random" behaviour of digits in the base-3 representation of the numbers 2^n.

Secondly, in Sect. 3 we consider representing natural numbers by arithmetical expressions that include also **subtraction**. Let us call this "representing numbers in basis $\{1, +, \cdot, -\}$".

Definition 2. *Let $\|n\|_-$ denote the **minimum** number of 1's in the expressions representing n in basis $\{1, +, \cdot, -\}$. The logarithmic complexity $\|n\|_{-\log}$ is defined as $\frac{\|n\|_-}{\log_3 n}$.*

$\|n\|_-$ corresponds to the sequence A091333 [19] in the OEIS.

We prove that almost all values of the logarithmic complexity $\|n\|_{-\log}$ are located in the segment $[3, 3.679]$. Having computed $\|n\|_-$ up to $n = 2 \cdot 10^{11}$, we present some of our observations.

2   Integer Complexity in Basis $\{1, +, \cdot\}$

2.1   Connections to the Sum-of-digits Problem

Throughout this subsection, we assume that p, q are multiplicatively independent positive integers, i.e., $p^a \neq q^b$ for all integers $a, b > 0$.

Definition 3. *Let $D_q(n, i)$ denote the i-th **digit** in the canonical base-q representation of the number n, and $S_q(n)$ – the **sum of digits** in this representation.*

Let us consider base-q representations of powers p^n. Imagine, for a moment (somewhat incorrectly), that, for fixed p, q, n, the digits $D_q(p^n, i)$ behave like statistically independent random variables taking the values $0, 1, ..., q-1$ with equal probabilities $\frac{1}{q}$. Then, the (pseudo) mean value and (pseudo) variance of $D_q(p^n, i)$ would be

$$E = \frac{q-1}{2}; V = \sum_{i=0}^{q-1} \frac{1}{q}\left(i - \frac{q-1}{2}\right)^2 = \frac{q^2-1}{12}.$$

The total number of digits in the base-q representation of p^n is $k_n \approx n \log_q p$. Hence, the (pseudo) mean value of the sum $S_q(p^n) = \sum\limits_{i=1}^{k_n} D_q(p^n, i)$ would be $E_n \approx n\frac{q-1}{2} \log_q p$ and, because of the assumed (pseudo) independence of digits, its (pseudo) variance would be $V_n \approx n\frac{q^2-1}{12} \log_q p$. As the final consequence, the corresponding centered and normed variable $\frac{S_q(p^n)-E_n}{\sqrt{V_n}}$ would behave like a standard normally distributed random variable with probability density $\frac{1}{\sqrt{2\pi}}e^{-\frac{x^2}{2}}$.

One can try verifying this conclusion experimentally. For example, let us compute $S_3(2^n)$ for n up to 100000, and let us draw the histogram of the corresponding centered and normed variable

$$s_3(2^n) = \frac{S_3(2^n) - n\log_3 2}{\sqrt{n\frac{2}{3}\log_3 2}}$$

(see Fig. 1). As we see, this variable behaves, indeed, almost exactly as a standard normally distributed random variable (the dashed curve).

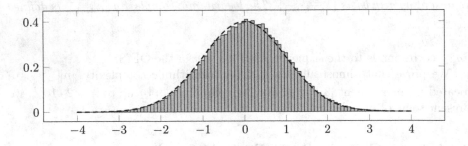

Fig. 1. Histogram of centered and normed variable $s_3(2^n)$

Observing such a phenomenon "out there", one could conjecture that $S_q(p^n)$, as a function of n, behaves almost like $n\frac{q-1}{2}\log_q p$, i.e., almost **linearly** in n. Let us try to estimate the amplitude of the possible deviations by "applying" the Law of the Iterated Logarithm (the idea proposed in [5,8]). Let us introduce centered and normed (pseudo) random variables:

$$d_q(p^n, i) = \frac{D_q(p^n, i) - \frac{q-1}{2}}{\sqrt{\frac{q^2-1}{12}}}.$$

By summing up these variables for i from 1 to k_n, we obtain a sequence of (pseudo) random variables:

$$\kappa_q(p, n) = \frac{S_q(p^n) - \frac{q-1}{2}k_n}{\sqrt{\frac{q^2-1}{12}}},$$

that "must obey" the Law of the Iterated Logarithm. Namely, if the sequence $S_q(p^n)$ behaves, indeed, like a "typical" sum of equally distributed random variables, then $\liminf_{n\to\infty}$ and $\limsup_{n\to\infty}$ of the fraction

$$\frac{\kappa_q(p,n)}{\sqrt{2k_n \log\log k_n}},$$

(log stands for the natural logarithm) "must be" -1 and $+1$ correspondingly. Therefore, it seems, we could conjecture that, if we denote

$$\sigma_q(p,n) = \frac{S_q(p^n) - (\frac{q-1}{2}\log_q p)n}{\sqrt{(\frac{q^2-1}{6}\log_q p)n \log\log n}},$$

then

$$\limsup_{n\to\infty}\sigma_q(p,n) = 1; \liminf_{n\to\infty}\sigma_q(p,n) = -1.$$

In particular, this would mean that

$$S_q(p^n) = \left(\frac{q-1}{2}\log_q p\right)n + O(\sqrt{n \log\log n}).$$

By setting $p = 2; q = 3$ (note that $\log_3 2 \approx 0.6309$):

$$S_3(2^n) = n \cdot \log_3 2 + O(\sqrt{n \log\log n});$$

$$\sigma_3(2,n) = \frac{S_3(2^n) - n\log_3 2}{\sqrt{(\frac{4}{3}\log_3 2)n \log\log n}} \approx \frac{S_3(2^n) - 0.6309n}{\sqrt{0.8412n \log\log n}},$$

$$\limsup_{n\to\infty}\sigma_3(2,n) = 1; \liminf_{n\to\infty}\sigma_3(2,n) = -1.$$

However, the behaviour of the expression $\sigma_3(2,n)$ up to $n = 10^7$ does not show convergence to the segment $[-1, +1]$ (see Fig. 2, obtained by Juris Čerņenoks). Although it is oscillating almost as required by the Law of the Iterated Logarithm, very many of its values lie outside the segment.

Could we hope to prove the above estimate of $S_q(p^n)$? To our knowledge, the best result on this problem is due to C. L. Stewart [17]. It follows from his Theorem 2 (put $\alpha = 0$), that

$$S_q(p^n) > \frac{\log n}{\log\log n + C_0} - 1,$$

where the constant $C_0 > 0$ can be effectively computed from q, p. Since then, no better than $\frac{\log n}{\log\log n}$ lower bounds of $S_q(p^n)$ have been proved.

However, it appears that from a well-known open conjecture about integer complexity in basis $\{1, +, \cdot\}$ (Hypothesis 1, attributed to J. L. Selfridge in [10]), one can derive a strong **linear** lower bound of $S_3(2^n)$.

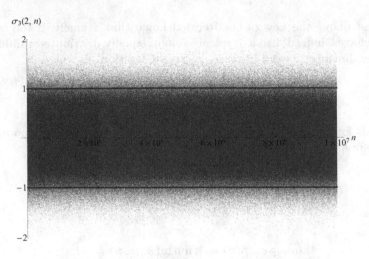

Fig. 2. Oscillating behaviour of the expression $\sigma_3(2, n)$

Proposition 1. *For any primes p, q, and all n, we have $S_q(p^n) \geq \|p^n\| - nq \log_q p$.*

Proof. Assume $a_m a_{m-1} \cdots a_0$ is a canonical base-q representation of the number p^n. One can derive from it a representation of p^n in basis $\{1, +, \cdot\}$, having length $\leq mq + S_q(p^n)$. Hence, $\|p^n\| \leq mq + S_q(p^n)$. Since $q^m \leq p^n < q^{m+1}$, we have $m \leq n \log_q p < m + 1$, and $\|p^n\| \leq nq \log_q p + S_q(p^n)$.   □

Theorem 1. *If, for a prime $p \neq 3$, $\epsilon > 0$, and $n > 0$, $\|p^n\|_{\log} \geq 3 + \epsilon$, then $S_3(p^n) \geq n\epsilon \log_3 p$.*

Proof. Since

$$3 + \epsilon \leq \|p^n\|_{\log} = \frac{\|p^n\|}{\log_3 p^n},$$

according to Proposition 1, we have

$$S_3(p^n) \geq (3 + \epsilon)n \log_3 p - 3n \log_3 p = n\epsilon \log_3 p.$$

□

Let us recall the well-known

Hypothesis 1. *For all $n \geq 1$, we have $\|2^n\| = 2n$. Moreover, the product of $1 + 1$'s is shorter than any other representation of 2^n.*

We consider proving or disproving Hypothesis 1 as a big challenge of number theory.

Hypothesis 1 has been verified as true up to $n = 39$ [13]. However, see [7, Sect. 4.3] for an argument against the general truth of it.

If $\|2^n\| = 2n$, then $\|2^n\|_{\log} = \frac{2}{\log_3 2}$, and thus, by taking $\epsilon = \frac{2}{\log_3 2} - 3$ in Theorem 1, we obtain

Corollary 1. *If Hypothesis 1 is true, then $S_3(2^n) > 0.107 \cdot n$ for all $n > 0$.*

Thus, proving Hypothesis 1 would yield a strong linear lower bound for $S_3(2^n)$. Does this mean that proving Hypothesis 1 is an extremely complicated task?

Similar considerations appear in [4] (see the discussion following Conjecture 1.3).

2.2   Compression of Powers

For a prime p, can the shortest expressions of powers p^n be obtained simply by multiplying the best expressions of p?

The answer "yes" can be proved easily for all powers of $p = 3$. For example, the shortest expression of $3^3 = 27$ is $(1 + 1 + 1) \cdot (1 + 1 + 1) \cdot (1 + 1 + 1)$, and $\|3^n\| = n \cdot \|3\| = 3n$ for all n. The same seems to be true for the powers of $p = 2$; see Hypothesis 1 above. For example, the shortest expression of $2^5 = 32$ is $(1+1) \cdot (1+1) \cdot (1+1) \cdot (1+1) \cdot (1+1)$. Thus, it seems that $\|2^n\| = n \cdot \|2\| = 2n$ for all n.

However, for $p = 5$ this is true only for $n = 1, 2, 3, 4, 5$, but the shortest expression of 5^6 is not $5 \cdot 5 \cdot 5 \cdot 5 \cdot 5 \cdot 5$, but

$$5^6 = 15625 = 1 + 2^3 \cdot 3^2 \cdot 217 = 1 + 2^3 \cdot 3^2(1 + 2^3 \cdot 3^3).$$

Thus, we have here a kind of "compression": $\|5^6\| = 29 < 6\,\|5\| = 30$.

Could we expect now that the shortest expression of 5^n can be obtained by multiplying the expressions of 5^1 and 5^6? This is true at least up to $n = 17$, as one can verify by using the online calculator [12] by Jānis Iraids. But, as observed by Juris Čerņenoks, $\|5^{36}\|$ is not $\|5^6\| \cdot 6 = 29 \cdot 6 = 174$ as one might expect. Namely:

$$5^{36} = 2^4 \cdot 3^3 \cdot 247 \cdot 244125001 \cdot 558633785731 + 1,$$

where

$$247 = 3 \cdot (3^4 + 1) + 1;$$

$$244125001 = 2^3 \cdot 3^2 \cdot (2^3 \cdot 3^3 + 1) \cdot (2^3 \cdot 3^2 \cdot (2^3 \cdot 3^3 + 1) + 1) + 1;$$

$$558633785731 = 2 \cdot 3 \cdot (2^3 \cdot 3^5 + 1) \cdot (2 \cdot 3^4 \cdot (2^6 \cdot 3^5 \cdot (2 \cdot 3^2 + 1) + 1) + 1) + 1.$$

In total, this expression of 5^{36} contains 173 ones.

Until now, no more "compression points" are known for powers of 5.

Let us define the corresponding general notion:

Definition 4. *Let us say that n is a **compression point** for powers of the prime p, if and only if for any numbers k_i such that $0 < k_i < n$ and $\sum k_i = n$:*

$$\|p^n\| < \sum \|p^{k_i}\|,$$

i.e., if the shortest expression of p^n is better than any product of expressions of smaller powers of p.

Question 1. *Which primes possess an infinite number of compression points, which ones a finite number, and which ones do not possess them at all?*

Powers of 3 (and, it seems, powers of 2 as well) do not possess compression points at all. Unfortunately this fact is a trivial consequence of Proposition 4 and thus does not generalize to powers of other primes. Powers of 5 possess at least two compression points. For more about compression of powers of particular primes, see our previous paper [13] (where compression is termed "collapse").

Proposition 2. *If a prime $p \neq 3$ possesses finite number of compression points, then there is an $\epsilon > 0$ such that $\|p^n\|_{\log} \geq 3 + \epsilon$ for all $n > 0$.*

Proof. If $p \neq 3$, then for any particular n, by Proposition 4 we have $\|p^n\|_{\log} > 3$.
    If n is not a compression point, then

$$\|p^n\| = \sum \|p^{k_i}\|$$

for some numbers k_i such that $0 < k_i < n$ and $\sum k_i = n$. Now, if some of the k_i are not compression points as well, then we can express $\|p^{k_i}\|$ as $\sum \|p^{l_j}\|$, where $0 < l_j < k_i$ and $\sum l_j = k_i$.
    In this way, if m is the last compression point of p, then, for any $n > m$, we can obtain numbers k_i such that $0 < k_i \leq m$, $\sum k_i = n$, and

$$\|p^n\| = \sum \|p^{k_i}\|.$$

Hence,

$$\|p^n\|_{\log} = \frac{\|p^n\|}{\log_3 p^n} = \frac{\sum \|p^{k_i}\|}{(\log_3 p) \sum k_i}.$$

Since, for any $a_i, b_i > 0$,

$$\frac{\sum a_i}{\sum b_i} \geq \min \frac{a_i}{b_i},$$

we obtain that

$$\|p^n\|_{\log} \geq \min \frac{\|p^{k_i}\|}{k_i \log_3 p} = \min \|p^{k_i}\|_{\log} = 3 + \epsilon,$$

for some $\epsilon > 0$.   □

As we established in Sect. 2.1, for any particular prime $p \neq 3$, proving $\|p^n\|_{\log} \geq 3 + \epsilon$ for some $\epsilon > 0$, and all sufficiently large $n > 0$, would yield a strong linear lower bound for $S_3(p^n)$. Therefore, for reasons explained in Sect. 2.1, proving the above inequality (even for a particular $p \neq 3$) seems to be an extremely complicated task. And hence, proving (even for a particular $p \neq 3$) that p possesses finite number of compression points seems to be an extremely complicated task as well.

Proposition 3. *For any number k, $\lim_{n \to \infty} \|k^n\|_{\log}$ exists, and does not exceed any particular $\|k^n\|_{\log}$.*

Proof. Function $A(n) = \|k^n\|$ is subadditive:

$$A(n + m) = \|k^{n+m}\| \leq \|k^n\| + \|k^m\| = A(n) + A(m).$$

The proposition follows from a general property of subadditive functions, see [1, Lemma 10.2.7].   □

For more about the spectrum of logarithmic complexity $\|n\|_{\log}$, see in our previous paper [13].

The following **weakest possible question** about the spectrum of logarithmic complexities appears as Q6 in [7]:

Question 2. *Is there an $\epsilon > 0$ such that $\|n\|_{\log} \geq 3 + \epsilon$ for infinitely many n?*

Theoretically, Question 2 should be easier to solve than Hypothesis 1 and other hypotheses from [13], but it remains still unsolved nevertheless.

On the other hand,

Question 3. *If, for all primes p, the limit $\lim_{n\to\infty} \|p^n\|_{\log} = 3$ holds, could this imply that $\lim_{N\to\infty} \|N\|_{\log} = 3$ for all N —, a negative solution to Question 2?*

3   Integer Complexity in the Basis $\{1, +, \cdot, -\}$

In this section, we consider representing natural numbers by arithmetical expressions using 1's, addition, multiplication, subtraction, and parentheses. According to Definition 2, $\|n\|_-$ denotes the number of 1's in the shortest expressions representing n in the basis $\{1, +, \cdot, -\}$.

Of course, for all n, we have $\|n\|_- \leq \|n\|$. The number 23 is the first one that possesses a better representation in the basis $\{1, +, \cdot, -\}$ than in the basis $\{1, +, \cdot\}$:

$$23 = 2^3 \cdot 3 - 1 = 2^2 \cdot 5 + 2; \|23\|_- = 10; \|23\| = 11.$$

Definition 5. *(a) Let $E(n)$ denote the **largest** m such that $\|m\| = n$.*
*(b) Let $E_-(n)$ denote the **largest** m such that $\|m\|_- = n$.*
*(c) Let $E_{-k}(n)$ denote the k-**th largest** m such that $\|m\|_- \leq n$ (if it exists). Thus, $E_-(n) = E_{-1}(n)$.*
*(d) Let $e_-(n)$ denote the **smallest** m such that $\|m\|_- = n$.*

According to Guy [10], it was J. L. Selfridge who first noticed that

Proposition 4. *For all $k \geq 0$:*

$$E(3k + 2) = 2 \cdot 3^k;$$

$$E(3k + 3) = 3 \cdot 3^k;$$

$$E(3k + 4) = 4 \cdot 3^k.$$

One can verify easily that $E_-(n) = E(n)$ for all $n > 0$, i.e., that the formulas discovered by J. L. Selfridge for $E(n)$ remain valid for $E_-(n)$ as well:

Proposition 5. *For all $n \geq 0$, we have $E(n) = E_-(n)$.*

One can verify also that $E_{-2}(n) = E_2(n)$ for $n \geq 5$; hence, the formula obtained by D. A. Rawsthorne [14] remains true for the basis $\{1, +, \cdot, -\}$: for all $n \geq 8$, we have $E_{-2}(n) = \frac{8}{9}E_-(n)$.

These formulas allow for building of dynamic programming algorithms for computing of $\|n\|_-$. Indeed, after filtering out all n with $\|n\|_- < k$, one can filter out all n with $\|n\|_- = k$ knowing that $n \leq E_-(k)$, and trying out representations of n as $A \cdot B, A + B, A - B$ for A, B with $\|A\|_-, \|B\|_- < k$. See [19] for a more sophisticated efficient computer program designed by Jānis Iraids. Juris Čerņenoks used another efficient program to compute $\|n\|_-$ up to $n = 2 \cdot 10^{11}$.

The values of $e_-(n)$ up to $n = 81$ are published in the OEIS [11]. 10 of the 81 values are composite numbers similarly to values of $e(n)$.

Does Fig. 3 provide some evidence that both kinds of logarithmic complexity of n do not tend to 3?

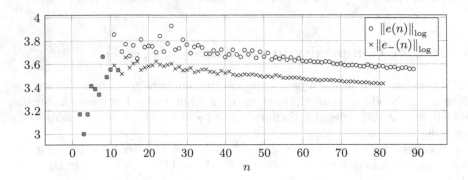

Fig. 3. Logarithmic complexities of the numbers $e(n)$ and $e_-(n)$

At least for all 2^n up to $2 \cdot 10^{11}$ Hypothesis 1 also remains true for the basis $\{1, +, \cdot, -\}$.

While observing the shortest expressions representing small numbers in basis $\{1, +, \cdot, -\}$, one might conclude that whenever subtraction is the **last** operation of a shortest expression, then it is subtraction of 1, for example, $23 = 2^3 \cdot 3 - 1$.

As established by Juris Čerņenoks, the first number for which this observation **fails** is larger than 55 billion:

$$\|n\|_- = 75; n = 55659409816 = (2^4 \cdot 3^3 - 1)(3^{17} - 1) - 2 \cdot 3.$$

Up to $2 \cdot 10^{11}$, there are only 3 numbers for which subtraction of 6 is necessary as the last operation of shortest expressions – the above one and the following two:

$$\|n\|_- = 77; n = 111534056696 = (2^5 \cdot 3^4 - 1)(3^{16} + 1) - 2 \cdot 3,$$

$$\|n\|_- = 78; n = 167494790108 = (2^4 \cdot 3^4 + 1)(3^{17} - 1) - 2 \cdot 3.$$

The need for subtraction by 8, 9, 12, or larger was not observed for numbers up to $2 \cdot 10^{11}$.

Theorem 2. *For all $n > 1$,*

$$3 \log_3 n \le \|n\|_- \le 6 \log_6 n + 5.890 < 3.679 \log_3 n + 5.890,$$

If n is a power of 3, then $\|n\|_- = 3 \log_3 n$, else $\|n\|_- > 3 \log_3 n$.

Proof. The lower bound follows from Proposition 5. Let us prove the upper bound.

If $n = 6k$, then we can start building the expression for n as $(1+1)(1+1+1)k$. Hence, by spending 5 ones, we reduce the problem to building the expression for the number $k \le \frac{n}{6}$.

Similarly, if $n = 6k + 1$, then, by spending 6 ones, we reduce the problem to building the expression for the number $k \le \frac{n-1}{6}$.

If $n = 6k + 2 = 2(3k + 1)$, then, by spending 6 ones, we reduce the problem to building the expression for the number $k \le \frac{n-2}{6}$.

If $n = 6k + 3 = 3(2k + 1)$, then, by spending 6 ones, we reduce the problem to building the expression for the number $k \le \frac{n-3}{6}$.

If $n = 6k+4 = 2(3k+2) = 2(3(k+1)-1)$, then, by spending 6 ones, we reduce the problem to building the expression for the number $k + 1 \le \frac{n+2}{6} = \frac{n}{6} + \frac{1}{3}$.

Finally, if $n = 6k + 5 = 6(k+1) - 1$, then, by spending 6 ones, we reduce the problem to building the expression for the number $k + 1 \le \frac{n+1}{6} = \frac{n}{6} + \frac{1}{6}$.

Thus, by spending no more than 6 ones, we can reduce building the expression for any number n to building the expression for some number $k \le \frac{n}{6} + \frac{1}{3}$. By applying this kind of operations 2 times to the number n, we will arrive at a number $k \le \frac{n}{6^2} + \frac{1}{6 \cdot 3} + \frac{1}{3}$. By applying them m times, we will arrive at a number

$$k < \frac{n}{6^m} + \frac{1}{3} \cdot \frac{1}{1 - \frac{1}{6}} = \frac{n}{6^m} + \frac{2}{5}.$$

Hence, if $\frac{n}{6^m} + \frac{2}{5} \le 5$, or, $6^m \ge \frac{5n}{23}$, or $m \ge \log_6 \frac{5n}{23}$, then, after m operations, spending $\le 6m$ ones, we will arrive at the number ≤ 5. Thus,

$$\|n\|_- \le 6 \left(\log_6 \frac{5n}{23} + 1 \right) + 5 = 6 \log_6 n + 5.890 < 3.679 \log_3 n + 5.890.$$

$\square$

According to Theorem 2, for all $n > 1$:

$$3 \le \|n\|_{-\log} \le 3.679 + \frac{5.890}{\log_3 n}.$$

It seems, the largest values of $\|n\|_{-\log}$ are taken by single numbers, see Table 1.

Table 1. Largest values of $\|n\|_{-\log}$

n	$\|n\|_-$	$\approx \|n\|_{-\log}$	$\|n\|$	Other properties
11	8	3.665	8	$e_-(8)$, prime
67	14	3.658	14	$e_-(14)$, prime
787	22	3.625	22	$e_-(22)$, prime
173	17	3.624	17	$e_-(17)$, prime
131	16	3.606	16	$e_-(16)$, prime
2767	26	3.604	26	$e_-(26)$, prime
2777	26	3.602	26	$e_{-2}(26)$, prime
823	22	3.600	22	$e_{-2}(22)$, prime
1123	23	3.598	23	$e_-(23)$, prime
2077	25	3.596	25	$e_-(25)$, $31 \cdot 67$
2083	25	3.594	25	$e_{-2}(25)$, prime
617	21	3.591	21	$e_-(21)$, prime
619	21	3.589	21	$e_{-2}(21)$, prime
29	11	3.589	11	$e_-(11)$, prime

4   Conclusion

Let us conclude with the summary of the most challenging **open problems**:

(1) The **Question of Questions** – prove or disprove Hypothesis 1: for all $n \geq 1$, $\|2^n\| = 2n$, moreover, the product of $1 + 1$'s is shorter than any other representation of 2^n, even in the basis with subtraction.

(2) Basis $\{1, +, \cdot\}$. Solve the **weakest possible** Question 2 (Q6 in [7]) about the spectrum of logarithmic complexity: is there an $\epsilon > 0$ such that for infinitely many numbers n: $\|n\|_{\log} \geq 3 + \epsilon$? An equivalent formulation: is there an $\epsilon > 0$ such that for infinitely many numbers n: $\log_3 e(n) \leq (\frac{1}{3} - \epsilon)n$. Proving Hypothesis 1 would solve Question 2 positively.

(3) Basis $\{1, +, \cdot, -\}$. Improve Theorem 2: for all $n > 1$,

$$\|n\|_- < 3.679 \log_3 n + 5.890.$$

References

1. Allouche, J.P., Shallit, J.: Automatic Sequences: Theory, Applications, Generalizations. Cambridge University Press, Cambridge (2003)
2. Altman, H.: Integer complexity and well-ordering. ArXiv e-prints, October 2013. http://arxiv.org/abs/1310.2894
3. Altman, H.: Integer complexity, addition chains, and well-ordering. Ph.D. thesis, University of Michigan (2014). http://www-personal.umich.edu/~haltman/thesis-FIXED.pdf

4. Altman, H., Zelinsky, J.: Numbers with integer complexity close to the lower bound. INTEGERS **12**(6), 1093–1125 (2012)
5. Aragón Artacho, F.J., Bailey, D.H., Borwein, J.M., Borwein, P.B.: Walking on real numbers. Math. Intell. **35**(1), 42–60 (2013). doi:10.1007/s00283-012-9340-x
6. Arias de Reyna, J., van de Lune, J.: Algorithms for determining integer complexity. ArXiv e-prints, April 2014. http://arxiv.org/abs/1404.2183
7. Arias de Reyna, J., van de Lune, J.: How many 1's are needed? revisited. ArXiv e-prints, April 2014. http://arxiv.org/abs/1404.1850
8. Belshaw, A., Borwein, P.: Champernowne's number, strong normality, and the X chromosome. In: Bailey, D.H., Bauschke, H.H., Borwein, P., Garvan, F., Théra, M., Vanderwerff, J.D., Wolkowicz, H. (eds.) Computational and Analytical Mathematics, Springer Proceedings in Mathematics & Statistics, vol. 50, pp. 29–44. Springer, New York (2013). doi:10.1007/978-1-4614-7621-4_3
9. Gnang, E.K., Radziwill, M., Sanna, C.: Counting arithmetic formulas. ArXiv e-prints (2014). http://arxiv.org/abs/1406.1704
10. Guy, R.K.: Some suspiciously simple sequences. Am. Math. Mon. **93**(3), 186–190 (1986)
11. Iraids, J.: The On-Line Encyclopedia of Integer Sequences, Smallest number requiring n 1's to build using $+$, $\cdot$ and $-$. http://oeis.org/A255641
12. Iraids, J.: Online calculator of integer complexity. http://expmath.lumii.lv/wiki/index.php/Special:Complexity. Accessed on 28 February 2015
13. Iraids, J., Balodis, K., Čerņenoks, J., Opmanis, M., Opmanis, R., Podnieks, K.: Integer complexity: Experimental and analytical results. Scientific Papers University of Latvia, Computer Science and Information Technologies 787, 153–179 (2012), arXiv preprint: http://arxiv.org/abs/1203.6462
14. Rawsthorne, D.A.: How many 1's are needed? Fibonacci Q. **27**(1), 14–17 (1989). http://www.fq.math.ca/Scanned/27-1/rawsthorne.pdf
15. Sloane, N.J.A.: The On-Line Encyclopedia of Integer Sequences, Complexity of n: number of 1's required to build n using $+$ and $\cdot$. http://oeis.org/A005245
16. Steinerberger, S.: A short note on integer complexity. Contributions to Discrete Mathematics 9(1) (2014). http://cdm.ucalgary.ca/cdm/index.php/cdm/article/view/361
17. Stewart, C.: On the representation of an integer in two different bases. Journal für die reine und angewandte Mathematik **319**, 63–72 (1980). http://eudml.org/doc/152278
18. Vatter, V.: Maximal independent sets and separating covers. Am. Math. Mon. **118**(5), 418–423 (2011). http://www.jstor.org/stable/pdfplus/10.4169/amer.math.monthly.118.05.418.pdf
19. Voß, J.: The On-Line Encyclopedia of Integer Sequences, Number of 1's required to build n using $+$, $-$, $\cdot$ and parentheses. http://oeis.org/A091333

Square on Ideal, Closed and Free Languages

Kristína Čevorová[✉]

Mathematical Institute, Slovak Academy of Sciences, Bratislava, Slovakia
cevorova@mat.savba.sk

Abstract. We study the deterministic state complexity of a language accepted by an n-state DFA concatenated with itself for languages from certain subregular classes. Tight upper bounds are obtained on optimal alphabets for prefix-closed, xsided-ideal and xfix-free languages, except for suffix-free, where a ternary alphabet is used.

1 Introduction

Janusz Brzozowski and his coauthors have recently published a series of articles concerning quotient complexity of basic operations such as concatenation, Kleene closure or boolean operations on ideal [1], free [2], and closed [3] languages. The results were usually significantly smaller compared to the state complexity of general languages, despite the fact that quotient and state complexity is numerically always the same.

None of these articles considered the operation square. Since it is a special case of concatenation – a product of language with itself – results for concatenation provide an immediate upper bound on state complexity of operation square. On the other hand, the state complexity of square on general languages has been studied by Rampersad [4]. He showed, that in the binary case it is $n2^n - 2^{n-1}$, whereas in unary $2n-1$. Our aim was to study the state complexity of a square on certain ideal, free or closed subclasses of regular languages and compare results with these upper bounds.

The study of these subregular classes is not isolated. Determination of many classes of NFAs was considered in [5], including free and closed languages. Syntactic complexity was studied for ideal and closed languages [6]. A more specialized study of suffix-free languages is in [7], of prefix-free languages in [8] and of prefix-closed languages in [9] and [10].

This paper is organized as follows. In next section we give the most important definitions. In Sect. 3 we study ideal languages, Sect. 4 is dedicated to prefix-closed languages and then, in Sect. 5, we discuss free languages. In the conclusion, we compare our results with results for concatenation on the same classes and for general regular languages.

2 Preliminaries

We assume, that the reader is familiar with basic notions from automata theory, for reference see [11]. Here we recall only the most important definitions.

© Springer International Publishing Switzerland 2015
J. Shallit and A. Okhotin (Eds.): DCFS 2015, LNCS 9118, pp. 70–80, 2015.
DOI: 10.1007/978-3-319-19225-3_6

Let $[c, d]$ denote the set $\{c, c+1, \ldots, d\}$ if $c \leq d$. A binary operation $\oplus$ called *symmetric difference* is defined on sets as $S \oplus S' = S \cup S' - (S \cap S')$. Cardinality of set S is denoted by $|S|$.

An *incomplete DFA* is an NFA $A = (Q, \Sigma, \delta, \{q_0\}, F)$ with single initial state and with the property that for every $q \in Q$ and $a \in \Sigma$ the inequality $|\delta(q, a)| \leq 1$ holds. Thus some transitions may be not defined, but this is the only trace of nondeterminism. It could be made deterministic by adding one nonfinal *dead state*, where all previously undefined transitions are incoming.

Let L be a regular language. The *state complexity* is denoted by $sc(L)$, and is defined as the number of states of its minimal DFA, whereas the *incomplete state complexity* $isc(L)$, is the number of states of its minimal incomplete DFA.

Let u, v, x, w be words. If $w = uvx$, then we call u a *prefix*, x a *suffix* and v a *factor* of word w. If $w = u_1 v_1 u_2 v_2 \cdots u_k v_k$, for some words u_i, v_i, then word $u = u_1 u_1 \cdots u_k$ is called *subword* of w.

We call L *xfix-free* for xfix $\in$ {prefix, suffix, factor, subword}, if whenever $u, v \in L$ and u is xfix of v, then $u = v$. Similarly, L is *xfix-closed*, if whenever u is xfix of v and $v \in L$, then also $u \in L$. And lastly we call L an *xsided* ideal for xsided $\in$ {left, right, 2-sided, all-sided}, when $L = \Sigma^* L$, $L = L\Sigma^*$, $L = \Sigma^* L\Sigma^*$, $L = L \shuffle \Sigma^*$ respectively, where $\shuffle$ is the shuffle operation. *Square* of L is language $L \cdot L$, denoted as L^2.

Sometimes we will think of letters as of a function $Q \longrightarrow Q$. In particular, this allows us to define for each letter a inverse function a^{-1} for all states with exactly one incoming transition on a. This can be naturally generalized for words.

Proposition 1. *Note that in the case $w^{-1}(q)$ exists, there is a unique state, from which state q could be reached on the word w.*

The standard construction of DFA for square of DFA $A = (Q, \Sigma, \delta, s, F)$. At first, we define an NFA accepting $L^2(A)$ as shown in Fig. 1. If A has a dead state, we remove it, to obtain an incomplete DFA. Then we take two exact copies of the DFA A with all transitions and change labels of states so they are unique and a state t becomes q_t. Other changes are, that s will not be initial and we change the finality of all states in the first copy to nonfinal. Lastly, we will add transitions. Whenever $f \in F$, then we add a transition $q_f \xrightarrow{\varepsilon} s$.

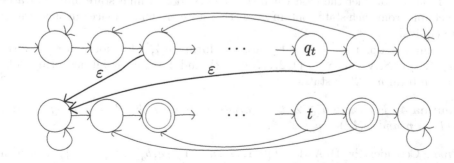

Fig. 1. NFA for square

Now we determine this NFA by the standard subset construction to obtain DFA accepting $L^2(A)$. Note that all reachable states contain at most one state from the first copy, because this copy is deterministic and there is no transition to it from the second copy.

3   Xsided Ideal Languages

A concatenation of left ideal languages has a quotient, i.e., also state complexity $m + n - 1$ [1]. Witness languages are unary $a^{n-1}a^*$ and $a^{m-1}a^*$; thus the upper bound on the state complexity of square of left ideal language is $2n - 1$ and is tight with the witness $a^{n-1}a^*$. Moreover, since 2-sided and all-sided ideals are also left ideals, this upper bound also holds for them. Because in the unary case the 2-sided or all-sided ideal is the same as left ideal, this bound is tight. Only nonunary right ideals remain unresolved by this reasoning.

Proposition 2. *A language L is right ideal if and only if the minimal DFA that recognizes L has exactly one final state with all outcoming transitions leading back to this final state.*

Lemma 3. *Let L be a right ideal language with $\mathrm{sc}(L) = n$. Then $\mathrm{sc}(L^2) \leq n + 2^{n-2}$.*

Proof. This is a direct consequence of Theorem 9 in [1], which states, that product (concatenation) of two right ideal languages with quotient complexities m and n has quotient complexity at most $m + 2^{n-2}$. We will also provide an elementary proof, since it provides the insight necessary for the following lemmata. Let $n - 1$ be the only final state of a minimal DFA for a right-ideal language. Consider the standard construction of an NFA for a square and then the subset determination of it as described in the preliminaries. What kind of subsets are reachable?

Since $n - 1$ loops to itself on all letters, states reachable from the state S containing q_{n-1} again contain q_{n-1}. The state $n - 1$ is a final state, thus, that $0 \in S$. Moreover, it is the *only* final state, thus if $i < n - 1$ and $q_i \in S$ is reachable, then $S = \{q_i\}$.

Finally, consider the case when $n - 1 \in S$. It is a final state and all states reachable from such state contain $n-1$, therefore are final. Hence, all such states are equivalent.

Summed up, that is $n - 1$ states in form $\{q_i\}$, 2^{n-2} for states in form $\{q_{n-1}, 0\} \cup S$, where $S \subseteq \{1, 2, \ldots, n - 2\}$ and 1 state for all accepting subsets, in total $n + 2^{n-2}$ states.   □

Lemma 4. *For every $n \geq 2$, there exists a binary right ideal language L with $\mathrm{sc}(L) = n$ and $\mathrm{sc}(L^2) = n + 2^{n-2}$.*

Proof. Consider the DFA $A = (\{0, 1, \ldots, n - 1\}, \{a, b\}, \delta, 0, \{n - 1\})$ shown in Fig. 2, where δ is defined as follows:

$$\delta(i,a) = \begin{cases} i+1, & \text{if } i < n-1; \\ n-1, & \text{if } i = n-1. \end{cases} \qquad \delta(i,b) = \begin{cases} 0, & \text{if } i = 0; \\ i+1, & \text{if } 1 \leq i \leq n-3; \\ 0, & \text{if } i = n-2; \\ n-1, & \text{if } i = n-1. \end{cases}$$

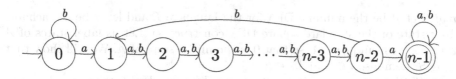

Fig. 2. Witness for right-ideal bound optimality

We will show that all subsets considered in Lemma 3 are reachable and distinguishable. We will start with reachability.

For $i < n-1$ states $\{q_i\}$ are reachable by word a^i. We have to accept word $a^{2(n-1)}$, thus final state could be also reached. The reachability of states of the form $\{q_{n-1}, 0, j_1, j_2, \ldots, j_k\}$, where $1 \leq j_1 < j_2 < \cdots < j_k$ is proven by induction on the size of maximal element of this set – the element j_k. The state $\{q_{n-1}, 0\}$ is reached by the word a^{n-1}.

The induction hypothesis is, that all states S with $\max S = j_k - 1$ are reachable. If $j_1 = 1$, then $\{q_{n-1}, 0, 1, j_2, \ldots, j_k\} = a(\{q_{n-1}, 0, j_2 - 1, \ldots, j_k - 1\})$; otherwise $\{q_{n-1}, 0, j_1, \ldots, j_k\} = a(\{q_{n-1}, 0, j_1 - 1, \ldots, j_k - 1\})$.

Proof of distinguishability is necessary only for nonfinal states. We will start by distinguishing states S and S', where $S \cap [1, n-1] \neq \varnothing$.

Let $m = \max(S \oplus S')$ – this is well-defined and $m \geq 0$, since $S \neq S'$. Without loss of generality let $m \in S$. Denote $B = \delta(S, b^{n-2-m})$ and $B' = \delta(S', b^{n-2-m})$. Note that the state $n-1$ is not reachable from the state 0 by any word shorter than $n-2$ and $m \leq n-2$ and the transition function on states other than $n-1$ is injective. Therefore a state reached by b^{n-2-m} contains $n-2$ iff we started in a state containing m and thus $n-2 \in B$, while $n-2 \in B'$. On the other hand, neither B nor B' does contain $n-1$, since $n-1$ has no incoming transition by b. And since the final state $n-1$ is reachable by a only from $n-1$ and $n-2$, the state $\delta(B, a)$ is final, while $\delta(B', a)$ is not. Therefore the word $b^{n-2-m}a$ distinguishes these two states.

Now we will distinguish states $\{q_i\}$ and $\{q_k\}$ for $0 \leq i < k \leq n-1$ (we treat $\{q_{n-1}, 0\}$ as $\{q_{n-1}\}$). The word a^{2k-2-k} is accepted from $\{q_k\}$, but not from $\{q_i\}$, since it is too short. $\qquad \square$

Combination of two previous lemmata yields the following result.

Theorem 5. *Let n be integer with $n \geq 2$ and L be a right ideal language with $\mathrm{sc}(L) = n$. Then $\mathrm{sc}(L^2) \leq n + 2^{n-2}$, and this bound is tight for an alphabet of size at least two.*

4   Prefix-Closed Languages

Since minimal DFA for prefix-closed has all states final [9], except for one dead state, it is much more convenient to use incomplete DFA. We will do so, and in the end, we will derive results for standard state complexity.

Lemma 6. *Let L be a prefix-closed language with* $\mathrm{isc}(L) = n$. *Then* $\mathrm{isc}(L^2) \leq (n+5)2^{n-2} - 2$.

Proof. Let A be the minimal DFA for the language L and let S be a reachable subset state of the standard square DFA construction. Let us label states of A with integers from $[0, n-1]$ so that 0 is the initial state of A. We will show that if $q_i \in S$, then also $i \in S$ and $s \in S$.

The initial subset state is $\{q_0, 0\}$. Let w be a word such that $\{q_0, 0\} \xrightarrow{w} S$. Then if $q_i \in S$, it means that $s \xrightarrow{w} i$ in A and therefore $\{q_i, i\} \subseteq S$. Since q_i is final, $0 \in S$. Moreover, state $\{0, 1, \ldots, n-1\}$ is unreachable, because it could be reached only from some state S' with some $q_i \in S$ on a letter a with an undefinied transition from i, but that results in a state with at most $n-1$ states.

Summing up, there are $(n-1)2^{n-2}$ states containing q_i other than q_0 plus 2^{n-1} for those with q_0 plus $2^n - 2$ for subsets of $[0, n-1]$, not counting the unreachable full state and the empty dead state uncounted in incomplete DFAs. □

Lemma 7. *Let $n \geq 2$. There exists a binary language L with* $\mathrm{isc}(L) = n$ *and* $\mathrm{isc}(L^2) = (n+5)2^{n-2} - 2$.

Proof. If $n = 2$, consider the DFA $(\{0, 1\}, \{a, b\}, \delta, 0, \{0, 1\})$ with transitions defined as $\delta(0, a) = 1$, $\delta(0, b) = \varnothing$ and $\delta(1, a) = \delta(1, b) = 0$.

For $n > 2$ consider the DFA $A = ([0, n-1], \{a, b\}, \delta, 0, [0, n-1])$ shown in Fig. 3 where

$$\delta(i, a) = \begin{cases} \varnothing, & \text{if } i = 0; \\ 0, & \text{if } i = 1; \\ i+1, & \text{if } 2 \leq i \leq n-2; \\ 2, & \text{if } i = n-1. \end{cases} \qquad \delta(i, b) = \begin{cases} i+1, & \text{if } 1 \leq i \leq n-2; \\ 0, & \text{if } i = n-1. \end{cases}$$

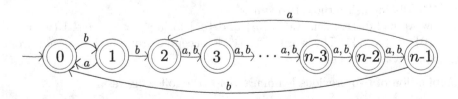

Fig. 3. The witness for prefix-closed bound optimality

Note that in the standard square DFA construction for this witness DFA (see preliminaries), if $n-1$ is a member of set S, then $|b(S)| = |S|$. Moreover,

$$|a(S)| = \begin{cases} |S|, & \text{if } 1 \notin S; \\ |S| - 1, & \text{if } 1 \in S. \end{cases}$$

To prove reachability, we will use a variation on an inductive proof. Largest sets will be used as the base of the induction and in an inductive step, we will show, how to reach smaller sets. As the base of the induction we will show that all sets in form $\{q_i, 0, 1, \ldots, n-2, n-1\}$ are reachable from the start state as follows:

$$\{q_0, 0\} \xrightarrow{b^{n-1}} \{q_{n-1}, 0, 1, \ldots, n-2, n-1\} \xrightarrow{b^{i+1}} \{q_i, 0, 1, \ldots, n-2, n-1\}.$$

The induction hypothesis is that all states S with $|S \cap [1, n-1]| \geq k+1$ with property $q_i \in S$ implies that $i \in S$ are reachable. We will show that all such states S' with $|S' \cap [1, n-1]| = k$ are also reachable. The proof is divided into four cases in all of them we suppose that $0 < j_r < j_{r+1}$ (r is used as an arbitrary index $1 \leq r$ in this proof).

1. $S = \{q_i, 0, j_1, \ldots, j_k\}$ **and** $j_1 > 1$ **and** $i \in S$. Since $j_r \neq 1$, necessarily also $i \neq 1$ (see the proof of Lemma 6). In that case is $a^{-1}(S)$ well defined and we have $a^{-1}(S) = \{q_{a^{-1}(i)}, 0, 1, a^{-1}(j_1), \ldots, a^{-1}(j_k)\} \xrightarrow{a} \{q_i, 0, j_1, \ldots, j_k\}$. Both $q_{a^{-1}(i)} \in S$ and $a^{-1}(i) \in S$. Moreover since a is bijection on $[2, n-1]$, set $a^{-1}(S)$ has $k+1$ elements from $[0, n-1]$.

2. $S = \{q_i, 0, 1, j_2, \ldots, j_k\}$. Let $m = \min\{b| \, b > 0 \text{ and } b \neq j_r\}$. Then, since b is a bijection, the set $B = \{q_{b^{-(m-1)}(i)}, b^{-(m-1)}(0), b^{-(m-1)}(j_1), \ldots, b^{-(m-1)}(j_k)\}$

   is well-defined. Then we have $B \xrightarrow{b^{m-1}} \{q_i, 0, 1, j_2, \ldots, j_k\}$.

   The choice of the exponent $m-1$ was deliberate so that $1 \notin B$. Moreover, since 0 is a member of each state $b^l(B)$ in this computing path, the state $b^{-1}(0) = n-1$ is also a member of $b^l(B)$. Therefore, as noted at the beginning of this proof, this implies that $|B| = |b(B)| = \cdots = |b^{m-1}(B)| = |S|$. This shows how S can be reached from a state of the same size containing 1, which has already been shown to be reachable in case 1.

3. $S = \{0, j_1, \ldots, j_k\}$. Then $\{q_0, 0, a^{-1}(j_1), \ldots, a^{-1}(j_k)\} \xrightarrow{a} \{j_1, \ldots, j_k\}$. Since letter a is an injection on states other than 0, the state on the left was shown to be reachable in 2.

4. $S = \{j_1, \ldots, j_k\}$. Then $\{0, j_2 - j_1, \ldots, j_k - j_1\} \xrightarrow{b^{j_1}} \{j_1, \ldots, j_k\}$. Since b is a bijection, the state on the left was shown to be reachable in 3.

The proof of distinguishability of states S and S' is divided into four cases. The empty state is the only nonfinal state; therefore in each of these cases, our aim is to find a word that leads to the empty state from one of these states, whereas from the other does not.

1. **Both S and S' are subsets of** $[0, n-1]$. Without loss of generality there exists $s \in S$ such that $s \notin S'$. The transition on b in states that are subsets

of $[0, n-1]$ never changes the size of a resulting state, while a transition on a changes it iff 1 is its member. We will call a state the successor of a given state on a word w, if it is reached on this word without using ε transition. State $i \in S'$ will have no successors on words with prefix $b^{n-i}a$. In this manner, we gradually construct word removing successors of all states in S' so it will have no successor and we reach the empty state. On the same word, we removed all states in S that are in $S \cap S'$. But we did not remove s, so the resulting state is not empty.

2. **S is a subset of** $[0, n-1]$. Let $q_i \in S'$. If $i \notin S$, then we just erase the state S as in case 1. The successor of q_i was not erased, so the result is an non-empty state. If $i \in S$, we erase everything except i. In the successor of S', there are still at least the states 0 and successor of i and q_i. We will not remove successors of states other than i while removing i. There always is some other state, unless we have states $\{q_0, 0\}$ and $\{0\}$. But the word $b^n a$ distinguishes these two.

3. $q_i \in S$, $q_j \in S'$ **and** $i < j$. Then $q_0 \in b^{n-j}(S')$ while $q_{i+n-j} \in b^{n-j}(S)$. The state q_0 has no successor on a, therefore $b^{n-j}a(S') \subseteq [0, n-1]$, while $b^{n-j}a(S) \nsubseteq [0, n-1]$. We distinguished these types of states in the case 2.

4. $q_i \in S$ **and** $q_i \in S'$ **for some** i. At first, suppose that $S \oplus S' \neq \{1\}$; we will resolve the opposite later. A transition on $a^{n-1-i}ba$ leads from both states to two different states in $[0, n-1]$. Since any difference other than 1 between S and S' is preserved by transitions on a, following a transition on b adds 1, but this leaves the difference (which is in the cycle $[2, n-1]$) untouched and so does the last transition on a. So we reduced this to case 1.

   Lastly, if $S \oplus S' \neq \{1\}$, that is $S = S' \cup \{1\}$, then $b(S) \oplus b(S') = \{2\}$ and this case was treated in previous paragraph. $\square$

Now we will combine previous results for incomplete state complexity to get a tight upper bound on the standard deterministic state complexity.

Theorem 8. *Let $n \geq 3$ and L be prefix-closed language with $\mathrm{sc}(L) = n$. Then $\mathrm{sc}(L^2) \leq (n+4)2^{n-3} - 1$, and this bound is tight for an alphabet of size at least two.*

Proof. Since all prefix-closed languages L with $\mathrm{sc}(L) = n \geq 2$ have a dead state, $\mathrm{isc}(L) = n - 1$. Prefix-closed languages are closed under the operation square. Therefore $\mathrm{sc}(L^2) = \mathrm{isc}(L^2) + 1 = (n - 1 + 5)2^{n-1-2} - 2 + 1 = (n+4)2^{n-3} - 1$. $\square$

5   Xfix-Free Languages

The state complexity of a concatenation of prefix-free languages is $m + n - 2$ [12]. In fact, this is not only a tight upper bound, but also a lower bound. Beside that witness languages are unary $\{a^{n-1}\}$ and $\{a^{m-1}\}$, so $\{a^{n-1}\}$ is also a unary witness for a prefix-free square. Moreover, bifix-, factor- and subword-free languages are also prefix-free and in the unary case all of these properties are the same, so the state complexity is the same for all of these classes. It remains to investigate suffix-free languages.

Lemma 9. *If L is a suffix-free regular language with $\mathrm{sc}(L) = n$, then $\mathrm{sc}(L^2) \leq n2^{n-3} + 1$.*

Proof. By [13], every DFA accepting L is nonreturning with a dead state. Let $A = ([0, n-3] \cup \{s, d\}, \Sigma, \delta, s, F)$ be the minimal DFA for L, where d denotes the dead state.

Let S be a reachable set in the standard square construction on A. Suffix-freeness imposes certain restrictions on S. Since A is nonreturning, $s \in S$ iff $S \cap F \neq \varnothing$ and, for the same reason, if $q_s \in S$, then $S = \{q_s\}$. Note that states S and $S \cup \{d\}$ are equivalent. Finally, we will show that if for index i with $i \neq d$ holds that if $q_i \in S$, then $i \notin S$.

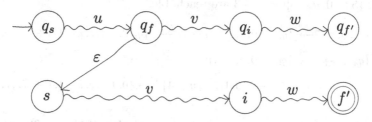

Fig. 4. Sketch of proof

Suppose that there exists reachable subset S and index i, such that both i and q_i are members of S. Consider a computation that shows reachability of S. Since i is in S, there was a step when the first copy of an DFA A was in state q_f corresponding to final state of A and from this step, the computation led from s to i and from q_f to q_i. Let u and v denote words corresponding to these two parts of the computation, respectively. Note that $u \neq \varepsilon$, because s is not final. If $i \neq d$, then i is a useful state and some final state f' is reachable from it on word w. So there are these two paths $q_s \xrightarrow{uvw} q_{f'}$ and $s \xrightarrow{vw} f'$. But this means that A accepts both uvw and vw, and that is impossible, because A accepts suffix-free language and $u \neq \varepsilon$.

This sums up to $(n-2)2^{n-3}$ equivalence classes of subset states when $q_d \notin S$ and $q_s \notin S$, plus 2^{n-2} classes of states such that $q_d \in S$ and plus one initial state $\{q_s\}$, in total $n2^{n-3} + 1$ states.   □

Lemma 10. *There exists a suffix-free language L with $\mathrm{sc}(L) = n$ and $\mathrm{sc}(L^2) = n2^{n-3} + 1$.*

Proof. Consider the DFA in Fig. 5, note that we did not draw transitions leading to the dead state d. This automaton satisfies requirements of Lemma 1 in [7], so it accepts a suffix-free language.

At first, we will show that all equivalence classes considered in the Lemma 9 are reachable. As presence of the dead state in a subset is unimportant and the presence of s determined, we will usually omit them, unless they are important. The initial state $\{q_s\}$ is reachable. For the reachability of S, we will use a variation of a mathematical induction on a size of S.

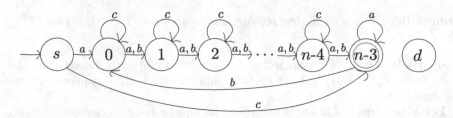

Fig. 5. Witness for suffix-free bound optimality

As the base of the induction we will show that the largest sets, that is sets such that $|S \cap [0, n-3]| = n - 3$ are reachable.

$$\{q_s\} \xrightarrow{c} \{q_{n-3}, s\} \xrightarrow{a^{n-3}} \{q_{n-3}, 0, 1, \ldots, n-4\} \xrightarrow{b^{i-1}} \{q_i, 0, 1, \ldots, i-1, i+1, \ldots, n-3\}$$

$$\{q_s\} \xrightarrow{c} \{q_{n-3}, s\} \xrightarrow{a^{n-3}} \{q_{n-3}, 0, 1, \ldots, n-4\} \xrightarrow{c}$$

$$\xrightarrow{c} \{0, 1, \ldots, n-4\} \xrightarrow{b^{i-1}} \{0, 1, \ldots, i-1, i+1, \ldots, n-3\}.$$

The inductive hypothesis now is that all sets S with $|S \cap [0, n-3]| = k+1$ are reachable. First, let it be the case that the set $S = \{j_1, j_2, \ldots, j_k\}$ and let $m = \min\{p \mid p \in [0, n-1]$ and $p \notin S\}$. Then $S = \{j_1, \ldots, j_{k-(n-3-m)}, m+1, m+2, \ldots, n-3\}$. Then the following computation shows reachability:

$$\{j_{m+1} - (m+1), \ldots, j_k - (m+1), n-3-m, \ldots, n-4, n-3\} \xrightarrow{c}$$

$$\xrightarrow{c} \{j_{m+1} - (m+1), \ldots, j_k - (m+1), n-3-m, \ldots, n-4\} \xrightarrow{b^{m+1}}$$

$$\xrightarrow{b^{m+1}} \{0, 1, \ldots, m-1, j_{m+1}, \ldots, j_k\}.$$

The proof for states of the form $\{q_i, j_1, \ldots, j_k\}$ is similar, with redefinition of $m = \min\{p \mid p \in [0, n-1]$ and $p \notin S$ and $p \neq i\}$. Just note that whenever we reach state q_{n-3} there is indeed 0 in the following state. The empty i.e., dead state, is reached from $\{n-1\}$ on c and lastly,

$$\{q_{n-3}, s, 0, 1, \ldots, n-4\} \xrightarrow{c} \{0, 1, \ldots, n-3\}.$$

Now we will prove distinguishability. Notice that the NFA for the square has the following properties:

1. the string ε is accepted only from state $n-1$;
2. the string b^{n-1-i} is accepted only from state i;
3. the string ab^{n-1} is accepted only from 0;
4. the string cab^{n-1} is accepted only from q_{n-3};
5. the string $a^{n-1-i}cab^{n-1}$ is accepted only from q_i.

Hence for each state q of this NFA, there is a string w_q which is accepted from q, but rejected from any other state. Now let S and S' be two distinct subsets in the subset automaton of this NFA. Then S and S' differ in a state q and the string w_q distinguishes them. □

Table 1. Comparison of results for square and concatenation.

		Square	$|\sum|$	Concatenation	$|\sum|$
Ideal	unary	$2n - 1$		$m + n - 1$	
	right	$n + 2^{n-2}$	2	$m + 2^{n-2}$	2
	left, 2-sided, all-sided	$2n - 1$	1	$m + n - 1$	1
Closed	unary	?		$m + n - 2$	
	suffix	?		$(m - 1)n + 1$	3
	prefix	$(n + 4)2^{n-3} - 1$	2	$(m + 1)2^{n-1}$	
	factor, subword	?		$m + n - 1$	2
Free	unary	$2n - 2$		$m + n - 2$	
	prefix, bifix, factor, subword	$2n - 2$	1	$m + n - 2$	1
	suffix	$n2^{n-3} + 1$	3	$(m - 1)2^{n-1} + 1$	3
Regular	unary	$2n - 1$		mn	if $(m, n) = 1$
	general	$n2^n - 2^{n-1}$	2	$m2^n - 2^{n-1}$	2

Theorem 11. *Let L be a suffix-free language with* $sc(L) = n$. *Then* $sc(L^2) \leq n2^{n-3} + 1$, *and this bound is tight for an alphabet of size at least three.*

Proof. This is a corollary of Lemmata 9 and 10. □

6   Conclusions

Table 1 is a summary of our results and comparison with catenation and regular languages [14].

The state complexity of square for all closed languages, except for prefix-closed, remains open.

References

1. Brzozowski, J.A., Jirásková, G., Li, B.: Quotient complexity of ideal languages. Theor. Comput. Sci. **470**, 36–52 (2013)
2. Brzozowski, J.A., Jirásková, G., Li, B., Smith, J.: Quotient complexity of bifix-, factor-, and subword-free regular languages. In: Dömösi, P., Iván, S. (eds.) AFL, pp. 123–137 (2011)
3. Brzozowski, J.A., Jirásková, G., Zou, C.: Quotient complexity of closed languages. Theor. Comp. Sys. **54**(2), 277–292 (2014)
4. Rampersad, N.: The state complexity of L^2 and L^k. Inf. Process. Lett. **98**(6), 231–234 (2006)
5. Bordihn, H., Holzer, M., Kutrib, M.: Determination of finite automata accepting subregular languages. Theor. Comput. Sci. **410**(35), 3209–3222 (2009). DCFS proceedings
6. Brzozowski, J., Ye, Y.: Syntactic complexity of ideal and closed languages. In: Mauri, G., Leporati, A. (eds.) DLT 2011. LNCS, vol. 6795, pp. 117–128. Springer, Heidelberg (2011)

7. Cmorik, R., Jirásková, G.: Basic operations on binary suffix-free languages. In: Kotásek, Z., Bouda, J., Černá, I., Sekanina, L., Vojnar, T., Antoš, D. (eds.) MEMICS 2011. LNCS, vol. 7119, pp. 94–102. Springer, Heidelberg (2012)
8. Jirásková, G., Krausová, M.: Complexity in prefix-free regular languages. In: McQuillan, I., Pighizzini, G. (eds.) DCFS. EPTCS, vol. 31, pp. 197–204 (2010)
9. Kao, J.Y., Rampersad, N., Shallit, J.: On NFAs where all states are final, initial, or both. Theor. Comput. Sci. **410**(4749), 5010–5021 (2009)
10. Čevorová, K., Jirásková, G., Mlynárčik, P., Palmovský, M., Šebej, J.: Operations on automata with all states final. In: Ésik, Z., Fülöp, Z. (eds.) Proceedings 14th International Conference on Automata and Formal Languages, AFL 2014, Szeged, Hungary, May 27–29, 2014. EPTCS, vol. 151, pp. 201–215 (2014)
11. Sipser, M.: Introduction to the Theory of Computation. PWS Publishing Company, Boston (1997)
12. Han, Y.S., Salomaa, K., Wood, D.: State complexity of prefix-free regular languages. In: Descriptional Complexity of Formal Systems, pp. 165–176 (2006)
13. Han, Y.-S., Salomaa, K.: State complexity of basic operations on suffix-free regular languages. In: Kučera, L., Kučera, A. (eds.) MFCS 2007. LNCS, vol. 4708, pp. 501–512. Springer, Heidelberg (2007)
14. Yu, S., Zhuang, Q., Salomaa, K.: The state complexities of some basic operations on regular languages. Theor. Comput. Sci. **125**(2), 315–328 (1994)

A Tentative Approach for the Wadge-Wagner Hierarchy of Regular Tree Languages of Index [0, 2]

Jacques Duparc[1] and Kevin Fournier[1,2](✉)

[1] Department of Information Systems Faculty of Business and Economics,
University of Lausanne, 1015 Lausanne, Switzerland
jacques.duparc@unil.ch
[2] Équipe de Logique Mathématique, Université Paris Diderot,
UFR de Mathématiques Case 7012, 75205 Paris Cedex 13, France
kevin.fournier@imj-prg.fr

Abstract. We provide a hierarchy of tree languages recognised by non-deterministic parity tree automata with priorities in $\{0, 1, 2\}$, whose length exceeds the first fixed point of the ε operation (that itself enumerates the fixed points of $x \mapsto \omega^x$). We conjecture that, up to Wadge equivalence, it exhibits all regular tree languages of index [0, 2].

1 Introduction

This paper contributes to the close investigation of regular tree languages of index [0, 2]. Our tool to measure and compare those languages is given by descriptive set theory through the notion of *topological complexity*. It is well known that deterministic parity tree automata recognize only languages in the $\mathbf{\Pi}_1^1$ class (coanalytic sets), whereas nondeterministic automata recognize languages that are neither analytic, nor coanalytic. The expressive power of nondeterministic automata is nonetheless bounded by the second level of the projective hierarchy, and, by Rabin's complementation result [7], all nondeterministic languages are in fact in the $\mathbf{\Delta}_2^1$ class. A more discriminating topological complexity measure than the Baire and the projective hierarchy is therefore needed: the Wadge hierarchy, which relies on the notion of reductions by continuous functions (Wadge-reducibility). Complexity classes, called Wadge degrees, consist of sets Wadge-reducible to each other, and constitute a hierarchy whose levels, called ranks, can be enumerated with ordinals. We describe a series of operations on automata that preserve the index and lift the Wadge degrees of the recognized languages[1]. These operations help us generate a hierarchy of regular tree languages of higher and higher topological complexity, one level higher than the first fixed point of the ordinal function[2] $x \mapsto \varepsilon_x$ which itself enumerates the fixed points of the exponentiation $x \mapsto \omega^x$.

---

[1] We emphasize that this is done without any determinacy principle. In particular, we do not require $\mathbf{\Delta}_2^1$-determinacy.

[2] Not to be mistaken with an ε-move.

© Springer International Publishing Switzerland 2015
J. Shallit and A. Okhotin (Eds.): DCFS 2015, LNCS 9118, pp. 81–92, 2015.
DOI: 10.1007/978-3-319-19225-3_7

2   Preliminaries

2.1   The Wadge Hierarchy and the Wadge Game

The Wadge theory is in essence the theory of *pointclasses*[3] (see [1]). For Γ a pointclass, we denote by $\check{\Gamma}$ its *dual* class containing all the subsets whose complements are in Γ, and by $\Delta(\Gamma)$ the ambiguous class $\Gamma \cap \check{\Gamma}$. If $\Gamma = \check{\Gamma}$, we say that Γ is *self-dual*.

Given any topological space X, the *Wadge preorder* $\leq_W$ on $\mathscr{P}(X)$ is defined for $A, B \subseteq X$ by $A \leq_W B$ if and only if there exists $f \colon X \longrightarrow X$ continuous such that $f^{-1}(B) = A$. It is merely by definition a preorder which induces an equivalence relation $\equiv_W$ whose equivalence classes – denoted by $[A]_W$ – are called the *Wadge degrees*. A set $A \subseteq X$ is *self-dual* if $[A]_W = [A^{\complement}]_W$, and *non-self-dual* otherwise. We use the same terminology for the Wadge degrees. We have a direct correspondence between $(\mathscr{P}(X), \leq_W)$ restricted to Γ and the pointclasses included in Γ with inclusion: the pointclasses are exactly the initial segments of the Wadge preorder. In particular, the Wadge hierarchy tremendously refines the Borel and the projective hierarchies.

The space T_Σ equipped with the standard Cantor topology is a Polish space, and is in fact homeomorphic to the Cantor space [2]. Let $L, M \subseteq T_\Sigma$, the Wadge game $W(L, M)$ is a two-player infinite game that provides a very useful characterization for the Wadge preorder. In this game, each player builds a tree, say t_I and t_{II}. At every round, player I plays first, and both players add a finite number of children to the terminal nodes of their tree. Player II is allowed to skip her turn, but has to produce a tree in T_Σ throughout a game. Player II wins the game if and only if $t_I \in L \Leftrightarrow t_{II} \in M$.

Lemma 1 ([9]). *Let $L, M \subseteq T_\Sigma$. Then $L \leq_W M$ if and only if player II has a winning strategy in the game $W(L, M)$.*

We write $A <_W B$ when II has a winning strategy in $W(A, B)$ *and* I has a winning strategy in $W(B, A)$[4]. Given a pointclass Γ of T_Σ with suitable closure properties, the assumption of the determinacy of Γ is sufficient to prove that Γ is semi-linearly ordered by $\leq_W$, denoted SLO(Γ), i.e., that for all $L, M \in \Gamma$,

$$L \leq_W M \qquad \text{or} \qquad M \leq_W L^{\complement},$$

and that $\leq_W$ is well founded when restricted to sets in Γ [1,8]. Under these conditions, the Wadge degrees of sets in Γ with the induced order is thus a hierarchy called the *Wadge hierarchy*. Therefore, there exists a unique ordinal, called the *height* of the Γ-Wadge hierarchy, and a mapping d_W^Γ from the Γ-Wadge hierarchy onto its height, called the *Wadge rank*, such that, for every L, M non-self-dual in Γ, $d_W^\Gamma(L) < d_W^\Gamma(M)$ if and only if $L <_W M$ and $d_W^\Gamma(L) = d_W^\Gamma(M)$

---

[3] A pointclass is a collections of subsets of a topological space that is closed under continuous preimages.

[4] This is in general stronger than the usual $A <_W B$ if and only if $A \leq_W B$ and $B \not\leq_W A$, but the two definitions coincide when the classes considered are determined.

if and only if $L \equiv_W M$ or $L \equiv_W M^{\complement}$. The wellfoundedness of the Γ-Wadge hierarchy ensures that the Wadge rank can be defined by induction as follows:

- $d_W^\Gamma(\emptyset) = d_W^\Gamma(\emptyset^{\complement}) = 1$.
- $d_W^\Gamma(L) = \sup\{d_W^\Gamma(M) + 1 : M \text{ is non-self-dual}, M <_W L\}$ for $L >_W \emptyset$.

Note that given two pointclasses Γ and Γ', for every $L \in \Gamma \cap \Gamma'$, we have $d_W^\Gamma(L) = d_W^{\Gamma'}(L)$. Under sufficient determinacy assumptions, we can therefore safely speak of *the* Wadge rank of a tree language, denoted by d_W, as its Wadge rank with respect to any topological class including it. However the main result of this article does not provide any Wadge rank for the canonical languages that are constructed, because we do not make use of any determinacy principle.

2.2   The Conciliatory Hierarchy

A *conciliatory* binary tree over a finite set Σ is a partial function $t : \{0,1\}^* \to \Sigma$ with a prefix-closed domain. Such trees can have both infinite and finite branches. A tree is called *full* if $\mathrm{dom}(t) = \{0,1\}^*$. Let $T_\Sigma^{\leq \omega}$ and T_Σ denote, respectively, the set of all conciliatory binary trees and the set of full binary trees over Σ. Given $x \in \mathrm{dom}(t)$, we denote by t_x the subtree of t rooted at x. Let $\{0,1\}^n$ denote the set of words over $\{0,1\}$ of length n, and let t be a conciliatory tree over Σ. We denote by $t[n]$ the finite initial binary tree of height $n+1$ given by the restriction of t to $\bigcup_{0 \leq i \leq n}\{0,1\}^i$.

For conciliatory languages L, M we define the *conciliatory* version of the Wadge game: $C(L, M)$ [4,5]. The rules are similar, except for the fact that both players are now allowed to skip and to produce trees with finite branches – or even finite trees. For conciliatory languages L, M we use the notation $L \leq_c M$ if and only if II has a winning strategy in the game $C(L, M)$. If $L \leq_c M$ and $M \leq_c L$, we will write $L \equiv_c M$. The conciliatory hierarchy is thus the partial order induced by $\leq_c$ on the equivalence classes given by $\equiv_c$. We write $A <_c B$ when II has a winning strategy in $C(A, B)$ *and* I has a winning strategy in $C(B, A)$.

From a conciliatory language L over Σ, one defines the corresponding language L^b of full trees over $\Sigma \cup \{b\}$ by

$$L^b = \{t \in T_{\Sigma \cup \{b\}} : t_{[\ /b]} \in L\},$$

where b is an extra symbol that stands for "blank", and $t_{[\ /b]}$, the *undressing* of t, is informally the conciliatory tree over Σ obtained once all the occurrences of b have been removed in a top-down manner. More precisely, if there is a node v such that $t(v) = b$, we ignore this node and replace it with $v0$. If, for each integer n, $t(v0^n) = b$, then $v \notin \mathrm{dom}(t_{[\ /b]})$. This process is illustrated by Fig. 1.

If Γ is a pointclass of full trees, we say that a conciliatory language L is in Γ if and only if L^b is in Γ.

Lemma 2. *Let L and M be conciliatory languages. Then*

$$L \leq_c M \text{ if and only if } L^b \leq_W M^b.$$

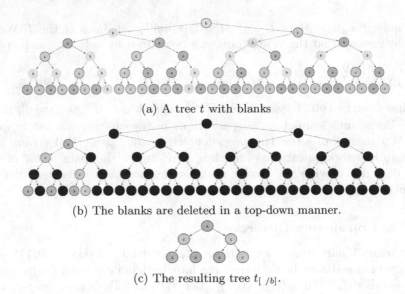

(a) A tree t with blanks

(b) The blanks are deleted in a top-down manner.

(c) The resulting tree $t_{[\ /b]}$.

Fig. 1. The undressing process.

The mapping $L \mapsto L^b$ gives thus a natural embedding of the preorder $\leq_c$ restricted to conciliatory sets in Γ into the Γ-Wadge hierarchy. Hence, for Γ with suitable closure and determinacy properties, the conciliatory degrees of sets in Γ with the induced order constitute a hierarchy called the *conciliatory hierarchy*. We define, by induction, the corresponding *conciliatory rank* of a language:

- $d_c^\Gamma(\emptyset) = d_c^\Gamma(\emptyset^{\mathbb{C}}) = 1$.
- $d_c^\Gamma(L) = \sup\{d_c^\Gamma(M) + 1 : M <_c L\}$ for $L >_c \emptyset$.

Similarly to the Wadge case, given two pointclasses Γ and Γ', for every conciliatory $L \in \Gamma \cap \Gamma'$, we have $d_c^\Gamma(L) = d_c^{\Gamma'}(L)$. Under sufficient determinacy assumptions, we can therefore speak safely of *the* conciliatory rank of a conciliatory tree language, denoted by d_c, as its conciliatory rank with respect to any topological class including it. Observe that the conciliatory hierarchy does not contain self-dual languages: a strategy for I in $C(L, L^{\mathbb{C}})$ is to skip in the first round, and then copy moves of II.

2.3   Automata and Conciliatory Trees

A *nondeterministic parity tree automaton* $\mathcal{A} = \langle \Sigma, Q, I, \delta, \mathrm{r} \rangle$ consists of a finite input alphabet Σ, a finite set Q of states, a set of initial states $I \subseteq Q$, a transition relation $\delta \subseteq Q \times \Sigma \times Q \times Q$ and a priority function $\mathrm{r} : Q \to \omega$. A run of automaton $\mathcal{A}$ on a binary conciliatory input tree $t \in \mathcal{T}_\Sigma^{\leq \omega}$ is a conciliatory tree $\rho_t \in \mathcal{T}_Q^{\leq \omega}$ with $\mathrm{dom}(\rho_t) = \{\varepsilon\} \cup \{va : v \in \mathrm{dom}(t) \wedge a \in \{0,1\}\}$ such that the root of this tree is labelled with a state $q \in I$, and for each $v \in \mathrm{dom}(t)$, transition $(\rho_t(v), t(v), \rho_t(v_1), \rho_t(v_1)) \in \delta$. The run ρ_t is *accepting* if parity condition is satisfied on each infinite branch of ρ_t, i.e., if the highest rank of a state occurring infinitely often on the branch is even, and if the rank of each leaf node in ρ_t is even. We say that a parity tree automaton A *accepts* a conciliatory tree t if it has an accepting run on t. The language *recognized* by A, denoted $L(A)$ is the set of trees accepted by A. We let $L^\omega(A)$ denote the set of full trees recognized by A, i.e., $L^\omega(A) = L(A) \cap T_\Sigma$. Notice that as the set of states is finite, the priority function is bounded. Moreover, shifting all ranks by an even number does not change the language recognized by a parity tree automaton. It is thus sufficient to consider parity tree automata whose priorities are restricted to intervals $[\iota, \kappa]$, for $\iota \in \{0, 1\}$. We say that an automaton is of index $[\iota, \kappa]$ if its priorities are restricted to intervals $[\iota, \kappa]$. A language is of index $[\iota, \kappa]$ if there is an automaton of index $[\iota, \kappa]$ that recognises it. This gives rise to the Mostowski-Rabin hierarchy [3]. Let $W_{[0,2]}$ be the game tree language of index $[0, 2]$. One can prove that $L \leq_W W_{[0,2]}$ holds for any regular tree language L of index $[0, 2]$, but fails for $L = W^{\mathsf{C}}_{[0,2]}$.

Corollary 1. *The mapping $L \mapsto L^b$ embeds the conciliatory hierarchy for Δ^1_2-sets restricted to languages of index $[0, 2]$ into the Δ^1_2-Wadge hierarchy restricted to languages of index $[0, 2]$.*

We use the following conventions in the diagrams. Nodes represent states of the automaton. Node labels correspond to state ranks. A red edge shows the state that is assigned to the left successor node of a transition, and a green edge goes to the right successor node. In order to lighten the notation, transitions that are not depicted on a diagram lead to some all-accepting state. Given automata $\mathcal{A}$ and $\mathcal{B}$, we write $\mathcal{A} \leq_c \mathcal{B}$ for $L(\mathcal{A}) \leq_c L(\mathcal{B})$, and same with $<_c, \leq_W, <_W$.

3 Operations on Languages and Their Automatic Counterparts

We present operations on conciliatory tree languages, which we then use to construct more and more complex languages. W.l.o.g. we assume the alphabet to be $\Sigma = \{a, c\}$.

3.1 The Sum

For $L, M \subseteq T_\Sigma^{\leq \omega}$, we define $L \oplus M$ (the *sum* of L and M) as the language formed of all those trees $t \in T_\Sigma^{\leq \omega}$ such that one of the following conditions holds:

- $t(10^n) = a$ for each integer n and $t_0 \in M$;
- the node 10^n is the first on the path 10^* labeled with c and either $t(10^n0) = a$ and $t_{10^n00} \in L$, or $t(10^n0) = c$ and $t_{10^n00} \in L^{\mathsf{C}}$.

This operation behaves well regarding the conciliatory hierarchy.

Facts 1 ([4,5]). *Given L, M, and M' any conciliatory tree languages over Σ,*

1. $(L \oplus M)^{\complement} \equiv_c L \oplus M^{\complement}$.
2. The operation $\oplus$ preserves the conciliatory ordering: if $M' \leq_c M$, then

$$L \oplus M' \leq_c L \oplus M.$$

3. Assuming enough determinacy:

$$d_c(L \oplus M) = d_c(L) + d_c(M).$$

Let $\mathcal{A}$ and $\mathcal{B}$ be two automata that recognize, respectively, the conciliatory languages M and L. Then the automaton $\mathcal{B}+\mathcal{A}$ depicted in Fig. 2 recognizes the sum of L and M. In this picture, $\mathcal{C}$ is any automaton of index $[0,2]$ that recognizes a language equivalent to $L^{\complement}$, and the parity i and j are defined as follows:

- $i = 0$ if and only if the empty tree is accepted by $\mathcal{A}$;
- $j = 1$ if and only if $L(\mathcal{A})$ is equivalent to $L(\mathcal{A}) \to \ominus$, where $\ominus$ denotes any automaton that rejects all trees.[5]

Notice that if $\mathcal{A}$ and $\mathcal{B}$ are parity tree automata of index $[0,2]$ such that $L(\mathcal{B})^{\complement}$ can be recognized by an automaton of index $[0,2]$, then $\mathcal{B}+\mathcal{A}$ is a parity tree automata of index $[0,2]$.

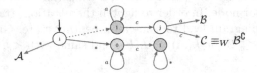

Fig. 2. The automaton $\mathcal{B}+\mathcal{A}$ that recognizes $L(\mathcal{B}) \oplus L(\mathcal{A})$. The values of i and j depend on properties of $\mathcal{A}$.

Lemma 3. *Let L, L', M and M' be conciliatory languages such that $L <_c L'$ and $M \leq_c M'$. Then the following hold.*

1. $M \oplus L <_c M' \oplus L'$;
2. $M <_c M \oplus L$.

3.2   Multiplication by a Countable Ordinal

In order to define the multiplication of a language by a countable ordinal, we first introduce the operation $\sup_{n<\omega}$. Let $(L_n)_{n\in\omega} \subseteq T_{\Sigma}^{\leq\omega}$ be a countable family of conciliatory languages. Define $\sup_{n<\omega} L_n$ as the conciliatory tree language containing all of those trees $t \in T_{\Sigma}^{\leq\omega}$ such that one of the following conditions holds:

---

[5] A player in charge of $L(\mathcal{A}) \to \ominus$ in a conciliatory game is like a player in charge of $L(\mathcal{A})$, but with the extra possibility at any moment of the play to reach a definitively rejecting position.

- $t(1^n) = a$ for all integer n;
- the node 1^n is the first on the path 1^* labeled with c and $t_{1^n 0} \in L_n$.

The multiplication by a countable ordinal is now defined as an iterated sum. Let $L \subseteq T_{\Sigma}^{\leq \omega}$, $L \odot 1 = L$, $L \odot (\alpha + 1) = (L \odot \alpha) \oplus L$, and $L \odot \lambda = \sup_{\alpha < \lambda} L \odot \alpha$, for λ limit.

Let $\mathcal{A}$ be an automaton that recognizes the conciliatory languages L. Then the automaton $\mathcal{A} \bullet \omega$ depicted in Fig. 3(a) recognizes a language equivalent to $L \odot \omega$. In this picture, $\mathcal{C}$ is any automaton that recognizes a language equivalent to $L^{\mathcal{C}}$. The automaton $\mathcal{A} \overline{\bullet} \omega$ that recognizes the complement of $L(\mathcal{A} \bullet \omega)$, and thus a language equivalent to the complement of $L \odot \omega$, is depicted in Fig. 3b. Notice that if $\mathcal{A}$ is of index $[0, 2]$, and if there exists an automaton that recognizes $L(\mathcal{A})^{\mathcal{C}}$ of index $[0, 2]$, then both $\mathcal{A} \bullet \omega$ and $\mathcal{A} \overline{\bullet} \omega$ are parity tree automata of index $[0, 2]$. Hence, for every ordinal $0 < \alpha < \omega^{\omega}$ and for every automaton $\mathcal{A}$, there exists an automaton $\mathcal{A} \bullet \alpha$ that recognizes $L(\mathcal{A}) \odot \alpha$. Moreover, if $\mathcal{A}$ is of index $[0, 2]$, and if there exists an automaton that recognizes $L(\mathcal{A})^{\mathcal{C}}$ of index $[0, 2]$, then $\mathcal{A} \bullet \alpha$ is a parity tree automaton of index $[0, 2]$.

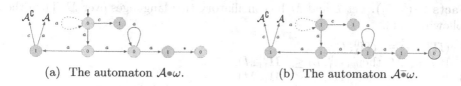

(a) The automaton $\mathcal{A} \bullet \omega$.   (b) The automaton $\mathcal{A} \overline{\bullet} \omega$.

Fig. 3. Automata that recognize respectively a language equivalent to $L \odot \omega$ and a language equivalent to its complement.

As a corollary of Lemma 3 and Facts 1, the multiplication by a countable ordinal behaves well regarding the conciliatory hierarchy.

Corollary 2. *Let L and M be conciliatory languages such that $L <_c M$. Then for every countable ordinals $0 < \alpha < \beta < \omega^{\omega}$:*

1. $L \odot \alpha <_c L \odot \beta$;
2. $L \odot \alpha <_c M \odot \alpha$.

3.3 The Pseudo-Exponentiation

Let $P \subseteq T_{\Sigma}^{\leq \omega}$ be a conciliatory tree language. For $t \in T_{\Sigma}^{\leq \omega}$, let:

$$i^P(t)(a_1, a_2, \ldots, a_n) = \begin{cases} t(a_1, 0, a_2, 0, \ldots, 0, a_n, 0), & \text{if } t_{a_1, 0, a_2, 0, \ldots, 0, a_n, 1} \in P; \\ b, & \text{otherwise.} \end{cases}$$

This process is illustrated in Fig. 4. The nodes in blue are called the *main run*. The blue arrows denote the dependency of a node of the main run on a subtree of auxiliary moves. If the auxiliary subtree of a main run node is not in P, then we say that the node is *killed*.

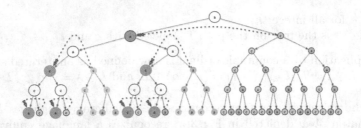

Fig. 4. Main run and auxiliary moves.

Let $L \subseteq \mathcal{T}_\Sigma^{\leq \omega}$, we define the *action* of P on L, in symbols (P, L), by

$$\left\{ t \in \mathcal{T}_\Sigma^{\leq \omega} : i^P(t)_{[\ /b]} \in L \right\}.$$

Let $P_{\mathbf{\Pi}_1^0}$ be the complete closed set of all full trees over Σ with all nodes on the leftmost branch 0^* labelled by a. For $L \subseteq \mathcal{T}_\Sigma^{\leq \omega}$, we denote by $(\mathbf{\Pi}_1^0, L)$ the action of $P_{\mathbf{\Pi}_1^0}$ on L. This operation $(\mathbf{\Pi}_1^0, \cdot)$ behaves well regarding the conciliatory hierarchy.

Facts 2 ([4,5]). *Let L and M be conciliatory tree languages over Σ. Then the following hold.*

1. $(\mathbf{\Pi}_1^0, L)^{\complement} \equiv_c (\mathbf{\Pi}_1^0, L^{\complement})$.
2. *If $L \leq_c M$, then $(\mathbf{\Pi}_1^0, L) \leq_c (\mathbf{\Pi}_1^0, M)$.*
3. *If $L <_c M$, then $(\mathbf{\Pi}_1^0, L) <_c (\mathbf{\Pi}_1^0, M)$.*
4. *Assuming enough determinacy, $d_c((\mathbf{\Pi}_1^0, L)) = \omega_1^{d_c(L)+\varepsilon}$, for* [6] *$\varepsilon \in \{-1, 0, 1\}$.*

Without assuming any determinacy hypothesis, we can nonetheless prove the following Proposition that links $(\mathbf{\Pi}_1^0, \cdot)$ to $\oplus$.

Proposition 1. *Let L, L' and M be conciliatory languages such that $L <_c (\mathbf{\Pi}_1^0, M)$ and $L' <_c (\mathbf{\Pi}_1^0, M)$. Then*

1. $L \oplus L' <_c (\mathbf{\Pi}_1^0, M)$;
2. $L \odot \alpha <_c (\mathbf{\Pi}_1^0, M)$, *for any $\alpha < \omega^\omega$.*

Given any automaton $\mathcal{A}$ recognizing $L \subseteq \mathcal{T}_\Sigma^{\leq \omega}$, the conciliatory language $(\mathbf{\Pi}_1^0, L)$ is recognized by the automaton $(\omega^\omega)^{\mathcal{A}}$ defined from $\mathcal{A}$ by replacing each state of $\mathcal{A}$ by a "gadget", as depicted in Fig. 5. By replacing a state by the gadget we mean that all transitions ending in this state should now end in the initial state of the gadget, and that all the transitions leaving this state should now start from the final state of the gadget. This sort of gadget first appeared in [5]. Notice that if $L \subseteq \mathcal{T}_\Sigma^{\leq \omega}$ is of index $[0, 2]$, then $(\mathbf{\Pi}_1^0, L)$ is also of index $[0, 2]$. Observe also that the game language $W_{[0,2]}$ is a fixed point for pseudo-exponentiation, i.e.,

$$(\mathbf{\Pi}_1^0, W_{[0,2]})^b \equiv_W W_{[0,2]}.$$

---

[6] $\varepsilon = \begin{cases} -1 & \text{if } d_c(L) < \omega; \\ 0 & \text{if } d_c(L) = \beta + n \text{ and } \operatorname{cof}(\beta) = \omega_1; \\ 1 & \text{if } d_c(L) = \beta + n \text{ and } \operatorname{cof}(\beta) = \omega. \end{cases}$

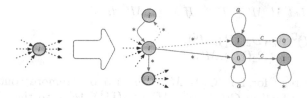

Fig. 5. The gadget to replace a state in $\mathcal{A}$.

4   Difference of Co-analytic Sets

The operations defined in Sect. 3 are *Borel* in the sense that when we apply them
to Borel languages, the resulting language is still Borel. In order to describe the
most of the Wadge hierarchy of languages recognized by parity tree automata of
index $[0, 2]$ we need to climb higher.

4.1   The Operation $(D_2(\Pi_1^1), \cdot)$

We define a conciliatory language of index $[0, 2]$ that is $D_2(\Pi_1^1)$-complete (Fig. 6a)
and such that its complement (Fig. 6b) is also of index $[0, 2]$, via the automata
that recognize each of them. We denote by $A_{D_2(\Pi_1^1)}$ and $A_{\check{D}_2(\Pi_1^1)}$ the conciliatory
languages recognized respectively by $\mathcal{A}_{D_2(\Pi_1^1)}$ and $\mathcal{A}_{\check{D}_2(\Pi_1^1)}$.

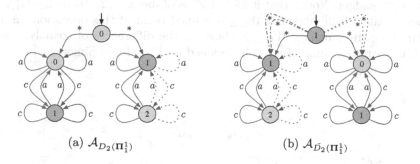

(a) $\mathcal{A}_{D_2(\Pi_1^1)}$ (b) $\mathcal{A}_{\check{D}_2(\Pi_1^1)}$

Fig. 6. Automata that recognize respectively a $D_2(\Pi_1^1)$-complete and a $\check{D}_2(\Pi_1^1)$-complete language.

For $M \subseteq \mathcal{T}_{\Sigma}^{\leq \omega}$, we denote by $(D_2(\Pi_1^1), M)$ the action of $L(\mathcal{A}_{D_2(\Pi_1^1)})$ on
M. Observe that this operation is highly non-Borel, since if we apply it to a
Σ_1^0-complete conciliatory language, the resulting language will be complete for
the pointclass of all the countable unions of $D_2(\Pi_1^1)$ languages. The operation
$(D_2(\Pi_1^1), \cdot)$ behaves well with respect to $\leq_c$.

Theorem 1. *Let* $M, M' \subseteq \mathcal{T}_\Sigma^{\leq\omega}$. *If* $M \leq_c M'$, *then*

1. $(D_2(\mathbf{\Pi}_1^1), M)^{\mathsf{C}} \equiv_c (D_2(\mathbf{\Pi}_1^1), M^{\mathsf{C}})$;
2. $(D_2(\mathbf{\Pi}_1^1), M) \leq_c (D_2(\mathbf{\Pi}_1^1), M')$.

A winning strategy for I in $C(M, M')$ can also be "remote controlled" to a winning strategy for I in $C((D_2(\mathbf{\Pi}_1^1), M), (D_2(\mathbf{\Pi}_1^1), M'))$, so that the following holds.

Corollary 3. *Let* M *and* M' *be conciliatory languages such that* $M <_c M'$. *Then*

$$(D_2(\mathbf{\Pi}_1^1), M) <_c (D_2(\mathbf{\Pi}_1^1), M')$$

The operation $(D_2(\mathbf{\Pi}_1^1), \cdot)$ is much stronger than $(\mathbf{\Pi}_1^0, \cdot)$, and is in fact a fixed point of it.

Proposition 2. *Let* $M \subseteq \mathcal{T}_\Sigma^{\leq\omega}$. *Then*

$$(\mathbf{\Pi}_1^0, (D_2(\mathbf{\Pi}_1^1), M)) \equiv_c (D_2(\mathbf{\Pi}_1^1), M).$$

Let $\mathcal{A}$ be an automaton that recognizes $M \subseteq \mathcal{T}_\Sigma^{\leq\omega}$. Then the conciliatory tree language $(D_2(\mathbf{\Pi}_1^1), M)$ is recognized by the automaton $\varepsilon_{\mathcal{A}}$ defined from $\mathcal{A}$ by replacing each state of $\mathcal{A}$ by a "gadget", as depicted in Fig. 7. As in the pseudo-exponentiation case, by replacing a state by the gadget we mean that all transitions ending in this state should now end in the initial state of the gadget, and that all the transitions starting from this state should now start from the final state of the gadget. Notice that if $M \subseteq \mathcal{T}_\Sigma^{\leq\omega}$ is of index $[0, 2]$, then $(D_2(\mathbf{\Pi}_1^1), M)$ is also of index $[0, 2]$, and that $W_{[0,2]}$ is a fixed point of this operation. In particular the game language $W_{[0,2]}$ is above all the differences of coanalytic sets, which is a strengthening of a result obtained by Finkel and Simonnet [6].

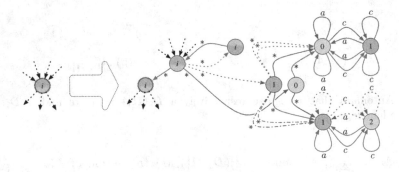

Fig. 7. The gadget to replace a state in $\mathcal{A}$.

5   A Fragment of the Wadge Hierarchy

Let $\varphi_2(0)$ denote the first fixed point[7] of the ordinal epsilon function, namely the one that enumerates the fixed points of the exponentiation of base ω:

$$\varepsilon_0 = \sup_{n<\omega} \underbrace{\omega^{\cdot^{\cdot^{\omega^0}}}}_{n} \; ; \; \varepsilon_{\alpha+1} = \sup_{n<\omega} \underbrace{\omega^{\cdot^{\cdot^{\omega^{(\varepsilon_\alpha+1)}}}}}_{n} \; ; \; \varepsilon_\lambda = \sup_{\alpha<\lambda} \varepsilon_\alpha, \text{ for } \lambda \text{ limit.}$$

Finally: $\varphi_2(0) = \sup_{n<\omega} \overbrace{\varepsilon_{\cdot_{\cdot_{\varepsilon_0}}}}^{n}$.

We recall that every ordinal $\alpha > 0$ admits a unique Cantor normal form of base ω^ω (CNF) which is an expression of the form $\alpha = (\omega^\omega)^{\alpha_k} \cdot \nu_k + \cdots + (\omega^\omega)^{\alpha_0} \cdot \nu_0$ where $k < \omega$, $0 < \nu_i < \omega^\omega$ for any $i \leq k$, and $\alpha_0 < \cdots < \alpha_k < \alpha$.

For every ordinal $0 < \alpha < \varphi_2(0)$, we inductively define a pair of automata $(\mathcal{A}_\alpha, \bar{\mathcal{A}}_\alpha)$ whose languages are incomparable through the conciliatory ordering. If the CNF of α is $\alpha = (\omega^\omega)^{\alpha_k} \cdot \nu_k + \cdots + (\omega^\omega)^{\alpha_0} \cdot \nu_0$ we set

$$\mathcal{A}_\alpha = \mathcal{A}_{(\omega^\omega)^{\alpha_k}} \bullet \nu_k + \cdots + \mathcal{A}_{(\omega^\omega)^{\alpha_0}} \bullet \nu_0, \quad \bar{\mathcal{A}}_\alpha = \mathcal{A}_{(\omega^\omega)^{\alpha_k}} \bullet \nu_k + \cdots + \bar{\mathcal{A}}_{(\omega^\omega)^{\alpha_0}} \bullet \nu_0,$$

where $\mathcal{A}_{(\omega^\omega)^{\alpha_i}}$ and $\bar{\mathcal{A}}_{(\omega^\omega)^{\alpha_i}}$ are respectively

- $\ominus$ and $\oplus$ if $\alpha_i = 0$;
- $(\omega^\omega)^{\mathcal{A}_{\alpha_i}}$ and $(\omega^\omega)^{\bar{\mathcal{A}}_{\alpha_i}}$ if $\alpha_i < (\omega^\omega)^{\alpha_i}$;
- $\varepsilon_{\mathcal{A}_{2+\beta}}$ and $\varepsilon_{\bar{\mathcal{A}}_{2+\beta}}$ if $\alpha_i = (\omega^\omega)^{\alpha_i}$ holds[8] and $\alpha_i = \varepsilon_\beta$ for some $\beta < \alpha_i$.

Lemma 4. *For $0 < \alpha < \beta < \varphi_2(0)$, we have*

1. $\mathcal{A}_\alpha \not\leq_c \bar{\mathcal{A}}_\alpha$ *and* $\bar{\mathcal{A}}_\alpha \not\leq_c \mathcal{A}_\alpha$.
2. $\mathcal{A}_\alpha <_c \mathcal{A}_\beta$; $\bar{\mathcal{A}}_\alpha <_c \mathcal{A}_\beta$; $\mathcal{A}_\alpha <_c \bar{\mathcal{A}}_\beta$ *and* $\bar{\mathcal{A}}_\alpha <_c \bar{\mathcal{A}}_\beta$.

Applying the embedding $L \mapsto L^b$, we have thus generated a family $\left(\mathcal{A}_\alpha{}^b\right)_{\alpha<\varphi_2(0)}$ of parity tree automata of index $[0, 2]$ that respects the strict Wadge ordering: $\alpha < \beta$ if and only if $\mathcal{A}_\alpha{}^b <_W \mathcal{A}_\beta{}^b$. Hence the main result follows.

Theorem 2. *There exists a family $\left(\mathcal{A}_\alpha{}^b\right)_{\alpha<\varphi_2(0)}$ of parity tree automata of index $[0, 2]$ such that*

1. *they recognize languages of full trees over the alphabet $\{a, b, c\}$;*
2. *$\alpha < \beta$ holds if and only if $\mathcal{A}_\alpha{}^b <_W \mathcal{A}_\beta{}^b$ holds as well.*

Let $\mathcal{A}_{\varphi_2(0)}{}^b$ be an automaton of index $[0, 2]$ over the alphabet $\{a, b, c\}$ that recognizes a language equivalent to $W_{[0,2]}$. We formulate the following conjecture.

Conjecture. Let L be a regular non-self-dual full language of index $[0, 2]$. Then either $L \equiv_W W_{[0,2]}$, or there exists $\alpha < \varphi_2(0)$ such that $L \equiv_W L(\mathcal{A}_\alpha{}^b)$ or $L^{\complement} \equiv_W L(\mathcal{A}_\alpha{}^b)$.

---

[7] Another way to characterise $\varphi_2(0)$ is to remember that an ordinal is the set of its predecessors and notice that a nonzero ordinal is of the form respectively ω^α iff it is closed under addition and ε_α iff it is closed under $x \longmapsto \omega^x$. Then $\varphi_2(0)$ is the first non null ordinal closed under $x \longmapsto \varepsilon_x$ as well as $x \longmapsto \omega^x$ and $x, y \longmapsto x + y$.

[8] Notice that we have $\alpha_i = (\omega^\omega)^{\alpha_i} \iff \alpha_i = \omega^{\alpha_i}$.

6 Conclusion

In this paper, we have produced a very long chain of parity tree automata of index $[0, 2]$ but of different Wadge degrees. Its length is $\varphi_2(0) + 1$, where $\varphi_2(0)$ is the first fixed point of the ordinal function that itself enumerates all fixed points of the ordinal exponentiation $x \mapsto \omega^x$. We conjecture that every regular non-self-dual language of index $[0, 2]$ is, up to Wadge equivalence, recognized by an automaton in $\left(\mathcal{A}_\alpha{}^b\right)_{\alpha < \varphi_2(0)+1}$. Since degrees of self-dual languages of index $[0, 2]$ are always immediately above and below two non-self-dual degrees of languages of index $[0, 2]$, this conjecture would imply that the height of the Wadge hierarchy of regular languages of index $[0, 2]$ is exactly $\varphi_2(0) + 1$.

The whole construction is effective, meaning that the mapping $\alpha \mapsto \mathcal{A}_\alpha{}^b$ (for $0 < \alpha < \varphi_2(0) + 1$) is recursive. It also means that, for any $0 < \alpha < \beta < \varphi_2(0) + 1$, the relation $\mathcal{A}_\alpha{}^b <_W \mathcal{A}_\beta{}^b$ which stipulates that there exist two strategies – one that is winning for player II in the game $W(\mathcal{A}_\alpha{}^b, \mathcal{A}_\beta{}^b)$ and another one that is winning for I in the game $W(\mathcal{A}_\beta{}^b, \mathcal{A}_\alpha{}^b)$ – can be established by recursively providing such strategies. However, we did not consider any decidability issue. It thus remains open whether one can decide, given any automaton $\mathcal{B}$ and any ordinal $0 < \alpha < \varphi_2(0) + 1$, whether $\mathcal{B} <_W \mathcal{A}_\alpha{}^b$ holds or not.

References

1. Andretta, A., Louveau, A.: Wadge degrees and pointclasses. In: Kechris, A.S., Löwe, B., Steel, J.R. (eds.) Wadge Degrees and Projective Ordinals: The Cabal Seminar, vol. II, pp. 3–23. Cambridge University Press, Cambridge (2012)
2. Arnold, A., Duparc, J., Murlak, F., Niwiński, D.: On the topological complexity of tree languages. Logic Automata: Hist. Perspect. **2**, 9–29 (2007)
3. Arnold, A., Niwiński, D.: Rudiments of μ-Calculus. Studies in Logic and the Foundations of Mathematics. Elseiver, Amsterdam (2001)
4. Duparc, J.: Wadge hierarchy and Veblen hierarchy, part I: Borel sets of finite rank. J. Symbolic Logic **66**(1), 56–86 (2001)
5. Duparc, J., Murlak, F.: On the topological complexity of weakly recognizable tree languages. In: Csuhaj-Varjú, E., Ésik, Z. (eds.) FCT 2007. LNCS, vol. 4639, pp. 261–273. Springer, Heidelberg (2007)
6. Finkel, O., Simonnet, P.: On recognizable tree languages beyond the Borel hierarchy. Fundam. Informaticae **95**(2–3), 287–303 (2009)
7. Rabin, M.O.: Decidability of second-order theories and automata on infinite trees. Trans. AMS **141**, 1–23 (1969)
8. Van Wesep, R.: Wadge degrees and descriptive set theory. In: Kechris, A.S., Moschovakis, Y.N. (eds.) Cabal Seminar 76–77, pp. 151–170. Springer, Heidelberg (1978)
9. Wadge, W.W.: Reducibility and determinateness on the Baire space. Ph.D. thesis, University of California, Berkeley (1984)

Compressibility of Finite Languages by Grammars

Sebastian Eberhard and Stefan Hetzl[✉]

Institute of Discrete Mathematics and Geometry, Vienna University of Technology,
Wiedner Hauptstraße 8-10, 1040 Wien, Austria
sebastian.eberhard84@gmail.com, stefan.hetzl@tuwien.ac.at

Abstract. We consider the problem of simultaneously compressing a
finite set of words by a single grammar. The central result of this paper is
the construction of an incompressible sequence of finite word languages.
This result is then shown to transfer to tree languages and (via a previously established connection between proof theory and formal language
theory) also to formal proofs in first-order predicate logic.

1 Introduction

In grammar-based compression, context-free grammars that generate exactly
one word are used for representing the input text. The smallest grammar problem asks for the smallest context-free grammar that generates a given word.
Its decision version is known to be NP-complete [28]. However, there is a number of fast algorithms which are practically useful [16,17,19] or achieve good
approximation ratios [5,14,15,23,24]. Grammar-based compression also has the
considerable practical advantage that many operations can be performed directly
on the compressed representation; see [18].

In this paper we consider the problem of simultaneously compressing a finite
set of words by a single grammar. Our motivation for investigating this problem
is rooted in proof theory and automated deduction: as shown in [7] there is
an intimate relationship between a certain class of formal proofs (those with
Π_1-cuts) in first-order predicate logic and a certain class of grammars (totally
rigid acyclic tree grammars). In particular, the number of production rules in
the grammar is a lower bound on the length of the proof. This relationship has
been exploited in a method for proof compression whose central combinatorial
step is a grammar-based compression of a finite tree language [8–10].

The proof-theoretic application of our work entails a shift of emphasis w.r.t.
traditional grammar-based compression in the following respects: first, we do
not have any freedom of choice regarding the type of grammar. Totally rigid
acyclic tree grammars have to be used because they can be translated to proofs
afterwards. Secondly, we are looking for a minimal grammar G s.t. $L(G) \supseteq L$
where L is the finite input language. This is the case because L describes a
disjunction which is required to be a tautology (a so-called Herbrand-disjunction,

Research supported by the Vienna Science Fund (WWTF) project VRG12-04.

© Springer International Publishing Switzerland 2015
J. Shallit and A. Okhotin (Eds.): DCFS 2015, LNCS 9118, pp. 93–104, 2015.
DOI: 10.1007/978-3-319-19225-3_8

see [2,6]) and if $L' \supseteq L$ then L' also describes a tautology. This condition is similar to (but different from) the one imposed on cover automata [3,4]: there an automaton A is sought s.t. $L(A) \supseteq L$, but in addition it is required that $L(A) \setminus L$ consists only of words longer than any word in L. And thirdly, the complexity measure we aim at minimising is not the number of symbols in the grammar, but the number of production rules of the grammar. This is due to the fact that the number of production rules corresponds to the number of certain logical inferences in a formal proof.

Along the lines of descriptional complexity measures such as automatic complexity [26] and automaticity [25] one can consider the size of the smallest grammar that covers a language L in the above sense as the complexity of L. Then the result of [7] shows that this complexity measure is a lower bound on the length of a proof with Π_1-cuts of the Herbrand-disjunction described by L.

The central result of this paper is the construction of an incompressible sequence of finite (word) languages. This result extends to tree languages in a straightforward way, and is then used to obtain an incompressibility result for proofs with Π_1-cuts in first-order predicate logic. The length of proofs with cuts is notoriously difficult to control (for propositional logic this is considered the central open problem in proof complexity [22]). Theorem 5 below is, to the best of our knowledge, the first such incompressiblity result in proof theory.

2   Grammar-Based Compression of Finite Languages

Definition 1. *A* context-free grammar (CFG) *is a 4-tuple* $G = (N, \Sigma, P, S)$ *where N is a finite set of nonterminals, Σ is a finite alphabet, $S \in N$ is the starting symbol and P is a finite set of productions of the form $A \to w$ where $A \in N$ and $w \in (\Sigma \cup N)^*$.*

As usual, the one-step derivation relation $\Longrightarrow_G$ of G is defined by $u \Longrightarrow_G v$ iff there is a production $A \to w$ in G s.t. v is obtained from u by replacing an occurrence of A by w. The derivation relation $\Longrightarrow_G^*$ is the reflexive and transitive closure of $\Longrightarrow_G$ and the language of G is $L(G) = \{w \in \Sigma^* \mid S \Longrightarrow_G^* w\}$. We omit the subscript G if the grammar is clear from the context.

Definition 2. *A* right-linear grammar *is a context-free grammar* (N, Σ, P, S) *s.t. all productions in P are of the form $A \to vB$ or $A \to v$ for $A, B \in N$ and $v \in \Sigma^*$.*

It is well-known, see e.g., [11], that the languages of right-linear grammars are exactly the regular languages.

Definition 3. *Let $G = (N, \Sigma, P, S)$ be a context-free grammar. The ordering $<_G^1$ of N is defined as follows: $A <_G^1 B$ iff there is a production $A \to w$ in P s.t. B occurs in w. The ordering $<_G$ is defined as the transitive closure of $<_G^1$. We say that G is cyclic (respectively acyclic) iff $<_G$ is.*

We abbreviate "right-linear acyclic grammar" as "RLAG". Let $A \in N$; then a production whose left hand side is A is called A-production. We write P_A for the set of A-productions in P. For $N' \subseteq N$ we define $P_{N'} = \bigcup_{A \in N'} P_A$. For a language L and a CFG G we say that G covers L if $L(G) \supseteq L$. The size of a CFG $G = (N, \Sigma, P, S)$ is defined as $|G| = |P|$. The length of a right-linear production rule $A \to wB$ or $A \to w$ for $w \in \Sigma^*$ is defined as $|w|$.

Definition 4. *A finite language L is called* compressible *if there is a RLAG G which covers L and satisfies $|G| < |L|$. It is called* incompressible *otherwise.*

A variant of this problem, the equality formulation, consists in asking for a grammar G with $L(G) = L$ and $|G| < |L|$. As explained in the introduction, the cover formulation is motivated by our proof-theoretic application, see Sect. 5. However, our main result on incompressibility also applies to the equality formulation; see Corollary 1.

The choice of RLAGs for the compression of finite languages is quite natural in view of the fact that right-linear grammars generate exactly the regular languages and the observation that a right-linear grammar where every nonterminal is accessible and which does not contain trivial productions generates a finite language iff it is acyclic.

Definition 5. *A sequence $(L_n)_{n \geq 1}$ of finite languages is called* incompressible *if there is an $M \in \mathbb{N}$ s.t. for all $n \geq M$ the language L_n is incompressible. A sequence $(L_n)_{n \geq 1}$ of finite languages is called* compressible *if for every $M \in \mathbb{N}$ there is an $n \geq M$ s.t. L_n is compressible.*

Note that it is trivial to construct an incompressible sequence of languages of small size, e.g., $L_n = \{a\}$ for a letter a. It is also trivial to construct a sequence of incompressible languages in an unbounded signature, e.g., $L_n = \{a_1, \ldots, a_n\}$ for letters $a_1, a_2, \ldots$. Consequently, in this paper we will construct an incompressible sequence of languages of unbounded size over a constant alphabet.

3 Incompressible Languages

3.1 Reduced Grammars

In this section we will make some preparatory observations on the structure of RLAGs which compress finite languages, leading to the notion of strong compressibility.

Definition 6. *Let $G = (N, \Sigma, P, S)$ be a RLAG. Then a rule $A \to w$ is called* trivial *if $A = S$ and $w \in \Sigma^*$. We define $G_t = (N, \Sigma, P_t, S)$ where P_t is the set of trivial rules of G.*

We say that a word u is a *subword* of a word w if there are words v_1, v_2 s.t. $w = v_1 u v_2$. We say that a word u is a *prefix* of a word w if there is a word v s.t. $w = uv$.

Definition 7. *Let L be a finite language and G be a RLAG that covers L. Then G is called* reduced w.r.t. *L if for every non-trivial production rule A → wB or A → w of G there are distinct u, v ∈ L \ L(G_t) s.t. w is a subword of both u and v.*

If the language is clear from the context we will say "reduced" instead of "reduced w.r.t. L". Intuitively, in a grammar which is reduced w.r.t. L all production rules are either trivial or useful for the compression of L. The following lemma shows that for questions of compressibility, it is sufficient to consider reduced RLAGs.

Lemma 1. *Let L be a finite language and G be a RLAG which covers L. Then there is a reduced RLAG G* which covers L and satisfies |G*| ≤ |G|.*

Proof Sketch. Replace each non-trivial production rule that is only used for deriving a single word w by the trivial production S → w.

Definition 8. *A language L is called* strongly compressible *if there is a reduced RLAG G without trivial rules s.t. G covers L and |G| < |L|. A sequence (L_n)_{n≥1} of finite languages is called* strongly compressible *if for every M ∈ ℕ there is an n ≥ M s.t. L_n is strongly compressible.*

Lemma 2. *Let L be a compressible language, then there is a language L' ⊆ L which is strongly compressible.*

Proof Sketch. After obtaining a reduced RLAG G' which compresses L from Lemma 1 let L' be all words in L derivable from the non-trivial part of G'.

3.2 Segmented Languages

From this section on we will often use the alphabet $\Sigma = \{0, 1, \mathbf{s}\}$. The letters **0** and **1** will be used for the binary representation of natural numbers, while the letter **s** will serve as a separator. The incompressible sequence of languages used for the main result of this paper will be a sequence of segmented languages, a notion which we define now and study in this section.

Definition 9. *Let $\Sigma = \{0, 1, \mathbf{s}\}$. A word $w \in \Sigma^*$ s.t. $w = (\mathbf{s}v)^k$ for some $k \geq 1$ and some $v \in \{0, 1\}^+$ is called* segmented word. *The word v is called the* building block *of w. Occurrences of v in w are called* segments. *A segmented word $(\mathbf{s}v)^k$ where $|v| = l$ is called a (k, l)-segmented word. A language consisting of (k, l)-segmented words is called a (k, l)-segmented language.*

The following lemma states the key property of segmented languages: long rules are not useful for compression.

Lemma 3. *Let L be a finite (k, l)-segmented language and G be a reduced RLAG that covers L. Then every non-trivial rule of G has length at most l.*

Proof Sketch. If a rule has length $l + 1$, it contains a whole building block and hence fixes a word of L.

Lemma 4. *Let L be a finite (k, l)-segmented language that is strongly compressible. Then $k < |L| - 1$.*

Proof. Note that we have $|w| = k(l+1)$ for all $w \in L$. Let G be a reduced RLAG which compresses L. Then by Lemma 3 every rule in G has length at most l. Hence for deriving a single $w \in L$ the grammar G needs at least $\frac{|w|}{l} > \frac{|w|}{l+1} = k$ rules. Since L is compressible, it is non-empty and hence $k < |G| < |L|$.

Lemma 5. *Let $(L_n)_{n \geq 1}$ be a compressible sequence of finite languages s.t. L_n is (k_n, l_n)-segmented and $(k_n)_{n \geq 1}$ is unbounded. Then there is a sequence of finite languages $(L'_n)_{n \geq 1}$ s.t. 1) $L'_n \subseteq L_n$ for all $n \geq 1$, 2) $(L'_n)_{n \geq 1}$ is strongly compressible, and 3) $(|L'_n|)_{n \geq 1}$ is unbounded.*

Proof. The pointwise application of Lemma 2 to an infinite subsequence of $(L_n)_{n \geq 1}$ that consists of compressible languages only yields a sequence $(L'_n)_{n \geq 1}$ satisfying 1 and 2. By Lemma 4 we have $k_i < |L'_i|$ for infinitely many $i \in \mathbb{N}$ which together with the assumption that $(k_n)_{n \geq 1}$ is unbounded entails 3.

The following Lemma 6 applies the uselessness of long rules for compression to provide an upper bound on the number of segments which a strongly compressing RLAG covers. This upper bound is a key ingredient of the proof of our main result.

Definition 10. *Let $G = (N, \Sigma, P, S)$ be a RLAG. Let $w \in L(G)$ be a (k, l)-segmented word with building block v and let $i \in \{1, \ldots, k\}$. Then $w = w_0 s v w_1$ for $w_0 = (sv)^{i-1}$ and $w_1 = (sv)^{k-i}$. Let δ be a derivation of w w.r.t. G; then it is of the form*

$$S \Longrightarrow^* w'_0 A_1 \Longrightarrow w_0 s v' A_2 \Longrightarrow \cdots w_0 s v'' A_n \Longrightarrow w_0 s v w'_1 A_{n+1} \Longrightarrow^* w$$

for some $A_1, \ldots, A_n, A_{n+1} \in N$ with v', v'' being prefixes of v, w'_0 a prefix of w_0 and w'_1 a prefix of w_1. We define $\mathrm{nonterms}(w, i, \delta) = \{A_j \mid 1 \leq j \leq n\}$.

Lemma 6. *Let L be a finite (k, l)-segmented language that is strongly compressed by a RLAG $G = (N, \Sigma, P, S)$. For each $w \in L$ fix a derivation δ_w of w w.r.t. G. Let $N_0 \subseteq N$, let $P_0 = P_{N_0}$ and let $S_0 = \{(w, i) \in L \times \{1, \ldots, k\} \mid \mathrm{nonterms}(w, i, \delta_w) \subseteq N_0\}$. Then we have $|S_0| \leq 2^{|P_0|} \cdot |P_0|$.*

Proof. For $w \in L$ define $S_{w,0} = \{i \in \{1, \ldots, k\} \mid \mathrm{nonterms}(w, i, \delta_w) \subseteq N_0\}$. By Lemma 3 every rule of G has length at most l. Due to acyclicity, each $A \in N_0$ can be used at most once in a derivation. Therefore by using all $A \in N_0$ in a derivation one can generate at most $|N_0| \cdot l$ terminal symbols, and hence at most $|N_0|$ segments. We thus obtain $|S_{w,0}| \leq |N_0|$.

Furthermore, define $L_0 = \{w \in L \mid \exists i \text{ s.t. } (w, i) \in S_0\}$. Let $P^* \subseteq P_0$ s.t. P^* contains exactly one production for each nonterminal of N_0 and note that there are at most $2^{|P_0|}$ such P^*. If P^* permits deriving a word that contains a subword $v \in \{0, 1\}^l$, then the choice of P^* uniquely determines a word $w \in L$. If P^* does not allow deriving such a word, then P^* may be used in the derivations δ_w of several $w \in L$; however, it does not contribute to any of its $S_{w,0}$. Therefore we have $|L_0| \leq 2^{|P_0|}$. Putting these two results together, we obtain $|S_0| = \sum_{w \in L_0} |S_{w,0}| \leq |L_0| \cdot |N_0| \leq 2^{|P_0|} \cdot |P_0|$.

3.3   Ordered Grammars

In a RLAG G the ordering $<_G$ is acyclic but, in general, not linear. For technical purposes it will be useful to fix a linearisation of $<_G$ and a corresponding linear order of the productions of G. To that aim we introduce the notion of ordered grammar.

Definition 11. *A* right-linear ordered grammar (RLOG) *is a tuple* $G = (N, \Sigma, P, A_1)$ *where N is a list* $A_1, \ldots, A_n$ *of nonterminals, P is a list* $p_1, \ldots, p_m$ *of productions s.t.*

1. $G' = (\{A_1, \ldots, A_n\}, \Sigma, \{p_1, \ldots, p_m\}, A_1)$ *is a RLAG,*
2. *if* $A_i <_{G'} A_j$ *then* $i < j$, *and*
3. $p_1, \ldots, p_m = q_{1,1}, \ldots, q_{1,k_1}, \ldots, q_{n,1}, \ldots, q_{n,k_n}$ *where* $\{q_{i,1}, \ldots, q_{i,k_i}\} = P_{A_i}$ *for all* $i \in \{1, \ldots, n\}$.

We say that an RLOG compresses a language L, is reduced w.r.t. L, etc., if the underlying RLAG fulfils the respective property.

Definition 12. *Let* $G = ((A_1, \ldots, A_n), \Sigma, P, S)$ *be a RLOG. Let* $w \in L(G)$ *be a (k, l)-segmented word, let* $i \in \{1, \ldots k\}$ *and let* δ *be a derivation of w w.r.t. G. Let* $m_1 = \min\{j \in \{1, \ldots, n\} \mid A_j \in \mathrm{nonterms}(w, i, \delta)\}$ *and* $m_2 = \max\{j \in \{1, \ldots, n\} \mid A_j \in \mathrm{nonterms}(w, i, \delta)\}$ *and define* $\mathrm{cost}(w, i, \delta) = \sum_{j=m_1}^{m_2} |P_{A_j}|$.

Note that the cost of the i-th segment of a word w also takes those nonterminals into account which are not used in the derivation δ of w. The following lemma shows that in a strongly compressed segmented language, many segments are cheap.

Lemma 7. *Let L be a finite (k, l)-segmented language and let G be a RLOG that strongly compresses L. Let $w \in L$ and δ be a derivation of w w.r.t. G. Then for at least half of the $i \in \{1, \ldots, k\}$, we have* $\mathrm{cost}(w, i, \delta) < \frac{4|L|}{k}$.

Proof. As G compresses L, it covers L, so by Lemma 3 every rule of G has length at most l. Hence each rule of G can contribute to the costs of at most two segments of w, so we have $2|G| \geq \sum_{i=1}^{k} \mathrm{cost}(w, i, \delta)$. Now suppose that $\lceil \frac{k}{2} \rceil$ segments of w have cost at least $\frac{4|L|}{k}$ each, then $\sum_{i=1}^{k} \mathrm{cost}(w, i, \delta) \geq \lceil \frac{k}{2} \rceil \cdot \frac{4|L|}{k} \geq 2|L|$, which is a contradiction to $|G| < |L|$.

Definition 13. *Let* $G = (N, \Sigma, (p_1, \ldots, p_m), A_1)$ *be a RLOG and let $s < m$. For* $A \in N$ *define* $\mathrm{pmin}(A) = \min\{j \mid p_j \in P_A\}$, *and* $\mathrm{pmax}(A) = \max\{j \mid p_j \in P_A\}$. *Furthermore, for* $j \in \{1, \ldots, \lceil \frac{m}{s} \rceil - 1\}$ *define* $N_j = \{A \in N \mid (j-1)s \leq \mathrm{pmin}(A) \ \text{and} \ \mathrm{pmax}(A) < (j+1)s\}$. *We say that* $(N_j)_{j=1}^{\lceil \frac{m}{s} \rceil - 1}$ *is the s-covering of G.*

Note that N_j and N_{j+1} can overlap, but N_j and N_{j+2} can not. Furthermore, note that $|P_{N_j}| \leq 2s$ for all $j \in \{1, \ldots, \lceil \frac{m}{s} \rceil - 1\}$. The following lemma applies Lemma 7 to obtain a lower bound on the number of segments covered by the productions of a single N_j.

Lemma 8. *Let L be a finite (k,l)-segmented language, let G be a RLOG which strongly compresses L and let $|G| > s \geq \frac{4|L|}{k}$. Let $N_1, \ldots, N_{\lceil \frac{|G|}{s} \rceil - 1}$ be the s-covering of G. Let $w \in L$ and δ be a G-derivation of w. Then for at least half of the $i \in \{1, \ldots, k\}$ there is a $j \in \{1, \ldots, \lceil \frac{|G|}{s} \rceil - 1\}$ s.t. $\mathrm{nonterms}(w, i, \delta) \subseteq N_j$.*

Proof. By Lemma 7 at least half of the $i \in \{1, \ldots, k\}$ have $\mathrm{cost}(w, i, \delta) < \frac{4|L|}{k}$. Let i be s.t. $\mathrm{cost}(w, i, \delta) < \frac{4|L|}{k}$, then $\mathrm{cost}(w, i, \delta) < s$. Let $A_0 \in N$ be the nonterminal used for entering the i-th segment of w in δ and let $j_0 = \max\{j \in \{1, \ldots, \lceil \frac{|G|}{s} \rceil - 1\} \mid A_0 \in N_j\}$. If $j_0 = \lceil \frac{|G|}{s} \rceil - 1$, then $\mathrm{nonterms}(w, i, \delta) \subseteq N_{j_0}$ because $(N_j)_{j=1}^{\lceil \frac{|G|}{s} \rceil - 1}$ covers all nonterminals and N_{j_0} is the last element of this list. If $j_0 < \lceil \frac{|G|}{s} \rceil - 1$, then $\mathrm{pmin}(A_0) < j_0 s$, for if $\mathrm{pmin}(A_0) \geq j_0 s$, then $A_0 \in N_{j_0+1}$. Therefore $\mathrm{pmin}(A_0) + \mathrm{cost}(w, i, \delta) < \mathrm{pmin}(A_0) + s < (j_0 + 1)s$; hence $\mathrm{nonterms}(w, i, \delta) \subseteq N_{j_0}$. ∎

3.4 The Main Result

For $n \geq 1$ and $k \in \{0, \ldots, 2^n - 1\}$ we write $\mathrm{b}_n k \in \{0, 1\}^n$ for the n-bit binary representation of k.

Definition 14 (Incompressible sequence). *For all $n \geq 1$ define $l(n) = \lceil \log(n) \rceil$, $k(n) = \lceil \frac{9n}{l(n)+1} \rceil$, and $L_n = \{(\mathrm{sb}_{l(n)}i)^{k(n)} \mid 0 \leq i \leq n - 1\}$.*

Note that $l(n)$ is the number of bits required to represent all elements of $\{0, \ldots, n-1\}$ in binary. Note furthermore that for every $n \geq 1$, we have $|L_n| = n$ and all words in L_n have the same length $k(n)(l(n) + 1)$. The number of segments $k(n)$ has been chosen s.t. $k(n)(l(n) + 1)$ is $9n$ padded up to the next multiple of $l(n) + 1$; hence the length of the words in L_n grows linearly in n.

Example 1. For $n = 5$ we have $l(5) = 3$, $k(5) = 12$ and $L_5 = \{(\mathrm{s000})^{12}, (\mathrm{s001})^{12}, (\mathrm{s010})^{12}, (\mathrm{s011})^{12}, (\mathrm{s100})^{12}\}$.

Theorem 1. *$(L_n)_{n \geq 1}$ is incompressible.*

The proof strategy for this theorem is as follows: both Lemmas 6 and 8 assume a strongly compressed segmented language. But while Lemma 6 states an upper bound on the number of segments covered by a certain part of a strongly compressing grammar, Lemma 8 provides a lower bound on the number of segments covered by the productions of a single N_j. The following proof will show these two bounds to be inconsistent, thus deriving the incompressibility of $(L_n)_{n \geq 1}$.

Proof. Suppose that $(L_n)_{n \geq 1}$ is compressible. Then by Lemma 5 there is a sequence $(L'_n)_{n \geq 1}$ which is strongly compressibly by a sequence $(G_n)_{n \geq 1}$ of RLAGs which we consider as RLOGs $(G'_n)_{n \geq 1}$ by fixing an arbitrary linear order satisfying Definition 11. Let us fix for every $n \geq 1$ and every $w \in L'_n$ a derivation δ_w of w w.r.t. G'_n. This is well-defined, since the L'_n are disjoint, and hence δ_w does not depend on n.

First note that for all $n \geq 1$ we have $k(n) = \lceil \frac{9n}{\lceil \log(n) \rceil + 1} \rceil \geq \frac{9n}{\log(n) + 2}$, and since $n \geq |L'_n|$ we have

$$k(n) \geq \frac{9|L'_n|}{\log(|L'_n|) + 2}. \tag{1}$$

Therefore $\frac{4|L'_n|}{k(n)} \leq \frac{4}{9}(\log(|L'_n|) + 2) =: s_n$. Let $N_1, \ldots N_{\lceil \frac{|G'_n|}{s_n} \rceil - 1}$ be the s_n-covering of G'_n and define $U_n := |\{(w, i) \in L'_n \times \{1, \ldots, k(n)\} \mid \exists j \text{ s.t. nonterms}(w, i, \delta_w) \subseteq N_j\}|$. By Lemma 8 we have $U_n \geq \frac{|L'_n| \cdot k(n)}{2}$, which, together with Theorem (1), entails

$$U_n \geq \frac{9|L'_n|^2}{2(\log(|L'_n|) + 2)}. \tag{2}$$

On the other hand, applying Lemma 6 to all N_j for $j = 1, \ldots, \lceil \frac{|G'_n|}{s_n} \rceil - 1$ and summing up yields $U_n \leq \sum_{j=1}^{\lceil \frac{|G'_n|}{s_n} \rceil - 1} 2^{|P_{N_j}|} \cdot |P_{N_j}| \leq (\lceil \frac{|G'_n|}{s_n} \rceil - 1) \cdot 2^{2s_n} \cdot 2s_n$. We have $2^{2s_n} \cdot 2s_n \leq C|L'_n|^{\frac{8}{9}}(\log(|L'_n|) + 2)$ for some $C \in \mathbb{N}$ and $\lceil \frac{|G'_n|}{s_n} \rceil - 1 \leq \frac{|L'_n|}{s_n} = \frac{9|L'_n|}{4(\log(|L'_n|) + 2)}$ and therefore

$$U_n \leq D|L'_n|^{\frac{17}{9}} \text{ for some } D \in \mathbb{N}. \tag{3}$$

Putting Theorem (2) and (3) together we obtain

$$|L'_n|^2 \leq E|L'_n|^{\frac{17}{9}}(\log(|L'_n|) + 2) \text{ for some } E \in \mathbb{N}. \tag{4}$$

But by Lemma 5 the function $n \mapsto |L'_n|$ is unbounded. Hence there is an $M \in \mathbb{N}$ s.t. for all $n \geq M$ the inequality Theorem (4) is not satisfied. This is a contradiction.

3.5   Remarks

Every sequence of languages which is incompressible in the cover formulation is also incompressible in the (more restricted) equality formulation. Therefore we immediately obtain the following corollary from Theorem 1.

Corollary 1. *There is no sequence $(G_n)_{n \geq 1}$ of RLAGs and $M \in \mathbb{N}$ s.t. $L(G_n) = L_n$ and $|G_n| < |L_n|$ for all $n \geq M$.*

On the other hand, the sequence $(L_n)_{n \geq 1}$ can be compressed by stronger formalisms:

Theorem 2. *There is a sequence $(G_n)_{n \geq 1}$ of acyclic CFGs which compresses $(L_n)_{n \geq 1}$.*

Proof. Let $G_n = (\{S, A_1, \ldots, A_{l(n)}\}, \{0, 1, s\}, P_n, S)$ where

$$P_n = \{S \to (sA_1)^{k(n)}, \; A_1 \to 0A_2 \mid 1A_2, \; \ldots, \; A_{l(n)} \to 0 \mid 1\}.$$

Then $L(G_n) \supseteq L_n$ for all $n \geq 1$ and $|G_n| = 2\lceil \log(n) \rceil + 1 < n = |L_n|$ for all $n \geq M$ for a certain M.

The length of the words in L_n grows linearly. Under the condition that $|L_n| = n$ this is the best possible:

Theorem 3. *Let $(L'_n)_{n \geq 1}$ be a sequence of finite languages over a finite alphabet $\Sigma = \{a_1, \ldots a_k\}$ s.t. $|L'_n| = n$ and s.t. there is a sublinear function that bounds the maximal length l_n of a word in L'_n. Then $(L'_n)_{n \geq 1}$ is compressible.*

Proof. Let $G_n = (\{A_1, \ldots, A_{l_n}\}, \Sigma, P_n, A_1)$ where

$$P_n = \{A_1 \to a_1 A_2 \mid \cdots \mid a_k A_2 \mid A_2, \ldots, A_{l_n} \to a_1 \mid \cdots \mid a_k \mid \varepsilon\}.$$

Then $L(G_n) = \Sigma^{\leq l_n} \supseteq L'_n$ and $|G_n| = (k+1) \cdot l_n$ which, from a certain $M \in \mathbb{N}$ on, is less than $|L'_n| = n$.

4 Application to Tree Languages

In this section we will transfer the main theorem to tree languages. The grammars we will be considering for the compression of finite tree languages are totally rigid acyclic tree grammars. Rigid tree languages were introduced in [12,13] with applications in verification in mind. A presentation of this class of languages via rigid grammars was given in [7].

For a ranked alphabet (i.e., a term signature) Σ we write $\mathcal{T}(\Sigma)$ for the set of all terms built from function and constant symbols of Σ.

Definition 15. *A regular tree grammar is a tuple $G = (N, \Sigma, P, S)$ where N is a set of nonterminals of arity 0, Σ is a term signature, $S \in N$ is the starting symbol and P is a finite set of productions of the form $A \to t$ where $A \in N$ and $t \in \mathcal{T}(\Sigma \cup N)$.*

The ordering $<_G$ of nonterminals is defined analogously to the case of word grammars. Hence we can speak about acyclic regular tree grammars. As usual for tree grammars, a derivation is a finite list of terms $t_1, \ldots, t_n$ s.t. t_{i+1} is obtained from t_i by applying a production rule to a single position. A derivation w.r.t. a grammar $G = (N, \Sigma, P, S)$ is said to satisfy the rigidity condition if for every nonterminal $A \in N$ it uses at most one A-production.

Definition 16. *A totally rigid acyclic tree grammar (TRATG) is an acyclic regular tree grammar $G = (N, \Sigma, P, S)$ whose language $L(G)$ is the set of all $t \in \mathcal{T}(\Sigma)$ that have a derivation from S satisfying the rigidity condition.*

(In)compressibility of tree languages and sequences thereof is defined analogously to the case of word languages replacing RLAGs with TRATGs.

Definition 17. *For an alphabet Σ define $\Sigma^{\mathrm{T}} = \{f_a/1 \mid a \in \Sigma\} \cup \{e\}$. For $w \in \Sigma^*$ define the term w^{T} recursively by $\varepsilon^{\mathrm{T}} = e$, and $(av)^{\mathrm{T}} = f_a(v^{\mathrm{T}})$. For $L \subseteq \Sigma^*$ define $L^{\mathrm{T}} = \{w^{\mathrm{T}} \mid w \in L\}$.*

Theorem 4. *The sequence of tree languages $(L_n^{\mathrm{T}})_{n \geq 1}$ is incompressible.*

Proof Sketch. Starting from the assumption that $(L_n^{\mathrm{T}})_{n \geq 1}$ is compressible, transform the compressing TRATGs into RLAGs compressing $(L_n)_{n \geq 1}$ hence arriving at a contradiction to Theorem 1.

5   Application to Proof Theory

A *sequent* is an expression of the form $\Gamma \vdash \Delta$ where Γ and Δ are finite sets of formulas in first-order predicate logic. The intended interpretation of a sequent $\Gamma \vdash \Delta$ is the formula $(\bigwedge_{\varphi \in \Gamma} \varphi) \to (\bigvee_{\psi \in \Delta} \psi)$. The logical complexity $\|\Gamma \vdash \Delta\|$ of a sequent is the number of logical connectives it contains.

A *proof* is a tree whose nodes are sequents which is built according to certain logical inference rules. The leaves are of the form $A \vdash A$. An important inference rule is the so-called *cut rule*:

$$\frac{\Gamma \vdash \Delta, A \quad A, \Pi \vdash \Lambda}{\Gamma, \Pi \vdash \Delta, \Lambda} \text{ cut}$$

For a complete list of inference rules of the sequent calculus, the interested reader is referred to [1]. The length of a proof π, written as $|\pi|$, is the number of inferences in π. The cut rule formalises the use of a lemma in mathematical practice and is of particular importance here because it allows compressing proofs in first-order logic non-elementarily [20,21,27]. A cut is said to be a Π_1-cut if its cut formula A is of the form $\forall x\, B$ for B quantifier-free.

The following result is a proof-theoretic corollary of the incompressiblity theorem proved in Sect. 3 and extended to tree languages in Sect. 4. It should be stressed that the proof-theoretic techniques for deriving it from Theorem 4 are simple standard techniques, the details of which are omitted from this paper for space reasons.

Theorem 5. *There is a sequences of sequents $(S_n)_{n \geq 1}$ such that $\|S_n\|$ is constant, there is a cut-free proof of S_n with $O(2^{2n})$ inferences and there is $M \in \mathbb{N}$ s.t. for all $n \geq M$: every proof with Π_1-cuts of S_n has at least 2^n inferences.*

Proof Sketch. For $i \geq 1$ define the sequent $R_i :=$

$\forall y \forall v\, P(0, y, e, v), \forall x \forall y \forall u \forall v\, (P(x, f_0(y), u, v) \to P(x, y, f_0(u), v)),$

$\forall x \forall y \forall u \forall v\, (P(x, f_1(y), u, v) \to P(x, y, f_1(u), v)),$

$\forall x \forall y \forall u \forall v\, (P(x, f_s(y), u, v) \to P(x, y, u, v)), \forall x \forall y \forall v\, (P(x, y, v, v) \to P(s(x), y, e, v)),$

$\forall v\, (P(\overline{k(n)}, e, e, v) \to Q(\overline{1}, v)), \forall x \forall v\, ((Q(x, f_0(v)) \wedge Q(x, f_1(v))) \to Q(s(x), v))$

$\vdash Q(\overline{l(n)}, e)$

where $\overline{n}$ for $n \in \mathbb{N}$ denotes the term $s^n(0)$. Define $(S_n)_{n \geq 1}$ by $S_n = R_{2^n}$. Then $\|S_n\|$ is constant and S_n has a straightforward cut-free proof with $O(2^{2n})$ inferences. Furthermore, every cut-free proof of S_n must instantiate the quantifier $\forall y$ in $\forall y \forall v\, P(0, y, e, v)$ with all terms in $L_{2^n}^T$. The lower bound on the proofs with cuts then follows immediately from Theorem 4 and a suitable (but straightforward) generalisation of Theorem 22 in [7].

6   Conclusion

We have investigated the problem of simultaneously compressing a finite set of words by a right-linear acyclic grammar. We have constructed an incompressible sequence of languages and applied it to obtain an incompressibility-result in proof theory.

This problem of simultaneous compression has received only little attention in the literature so far and consequently there is a number of interesting open questions, for example: what is the complexity of the smallest grammar problem in this setting? How difficult is the approximation of the smallest grammar? Can approximation algorithms and techniques be carried over from the case of one word to this setting? How does the situation change when we do not minimise the number of production rules but the symbol complexity of the grammar?

Fast approximation algorithms for computing a minimal RLAG (or TRATG) that covers a given finite input language would also be of high practical value in the cut-introduction method [9,10] and its implementation [8].

Acknowledgments. The authors would like to thank Manfred Schmidt-Schauß for several helpful conversations about the topic of this paper, Werner Kuich for a number of remarks that improved the presentation of the results, and the anonymous reviewers for numerous important comments and suggestions.

References

1. Buss, S.: An Introduction to proof theory. In: Buss, S. (ed.) The Handbook of Proof Theory, pp. 2–78. Elsevier, Amsterdam (1998)
2. Buss, S.R.: On Herbrand's theorem. In: Leivant, D. (ed.) LCC'94. LNCS, vol. 960, pp. 195–209. Springer, Heidelberg (1995)
3. Câmpeanu, C., Sântean, N., Yu, S.: Minimal cover-automata for finite languages. In: Champarnaud, J.-M., Maurel, D., Ziadi, D. (eds.) WIA 1998. LNCS, vol. 1660, pp. 43–56. Springer, Heidelberg (1999)
4. Câmpeanu, C., Santean, N., Yu, S.: Minimal cover-automata for finite languages. Theoret. Comput. Sci. **267**(1–2), 3–16 (2001)
5. Charikar, M., Lehman, E., Liu, D., Panigrahy, R., Prabhakaran, M., Sahai, A., Shelat, A.: The smallest grammar problem. IEEE Trans. Inf. Theory **51**(7), 2554–2576 (2005)
6. Herbrand, J.: Recherches sur la théorie de la démonstration. Ph.D. thesis, Université de Paris (1930)
7. Hetzl, S.: Applying tree languages in proof theory. In: Dediu, A.-H., Martín-Vide, C. (eds.) LATA 2012. LNCS, vol. 7183, pp. 301–312. Springer, Heidelberg (2012)
8. Hetzl, S., Leitsch, A., Reis, G., Tapolczai, J., Weller, D.: Introducing quantified cuts in logic with equality. In: Demri, S., Kapur, D., Weidenbach, C. (eds.) IJCAR 2014. LNCS, vol. 8562, pp. 240–254. Springer, Heidelberg (2014)
9. Hetzl, S., Leitsch, A., Reis, G., Weller, D.: Algorithmic introduction of quantified cuts. Theoret. Comput. Sci. **549**, 1–16 (2014)
10. Hetzl, S., Leitsch, A., Weller, D.: Towards algorithmic cut-introduction. In: Bjørner, N., Voronkov, A. (eds.) LPAR-18 2012. LNCS, vol. 7180, pp. 228–242. Springer, Heidelberg (2012)
11. Hopcroft, J.E., Ullman, J.D.: Introduction to Automata Theory, Languages, and Computation. Addison-Wesley, Cambridge (1979)

12. Jacquemard, F., Klay, F., Vacher, C.: Rigid tree automata. In: Dediu, A.H., Ionescu, A.M., Martín-Vide, C. (eds.) LATA 2009. LNCS, vol. 5457, pp. 446–457. Springer, Heidelberg (2009)

13. Jacquemard, F., Klay, F., Vacher, C.: Rigid tree automata and applications. Inf. Comput. **209**, 486–512 (2011)

14. Jeż, A.: Approximation of grammar-based compression via recompression. In: Fischer, J., Sanders, P. (eds.) CPM 2013. LNCS, vol. 7922, pp. 165–176. Springer, Heidelberg (2013)

15. Jeż, A.: A *really* simple approximation of smallest grammar. In: Kulikov, A.S., Kuznetsov, S.O., Pevzner, P. (eds.) CPM 2014. LNCS, vol. 8486, pp. 182–191. Springer, Heidelberg (2014)

16. Kieffer, J.C., Yang, E.H.: Grammar-based codes: a new class of universal lossless source codes. IEEE Trans. Inf. Theory **46**(3), 737–754 (2000)

17. Larsson, N.J., Moffat, A.: Offline dictionary-based compression. In: Data Compression Conference (DCC 1999). pp. 296–305. IEEE Computer Society (1999)

18. Lohrey, M.: Algorithmics on SLP-compressed strings: a survey. Groups Complex. Cryptol. **4**(2), 241–299 (2012)

19. Nevill-Manning, C.G., Witten, I.H.: Identifying hierarchical structure in sequences: a linear-time algorithm. J. Artif. Intell. Res. **7**, 67–82 (1997)

20. Orevkov, V.: Lower bounds for increasing complexity of derivations after cut elimination. Zapiski Nauchnykh Seminarov Leningradskogo Otdeleniya Matematicheskogo Instituta **88**, 137–161 (1979)

21. Pudlák, P.: The Lengths of proofs. In: Buss, S. (ed.) Handbook of Proof Theory, pp. 547–637. Elsevier, Amsterdam (1998)

22. Pudlák, P.: Twelve problems in proof complexity. In: Hirsch, E.A., Razborov, A.A., Semenov, A., Slissenko, A. (eds.) Computer Science – Theory and Applications. LNCS, vol. 5010, pp. 13–27. Springer, Heidelberg (2008)

23. Rytter, W.: Application of Lempel-Ziv factorization to the approximation of grammar-based compression. Theoret. Comput. Sci. **302**(1–3), 211–222 (2003)

24. Sakamoto, H.: A fully linear-time approximation algorithm for grammar-based compression. J. Discrete Algorithms **3**(2–4), 416–430 (2005)

25. Shallit, J., Breitbart, Y.: Automaticity I: properties of a measure of descriptional complexity. J. Comput. Syst. Sci. **53**, 10–25 (1996)

26. Shallit, J., Wang, M.W.: Automatic complexity of strings. J. Automata, Lang. Comb. **6**(4), 537–554 (2001)

27. Statman, R.: Lower bounds on Herbrand's theorem. Proc. Am. Math. Soc. **75**, 104–107 (1979)

28. Storer, J.A., Szymanski, T.G.: The macro model for data compression (extended abstract). In: Proceedings of the Tenth Annual ACM Symposium on Theory of Computing (STOC '78). pp. 30–39. ACM, New York (1978)

On the Complexity and Decidability of Some Problems Involving Shuffle

Joey Eremondi[1], Oscar H. Ibarra[2], and Ian McQuillan[3]($\boxtimes$)

[1] Department of Information and Computing Sciences, Utrecht University,
P.O. Box 80.089, 3508 TB Utrecht, The Netherlands
`j.s.eremondi@students.uu.nl`
[2] Department of Computer Science, University of California,
Santa Barbara, CA 93106, USA
`ibarra@cs.ucsb.edu`
[3] Department of Computer Science, University of Saskatchewan,
Saskatoon, SK S7N 5A9, Canada
`mcquillan@cs.usask.ca`

Abstract. The complexity and decidability of various decision problems involving the shuffle operation (denoted by $ш$) are studied. The following three problems are all shown to be NP-complete: given a non-deterministic finite automaton (NFA) M, and two words u and v, is $L(M) \not\subseteq u ш v$, is $u ш v \not\subseteq L(M)$, and is $L(M) \neq u ш v$? It is also shown that there is a polynomial-time algorithm to determine, for NFAs M_1, M_2 and a deterministic pushdown automaton M_3, whether $L(M_1)шL(M_2) \subseteq L(M_3)$. The same is true when M_1, M_2, M_3 are one-way nondeterministic l-reversal-bounded k-counter machines, with M_3 being deterministic. Other decidability and complexity results are presented for testing whether given languages L_1, L_2 and L from various languages families satisfy $L_1 ш L_2 \subseteq L$.

1 Introduction

The shuffle operator models the natural interleaving between strings. It was introduced by Ginsburg and Spanier [14], where it was shown that context-free languages are closed under shuffle with regular languages, but not context-free languages. It has since been applied in a number of areas such as concurrency [25], coding theory [9], and biocomputing [9,21], and has also received considerable study in the area of formal languages. However, there remains a number of open questions, such as the long-standing problem as to whether it is decidable, given a regular language R to tell if R has a non-trivial decomposition; that is, $R = L_1 ш L_2$, for some L_1, L_2 that are not $\{\lambda\}$ [7].

This paper addresses several complexity-theoretic and decidability questions involving shuffle. In the past, similar questions have been studied by Ogden,

The research of O. H. Ibarra was supported, in part, by NSF Grant CCF-1117708.
The research of I. McQuillan was supported, in part, by the Natural Sciences and Engineering Research Council of Canada.

J. Shallit and A. Okhotin (Eds.): DCFS 2015, LNCS 9118, pp. 105–116, 2015.
DOI: 10.1007/978-3-319-19225-3_9

Riddle, and Round [25], who showed that there exists deterministic context-free languages L_1, L_2 where $L_1 \shuffle L_2$ is NP-complete. More recently, L. Kari studied problems involving solutions to language equations of the form $R = L_1 \shuffle L_2$, where some of R, L_1, L_2 are given, and the goal is to determine a procedure, or determine that none exists, to solve for the variable(s) [20]. Also, there has been similar decidability problems investigated involving shuffle on trajectories [22], where the patterns of interleaving are restricting according to another language $T \subseteq \{0, 1\}^*$ (a zero indicates that a letter from the first operand will be chosen next, and a one indicates a letter from the second operand is chosen). L. Kari and Sosík show that it is decidable, given L_1, L_2, R as regular languages with a regular trajectory set T, whether $R = L_1 \shuffle_T L_2$ (the shuffle of L_1 and L_2 with trajectory set T). Furthermore, if L_1 is allowed to be context-free, then the problem becomes undecidable as long as, for every $n \in \mathbb{N}$, there is some word of T with more than n 0's (with a symmetric result if there is a context-free language on the right). This implies that it is undecidable whether $L_1 \shuffle L_2 = R$, where R and one of L_1, L_2 are regular, and the other is context-free. In [5], it is demonstrated that given two linear context-free languages, it is not semi-decidable whether their shuffle is linear context-free, and given two deterministic context-free languages, it is not semi-decidable whether their shuffle is deterministic context-free. Complexity questions involving so-called *shuffle languages*, which are augmented from regular expressions by shuffle and iterated shuffle, have also been studied [19]. It has also been determined that it is NP-hard to determine if a given string is the shuffle of two identical strings [6].

Recently, there have been several papers involving the shuffle of two words. It was shown that the shuffle of two words with at least two letters has a unique decomposition into the shuffle of words [2]. Also, a polynomial-time algorithm has been developed that, given a deterministic finite automaton (DFA) M and two words u, v, can test if $u \shuffle v \subseteq L(M)$ [3]. In the same work, an algorithm was presented that takes a DFA M as input and outputs a "candidate solution" u, v; this means, if $L(M)$ has a decomposition into the shuffle of two words, u and v must be those two unique words. But the algorithm cannot guarantee that $L(M)$ has a decomposition. This algorithm runs in $O(|u| + |v|)$ time, which is often far less than the size of the input DFA, as DFAs accepting the shuffle of two words can be exponentially larger than the words [8]. It has also been shown [4] that the following problem is NP-complete: given a DFA M and two words u, v, is it true that $L(M) \not\subseteq u \shuffle v$?

In this paper, problems are investigated involving three given languages R, L_1, L_2, and the goal is to determine decidability and complexity of testing if $R \not\subseteq L_1 \shuffle L_2, L_1 \shuffle L_2 \not\subseteq R$, and $L_1 \shuffle L_2 = R$, depending on the language families that L_1, L_2 and R are from. In Sect. 3, it is demonstrated that the following three problems are NP-complete: to determine, given an NFA M and two words u, v whether $u \shuffle v \not\subseteq L(M)$ is true, $L(M) \not\subseteq u \shuffle v$ is true, and $u \shuffle v \neq L(M)$ is true. Then, the DFA algorithm from [3] that can output a "candidate solution" is extended to an algorithm on NFAs that operates in polynomial time, and outputs two words u, v such that if the NFA is decomposable into the shuffle of words, then $u \sqcup v$ is the unique solution. And in Sect. 4, decidability and complexity of these shuffle problems are investigated involving more general language families. In particular, it is shown that it is decidable in polynomial time, given NFAs M_1, M_2 and a deterministic pushdown automaton M_3, whether $L(M_1) \sqcup L(M_2) \subseteq L(M_3)$. The same is true given M_1, M_2 that are one-way nondeterministic l-reversal-bounded k-counter machines, and M_3, a one-way deterministic l-reversal-bounded k-counter machine. However, if M_3 is a nondeterministic 1-counter machine that makes only one reversal on the counter, and M_1 and M_2 are fixed DFAs accepting a^* and b^* respectively, then the question is undecidable. Also, if we have fixed languages $L_1 = (a + b)^*$ and $L_2 = \{\lambda\}$, and M_3 is an NFA, then testing whether $L_1 \sqcup L_2 \not\subseteq L(M_3)$ is PSPACE-complete. Also, testing whether $a^* \sqcup \{\lambda\} \not\subseteq L$ is NP-complete for L accepted by an NFA. For finite languages L_1, L_2, and L_3 accepted by an NPDA, it is NP-complete to determine if $L_1 \sqcup L_2 \not\subseteq L_3$. Results on unary languages are also provided.

2   Preliminaries

We assume an introductory background in formal language theory and automata [16], as well as computational complexity [12].

We will use the notation below to represent classes of automata: NPDA for nondeterministic pushdown automata; DPDA for deterministic pushdown automata; NCA for NPDAs that uses only one stack symbol in addition to the bottom of stack symbol, which is never altered; DCA for deterministic NCAs; NFA for nondeterministic finite automata; DFA for deterministic finite automata; and DTM for deterministic Turing machines. We will also use these same notations to represent the language families accepted by the respective automata classes. As is well-known, NFAs, NPDAs, halting DTMs, and DTMs, accept exactly the regular languages, context-free languages, recursive languages, and recursively enumerable languages, respectively. We refer the reader to [16] for the formal definitions of these devices.

A *counter* is an integer variable that can be incremented by 1, decremented by 1, left unchanged, and tested for zero. It starts at zero and cannot store negative values. Thus, a counter is a pushdown stack on a unary alphabet, in addition to the bottom of the stack symbol which is never altered.

An automaton (NFA, NPDA, etc.) can be augmented with a finite number of counters, where the "move" of the machine also now depends on the status (zero or non-zero) of each counter, and the move can update the counters. It is well known that a DFA augmented with two counters is equivalent to a DTM [24].

In this paper, we will restrict the augmented counter(s) to be reversal-bounded in the sense that each counter can only reverse (i.e., change mode from non-decreasing to non-increasing and vice-versa) at most r times for some given r. In particular, when $r = 1$, the counter reverses only once, i.e., once it decrements, it can no longer increment. Note that a counter that makes r reversals can be simulated by $\lceil \frac{r+1}{2} \rceil$ 1-reversal counters. Closure and decidable properties of various machines augmented with reversal-bounded counters have been studied in the literature (see, e.g., [10,11,17,18]). We will use the notation NCM, NPCM, respectively, to denote NFAs, NPDAs, augmented with reversal-bounded counters. Also, $NCM(k,r)$, $NPCM(k,r)$, respectively will denote the machines with k r-reversal counters. In particular, $NCM(k,1)$, $NPCM(k,1)$, etc. are machines with k 1-reversal counters. We use 'D' in place of 'N' for the deterministic versions, e.g., DCM, $DCM(k,r)$, $DPCM(k,r)$, $DPCM(k,1)$, etc.

Let Σ be a finite alphabet. Then Σ^* (Σ^+) is the set of all (non-empty) words over Σ. A language over Σ is any $L \subseteq \Sigma^*$. Given a language $L \subseteq \Sigma^*$, the complement of L, $\overline{L} = \Sigma^* - L$. The length of a word $w \in \Sigma^*$ is $|w|$, and for $a \in \Sigma$, $|w|_a$ is the number of a's in w.

Let $u, v \in \Sigma^*$. The *shuffle* of u and v, denoted $u \shuffle v$ is the set

$$\{u_1 v_1 u_2 v_2 \cdots u_n v_n \mid u_i, v_i \in \Sigma^*, 1 \leq i \leq n, u = u_1 \cdots u_n, v = v_1 \cdots v_n\}.$$

This can be extended to languages $L_1, L_2 \subseteq \Sigma^*$ by $L_1 \shuffle L_2 = \bigcup_{u \in L_1, v \in L_2} u \shuffle v$. Given $u, v \in \Sigma^*$, there is an obvious NFA with $(|u|+1)(|v|+1)$ states accepting $u \shuffle v$, where each state stores a position within both u and v. This has been called the *naive* NFA for $u \shuffle v$ [8]. It was also mentioned in [8] that if u and v are over disjoint alphabets, then the naive NFA is a DFA.

An NFA $M = (Q, \Sigma, q_0, F, \delta)$ is *accessible* if, for each $q \in Q$, there exists $u \in \Sigma^*$ such that $q \in \delta(q_0, u)$. Also, M is *co-accessible* if, for each $q \in Q$, there exists $u \in \Sigma^*$ such that $\delta(q, u) \cap F \neq \emptyset$. Lastly, M is *trim* if it is both accessible and co-accessible, and M is *acyclic* if $q \notin \delta(q, u)$ for every $q \in Q, u \in \Sigma^+$.

3   Comparing Shuffle on Words to NFAs

The results to follow in this section depend on a result from [4], which is restated here.

Proposition 1. *It is NP-complete to determine, given a DFA M and words u, v over an alphabet of at least two letters, if $L(M) \not\subseteq u \shuffle v$.*

First, it is noted here that this NP-completeness extends to NFAs.

Corollary 1. *It is NP-complete to determine, given an NFA M and words u, v over an alphabet of at least two letters, if $L(M) \not\subseteq u \shuffle v$.*

Proof. NP-hardness follows from Proposition 1.

To show it is in NP, let M be an NFA with state set Q. Create a nondeterministic Turing machine that guesses a word w of length at most $|uv| + |Q|$, and verify that $w \in L(M)$ and that $w \notin u \shuffle v$ in polynomial time [4]. And indeed, $L(M) \not\subseteq u \shuffle v$ if and only if $L(M) \cap \{w \mid |w| \leq |uv| + |Q|, w \in \Sigma^*\} \not\subseteq u \shuffle v$, since any word longer than $|uv| + |Q|$ that is in $L(M)$ implies there is another one in $L(M)$ with length between $|uv| + 1$ and $|uv| + |Q|$, which is therefore not in $u \shuffle v$ (all words in $u \shuffle v$ are of length $|uv|$).   □

Next, the reverse inclusion of Corollary 1 will be examined. In contrast to the polynomial-time testability of $u \shuffle v \subseteq L(M)$ when M is a DFA ([3], with an alternate shorter proof appearing in Proposition 6 of this paper), testing $u \shuffle v \not\subseteq L(M)$ is NP-complete for NFAs.

Proposition 2. *It is* NP-*complete to determine, given an* NFA *M and u, v over an alphabet of at least two letters, whether $u \shuffle v \not\subseteq L(M)$.*

Proof. First, it is in NP, since all words in $u \shuffle v$ are of length $|uv|$, and so a nondeterministic Turing machine can be built that nondeterministically guesses one and tests if it is not in $L(M)$ in polynomial time.

For NP-hardness, let F be an instance of the 3SAT problem (a known NP-complete problem [12]) with a set of Boolean variables $X = \{x_1, \ldots, x_p\}$, and a set of clauses $\{c_1, \ldots, c_q\}$, where each clause has three literals.

If d is a truth assignment, then d is a function from X to $\{+, -\}$ (true or false). For a variable x, then x^+ and x^- are literals. In particular, the literal x^+ is true under d if and only if the variable x is true under d. And, the literal x^- is true under d if and only if the variable x is false [12]. Let $y = \lceil \log_2 p \rceil + 1$, which is enough to hold the binary representation of any of $1, \ldots, p$. For an integer i, $1 \le i \le p$, let $b(i)$ be the string 1 followed by the y-bit binary representation of i, followed by 1 again.

For $1 \le i \le p, 1 \le j \le q$, let $f(i, j)$ be defined as follows, where each element is a set of strings over $\{0, 1\}$:

$$f(i, j) = \begin{cases} \{01b(i)\}, & \text{if } x_i^+ \in c_j; \\ \{10b(i)\}, & \text{if } x_i^- \in c_j; \\ \{10b(i), 01b(i)\}, & \text{otherwise.} \end{cases}$$

For $1 \le j \le q$, let $F_j = f(1, j)f(2, j) \cdots f(p, j)$.

We will next give the construction. Let $u = 1b(1)1b(2) \cdots 1b(p)$, and let $v = 0^p$.

Let $T = \{e_1 b(1)e_2 b(2) \cdots e_p b(p) \mid e_i \in \{10, 01\}, 1 \le i \le p\}$. Clearly $T \subseteq u \shuffle v$, and also T is a regular language, and a DFA M_T can be built accepting this language in polynomial time, as with a DFA $\overline{M_T}$ accepting $\overline{L(M_T)}$.

Then, make an NFA M' accepting $\bigcup_{1 \le j \le q} F_j$. It is clear that this NFA is of polynomial size. Then, make another NFA M'' accepting $L(M') \cup L(\overline{M_T})$. The following claim shows that deciding $u \shuffle v \not\subseteq L(M'')$ is equivalent to deciding if there is a solution to the 3SAT instance.

Claim. The following three conditions are equivalent:

1. $u \shuffle v \cap \overline{L(M'')} \ne \emptyset$,
2. $T \cap \overline{L(M'')} \ne \emptyset$,
3. F has a solution.

Proof. "1 $\Rightarrow$ 2". Let $w \in u \shuffle v \cap \overline{L(M'')}$. Then $w \notin L(M'')$, and since $L(\overline{M_T}) \subseteq L(M'')$, necessarily $w \in L(T)$.

"2 ⇒ 1". Let $w \in T \cap \overline{L(M'')}$. But, $T \subseteq u \shuffle v$; and so $w \in u \shuffle v \cap \overline{L(M'')}$.

"2 ⇒ 3". Assume $w \in T \cap \overline{L(M'')}$. Thus, $w = e_1 b(1) e_2 b(2) \cdots e_p b(p), e_i \in \{10, 01\}$, but $w \notin \bigcup_{1 \leq j \leq q} F_j$. Let d be the truth assignment obtained from w where

$$d(x_i) = \begin{cases} +, & \text{if } e_i = 10; \\ -, & \text{if } e_i = 01; \end{cases}$$

for all $i, 1 \leq i \leq p$. Thus, for every j, $1 \leq j \leq q$, $w \notin F_j$, but for all variables x_i not in c_j, $e_i b(i)$ must be an infix of words in F_j since $10b(i)$ and $01b(i)$ are both in $f(i, j)$ when x_i is not in c_j. So one of the words encoding the (three) variables in c_j, must have $10b(i)$ as an infix of words in F_j where $d(x_i) = +$, or $01b(i)$ as an infix of words in F_j where $d(x_i) = -$, since otherwise F_j would have as infix, for each x_i that is a variable of c_j, $01b(i)$ if $x_i^+ \in c_j$, and $10b(i)$ if $x_i^- \in c_j$, and so w would be in F_j, a contradiction. Thus, d makes clause c_j true, as is the case with every clause. Hence, d is a satisfying truth assignment, and F is satisfiable.

"3 ⇒ 2". Assume F is satisfiable, hence d is a satisfying truth assignment. Create

$$w = e_1 b(1) e_2 b(2) \cdots e_p b(p),$$

where

$$e_i = \begin{cases} 10, & \text{if } d(x_i) = +; \\ 01, & \text{if } d(x_i) = -. \end{cases}$$

Then $w \in T$. Also, for each j, d applied to some variable, say x_i, must be in c_j, but then by the construction of F_j, $e_i b(i)$ must not be an infix of any word in F_j. Hence, $w \notin \bigcup_{1 \leq j \leq q} F_j$, $w \notin L(M')$, and $w \notin L(M'')$. Hence, $w \in T \cap \overline{L(M'')}$. □

Next, we examine the complexity of testing inequality between languages accepted by NFAs and words of a very simple form.

Proposition 3. *It is* NP-*complete to test, given $a^p, b^q \in \Sigma^*, p, q \in \mathbb{N}_0$, and M an NFA over $\Sigma = \{a, b\}$, whether $L(M) \neq a^p \shuffle b^q$.*

Proof. First, it is immediate that the problem is in NP, by Corollary 1 and Proposition 2.

To show NP-hardness, the problem in Proposition 1 is used.

Given M, a DFA, and words u, v, we can construct the naive shuffle NFA N for $u \shuffle v$. The naive NFA is of polynomial size in the length of u and v. Let $(p, q) = (|uv|_a, |uv|_b)$. Then construct the naive NFA A accepting $a^p \shuffle b^q$, which is a polynomially sized DFA since a^p, b^q are over disjoint alphabets. Thus, another DFA can be built accepting $\overline{L(A)}$. We can then construct an NFA M' in polynomial time which accepts $(a^p \shuffle b^q \cap \overline{L(M)}) \cup L(N) \cup (L(M) \cap (\overline{a^p \shuffle b^q}))$ as M is already a DFA. Also, $u \shuffle v \subseteq a^p \shuffle b^q$ since the latter contains all words with p a's and q b's.

We will show $L(M) \subseteq u \shuffle v$ if and only if $L(M') = a^p \shuffle b^q$.

Assume $L(M) \subseteq L(N) (= u \shuffle v)$. Then $L(M) \cap (\overline{a^p \shuffle b^q}) = \emptyset$ since $u \shuffle v \subseteq a^p \shuffle b^q$. All other words in $L(M')$ are in $a^p \shuffle b^q$. Thus, $L(M') \subseteq a^p \shuffle b^q$. Let $w \in a^p \shuffle b^q$.

If $w \notin L(M)$, then $w \in L(M')$. If $w \in L(M)$, then $w \in L(N) \subseteq L(M')$, by the assumption.

Assume $L(M') = a^p \sqcup\!\!\sqcup b^q$. Let $w \in L(M)$. Then $L(M) \cap (\overline{a^p \sqcup\!\!\sqcup b^q}) = \emptyset$ by the assumption. So, $L(M) \subseteq a^p \sqcup\!\!\sqcup b^q$. Assume $w \in L(M)$ but $w \notin L(N)$. However, $w \in L(M')$ by the assumption, a contradiction, as $w \notin a^p \sqcup\!\!\sqcup b^q \cap \overline{L(M)}$, and $w \notin L(M) \cap \overline{a^p \sqcup\!\!\sqcup b^q}$, implying $w \in L(N)$.

Hence, the problem is NP-complete. $\qquad\qquad\qquad\qquad\qquad\qquad\qquad\qquad\square$

To obtain the result of the following corollary, it only needs to be shown that the problem is in NP, which again follows from Corollary 1 and Proposition 2.

Corollary 2. *It is NP-complete to determine, given an NFA M and words u, v over an alphabet of size at least two, if $L(M) \neq u \sqcup\!\!\sqcup v$.*

It is known that there is a polynomial-time algorithm that, given a DFA, will output two words u and v such that, if $L(M)$ is decomposable into the shuffle of two words, then this implies $L(M) = u \sqcup\!\!\sqcup v$ [3]. Moreover, this algorithm runs in time $O(|u| + |v|)$, which is sublinear. This main result from [3] is as follows:

Proposition 4. *Let M be an acyclic, trim, non-unary DFA over Σ. Then we can determine words $u, v \in \Sigma^+$ such that, $L(M)$ has a shuffle decomposition into two words implies $L(M) = u \sqcup\!\!\sqcup v$ is the unique decomposition. This can be calculated in $O(|u| + |v|)$ time.*

However, the downside to this algorithm is that it can output two strings u and v, when $L(M)$ is not decomposable. Thus, the algorithm does not check whether $L(M)$ is decomposable, but if it is, it can find the decomposition in time usually far less than the number of states of the DFA. The decomposition also must be unique over words (this is always true when there are at least two combined letters) [2].

It is now shown that this result scales to NFAs, while remaining polynomial time complexity. The algorithm in [3] scans at most $O(|u| + |v|)$ transitions and states of the DFA from initial state towards final state. From an NFA, it becomes possible to apply the standard subset construction [16] on the NFA only by creating states and transitions for the transitions and states examined by this algorithm (thus, the NFA is never fully determinized, and only a subset of the transitions and states of the DFA are created and traversed). Because the algorithm essentially follows one "main" path from initial state to final state in the DFA, the amount of work required for NFAs is still polynomial.

Proposition 5. *There is a polynomial-time algorithm that, given an acyclic, non-unary NFA $M = (Q_N, \Sigma, q_{N0}, F_N, \delta_N)$, can find strings $u, v \in \Sigma^+$, such that, $L(M)$ has a decomposition into two words implies $L(M) = u \sqcup\!\!\sqcup v$ is the unique decomposition. Moreover, this algorithm runs in time $O((|u|+|v|)\,|Q_N|^2)$.*

Proof sketch. Uniqueness again follows from [2].

The algorithm outputs words $u, v \in \Sigma^+$ such that either $L(M) = u \sqcup\!\!\sqcup v$ or M is not shuffle decomposable. It is based off the one described in [3], which is quite detailed, and thus not reproduced here, although we will refer to it.

In order to use the algorithm in Proposition 4, first all states that are not accessible or not co-accessible are removed. For this, a breadth-first graph search algorithm can be used to detect which states can be reached from q_0 in $O(|Q_N|^2)$ time. It also verifies that all final states reached are the same distance from the initial state, and if not, M is not decomposable. Then, collapse these final states down to one state q_f and remove all outgoing transitions, which does not change the language accepted since M is acyclic. Then, check which states can be reached from q_f following transitions in reverse using the graph search, and remove all states that cannot be reached. This results in an NFA $M_1 = (Q_1, \Sigma, q_1, \{q_f\}, \delta_1)$ that is trim and accepts $L(M)$.

Let $M_D = (Q_D, \Sigma, q_{D0}, F_D, \delta_D)$ be the DFA obtained from M_1 via the subset construction (we do not compute this, but will refer to it). Necessarily M_D is trim and acyclic, since M_1 is as well. Then $q_{D0} = \{q_1\}, F_D = \{P \mid P \in Q_D, q_f \in P\}$.

We modify the algorithm of Proposition 4 as follows: In place of DFA states, we use subsets of Q_1 from Q_D [16]. However, states and transitions are only computed as needed in the algorithm. Any time $\delta(P, a)$ is referenced in the algorithm, we first compute the deterministic transition as follows: $\delta_D(P, a) = \bigcup_{p \in P} \delta_1(p, a)$, and then use this transition. Since there are at most $|Q_1|$ states in a subset of Q_1, any transition of δ_D defined on a given state and a given letter can transition to at most $|Q_1|$ states. Then, we can compute $\delta_D(P, a)$ in $O(|Q_1|^2)$ time (for each state $p \in P$, add $\delta_1(p, a)$ into a sorted list without duplicates). As it is making the list, it can test if this state is final by testing if $q_f \in \delta_D(P, a)$. Therefore, this algorithm inspects $O(|u| + |v|)$ states and transitions of M_D, which takes $O(|Q_1|^2(|u| + |v|))$ time to compute using the subset construction. □

4   Shuffle on Languages

A known result involving shuffle on words is that there is a polynomial-time test to determine, given words $u, v \in \Sigma^+$ and a DFA M, whether $u \shuffle v \subseteq L(M)$ [3]. An alternate simpler proof of this result is demonstrated next, and then this proof technique will be used to extend to more general decision problems.

Proposition 6. *There is a polynomial-time algorithm to determine, given $u, v \in \Sigma^+$, and a DFA M, whether or not $u \shuffle v \subseteq L(M)$.*

Proof. Clearly, $u \shuffle v$ is a subset of $L(M)$ if and only if $L(A) \cap \overline{L(M)} = \emptyset$, where A is the naive NFA accepting $u \shuffle v$. A DFA accepting $\overline{L(M)}$ can be built in polynomial time, and the intersection is accepted by an NFA using the standard construction [16] whose emptiness can be checked in polynomial time [16]. □

This result will be generalized in two ways. First, instead of individual words u and v, languages from NCM(k, r), for some fixed k, r will be used. Moreover, instead of a DFA for the right side of the inclusion, a DCM(k, r) machine will be used.

Proposition 7. *Let k, r be any fixed integers. It is decidable, given $M_1, M_2 \in$ NCM(k, r) and $M_3 \in$ DCM(k, r), whether $L(M_1) \uplus L(M_2) \subseteq L(M_3)$. Moreover, the decision procedure is polynomial in $n_1 + n_2 + n_3$, where n_i is the size of M_i.*

Proof. First, construct from M_1 and M_2, an NCM M_4 that accepts $L(M_1) \uplus L(M_2)$. Clearly M_4 is an NCM$(2k, r)$, and the size of M_4 is polynomial in $n_1 + n_2$.

Then, construct from M_3 a DCM(k, r) machine M_5 accepting the complement of $L(M_3)$, which can be done in polynomial time [17].

Lastly, construct from M_4 and M_5 an NCM$(3k, r)$ machine M_6 accepting $L(M_4) \cap L(M_5)$ by simulating the machines in parallel.

It is immediate that $L(M_1) \uplus L(M_2) \subseteq L(M_3)$ if and only if $L(M_6) = \emptyset$. Further, it has been shown that for any fixed t, s, it is decidable in polynomial time, given M in NCM(t, s), whether $L(M) = \emptyset$ [15]. □

Actually, the above proposition can be made stronger. For any fixed k, r, the decidability of non-emptiness of $L(M)$ for an NCM(k, r) is in NLOG, the class of languages accepted by nondeterministic Turing machines in logarithmic space [15]. It is known that NLOG is contained in the class of languages accepted by deterministic Turing machines in polynomial time (whether or not the containment is proper is open). By careful analysis of the constructions in the proof of the above proposition, M_6, could be constructed by a logarithmic space deterministic Turing machine. Hence:

Corollary 3. *Let k, r be any fixed integers. The problem of deciding, given $M_1, M_2 \in$ NCM(k, r) and $M_3 \in$ DCM(k, r), whether $L(M_1) \uplus L(M_2) \subseteq L(M_3)$, is in NLOG.*

Proposition 7 also holds if M_1 and M_2 are NFAs and M_3 is a deterministic pushdown automaton.

Proposition 8. *It is decidable, given NFAs M_1, M_2 and $M_3 \in$ DPDA, whether $L(M_1) \uplus L(M_2) \subseteq L(M_3)$. Moreover, the decision procedure is polynomial in $n_1 + n_2 + n_3$, where n_i is the size of M_i.*

Proof. The proof and algorithm proceeds much like the proof of Proposition 7. Given two NFAs M_1, M_2, another NFA M_4 that accepts $L(M_1) \uplus L(M_2)$ can be constructed in polynomial time. Then, for a given DPDA M_3, a DPDA M_5 can be constructed accepting its complement in polynomial time (and is of polynomial size) [13]. Also, given an NFA M_4 and a DPDA, a PDA M_6 can be built in polynomial time accepting $L(M_4) \cap L(M_5)$. As above, $L(M_1) \uplus L(M_2) \subseteq L(M_3)$ if and only if $L(M) = \emptyset$, and emptiness is decidable in polynomial time for NPDAs [16]. □

In contrast to Proposition 7, the following is shown:

Proposition 9. *It is undecidable, given one-state DFAs M_1 accepting a^* and M_2 accepting b^*, and an NCM$(1, 1)$ machine M_3 over $\{a, b\}$, whether $L(M_1) \uplus L(M_2) \subseteq L(M_3)$.*

Proof. Let $\Sigma = \{a, b\}$. Then $L_1 \shuffle L_2 = \Sigma^*$. Let $M_3 \subseteq \Sigma^*$ be an $\mathsf{NCM}(1,1)$ machine. Then $L_1 \shuffle L_2 \subseteq L_3$ if and only if $L_3 = \Sigma^*$. The result follows, since the universality problem for $\mathsf{NCM}(1,1)$ is undecidable. The idea is the following: Given a single-tape deterministic Turing machine Z, we construct M_3 which, when given any input w, accepts if and only if w does not represent a halting sequence of configurations of Z on an initially blank tape (by guessing a configuration ID_i, and extracting the symbol at a nondeterministically chosen position j within this configuration, storing j in the counter, and then checking that the symbol in position j in the next configuration ID_{i+1} determined by decrementing the counter is not compatible with the next move of the DTM from ID_i; see [1]). Hence, $L(M_3)$ accepts the universe if and only if Z does not halt. By appropriate coding, the universe can be reduced to $\{a, b\}^*$.   □

Note that the proof of Proposition 9 shows: Let G be a language family such that universality is undecidable. Then it is undecidable, given one-state DFAs M_1 accepting a^* and M_2 accepting b^*, and L in G, whether $L(M_1) \shuffle L(M_2) \subseteq L$.

Proposition 10. *Let $L_1 = (a + b)^*$ and $L_2 = \{\lambda\}$. It is PSPACE-complete, given an NFA M with input alphabet $\{a, b\}$, whether $L_1 \shuffle L_2 \not\subseteq L(M)$.*

Proof. Clearly, $(a+b)^* \shuffle \{\lambda\} \not\subseteq L$ if and only if $L \neq (a+b)^*$. The result follows, since it is known that this question is PSPACE-complete (see, e.g., [12]).   □

Remark. In Proposition 9, if M_1 and M_2 are DFAs accepting finite languages, and L is a language in any family with a decidable membership problem, then it is decidable whether $L(M_1) \shuffle L(M_2) \subseteq L$. This is clearly true by enumerating all strings in $L(M_1) \shuffle L(M_2)$ and testing membership in L.

Next, shuffle over unary alphabets is considered.

Proposition 11. *It is decidable, given languages L_1, L_2, L_3 over alphabet $\{a\}$ accepted by NPCMs, whether:*

1. $L_1 \shuffle L_2 \subseteq L_3$
2. $L_3 \subseteq L_1 \shuffle L_2$

Proof. It is known that the Parikh map of the language accepted by any NPCM is an effectively computable semilinear set (in this case over $\mathbb{N}$) [17] and, hence, the languages L_1, L_2, L_3 can be accepted by DFAs over a unary alphabet.   □

Proposition 12. *It is NP-complete to decide, for an NFA M over alphabet $\{a\}$, whether $a^* \shuffle \{\lambda\} \not\subseteq L(M)$.*

Proof. Clearly, $a^* \shuffle \{\lambda\} \not\subseteq L$ if and only if $L \neq a^*$. The result follows, since it is known that this question is NP-complete (see, e.g., [12]).   □

For the case when the unary languages L_1 and L_2 are finite:

Proposition 13. *It is polynomial-time decidable, given two finite unary languages L_1 and L_2 (where the lengths of the strings in L_1 and L_2 are represented in binary) and a unary language L_3 accepted by an NFA M, all over the same letter, whether $L_1 \shuffle L_2 \subseteq L_3$.*

Proof. Let r be the sum of the cardinalities of L_1 and L_2, s be the length of the binary representation of the longest string in $L_1 \cup L_2$, and t be the length of binary representation of M.

We represent the NFA M by an $n \times n$ Boolean matrix A_M, where n is the number of states of M, and $A_M(i, j) = 1$ if there is a transition from state i to state j; 0 otherwise.

Let x be the binary representation of a unary string a^d, where $d = d_1 + d_2$, $a^{d_1} \in L_1$, and $a^{d_2} \in L_2$. To determine if a^d is in L_3, we compute A_M^d and check that for some accepting state p, the $(1, p)$ entry is 1. Since the computation of A_M^d can be accomplished in $O(\log d)$ Boolean matrix multiplications (using the "right-to-left binary method for exponentiation" technique used to compute x^m where m is a positive integer in $O(\log m)$ multiplications, described in Sect. 4.6.3 of [23]), and since matrix multiplication can be calculated in polynomial time, it follows that we can decide whether $L_1 \sqcup L_2 \subseteq L_3$ in time polynomial in $r + s + t$.
□

However, when the alphabet of the finite languages L_1, L_2 is at least binary:

Proposition 14. *It is* NP-*complete to determine, given finite language L_1 and L_2, and an* NPDA *M accepting L_3, whether $L_1 \sqcup L_2 \not\subseteq L_3$.*

Proof. NP-hardness follows from Proposition 2. To show that it is in NP, guess a word $u \in L_1$, and $v \in L_2$, guess a word w of length $|u| + |v|$, and verify that it is in $u \sqcup v$ [4]. Then, verify that $w \notin L_3$, which can be done in polynomial time since the membership problem for NPDAs can be solved in polynomial time.

Finally, the following proposition follows from the proof of Theorem 6 in [5].

Proposition 15. *It is undecidable, given two languages accepted by 1-reversal-bounded* DPDAs *(resp.,* DCAs*), whether their shuffle is accepted by a 1-reversal-bounded* DPDA *(resp.,* DCA*).*

References

1. Baker, B.S., Book, R.V.: Reversal-bounded multipushdown machines. J. Comput. Syst. Sci. **8**(3), 315–332 (1974)
2. Berstel, J., Boasson, L.: Shuffle factorization is unique. Theoret. Comput. Sci. **273**, 47–67 (2002)
3. Biegler, F., Daley, M., McQuillan, I.: Algorithmic decomposition of shuffle on words. Theoret. Comput. Sci. **454**, 38–50 (2012)
4. Biegler, F., McQuillan, I.: On comparing deterministic finite automata and the shuffle of words. In: Holzer, M., Kutrib, M. (eds.) CIAA 2014. LNCS, vol. 8587, pp. 98–109. Springer, Heidelberg (2014)
5. Bordihn, H., Holzer, M., Kutrib, M.: Some non-semi-decidability problems for linear and deterministic context-free languages. In: Domaratzki, M., Okhotin, A., Salomaa, K., Yu, S. (eds.) CIAA 2004. LNCS, vol. 3317, pp. 68–79. Springer, Heidelberg (2005)
6. Buss, S., Soltys, M.: Unshuffling a square is NP-hard. J. Comput. Syst. Sci. **80**(4), 766–776 (2014)

7. Câmpeanu, C., Salomaa, K., Vágvölgyi, S.: Shuffle quotient and decompositions. In: Kuich, W., Rozenberg, G., Salomaa, A. (eds.) DLT 2001. LNCS, vol. 2295, pp. 186–196. Springer, Heidelberg (2002)
8. Daley, M., Biegler, F., McQuillan, I.: On the shuffle automaton size for words. J. Autom. Lang. Comb. **15**, 53–70 (2010)
9. Domaratzki, M.: More words on trajectories. Bull. EATCS **86**, 107–145 (2005)
10. Eremondi, J., Ibarra, O.H., McQuillan, I.: Deletion operations on deterministic families of automata. In: Jain, R., Jain, S., Stephan, F. (eds.) TAMC 2015. LNCS, vol. 9076, pp. 388–399. Springer, Heidelberg (2015)
11. Eremondi, J., Ibarra, O.H., McQuillan, I.: Insertion operations on deterministic reversal-bounded counter machines. In: Dediu, A.-H., Formenti, E., Martín-Vide, C., Truthe, B. (eds.) LATA 2015. LNCS, vol. 8977, pp. 200–211. Springer, Heidelberg (2015)
12. Garey, M.R., Johnson, D.S.: Computers and Intractability: A Guide to the Theory of NP-Completeness. Series of Books in the Mathematical Sciences. W. H. Freeman and Company, New York (1979)
13. Geller, M.M., Hunt III, H.B., Szymanski, T.G., Ullman, J.D.: Economy of description by parsers, dpda's, and pda's. Theoret. Comput. Sci. **4**, 143–153 (1977)
14. Ginsburg, S., Spanier, E.H.: Mappings of languages by two-tape devices. J. ACM **12**(3), 423–434 (1965)
15. Gurari, E.M., Ibarra, O.H.: The complexity of decision problems for finite-turn multicounter machines. J. Comput. Syst. Sci. **22**(2), 220–229 (1981)
16. Hopcroft, J.E., Ullman, J.D.: Introduction to Automata Theory, Languages, and Computation. Addison-Wesley, Reading (1979)
17. Ibarra, O.H.: Reversal-bounded multicounter machines and their decision problems. J. ACM **25**(1), 116–133 (1978)
18. Ibarra, O.H.: Automata with reversal-bounded counters: a survey. In: Jürgensen, H., Karhumäki, J., Okhotin, A. (eds.) DCFS 2014. LNCS, vol. 8614, pp. 5–22. Springer, Heidelberg (2014)
19. Jędrzejowicz, J., Szepietowski, A.: Shuffle languages are in P. Theoret. Comput. Sci. **250**, 31–53 (2001)
20. Kari, L.: On language equations with invertible operations. Theoret. Comput. Sci. **132**(1–2), 129–150 (1994)
21. Kari, L., Konstandtinidis, S., Sosík, P.: On properties of bond-free DNA languages. Theoret. Comput. Sci. **334**, 131–159 (2005)
22. Kari, L., Sosík, P.: Aspects of shuffle and deletion on trajectories. Theoret. Comput. Sci. **332**(1–3), 47–61 (2005)
23. Knuth, D.E.: Seminumerical Algorithms, The Art of Computer Programming, 3rd edn. Addison-Wesley, Reading (1998)
24. Minsky, M.L.: Recursive unsolvability of Post's problem of "tag" and other topics in theory of Turing machines. Ann. Math. **74**(3), 437–455 (1961)
25. Ogden, W.F., Riddle, W.E., Round, W.C.: Complexity of expressions allowing concurrency. In: Proceedings of the 5th ACM SIGACT-SIGPLAN Symposium on Principles of Programming Languages, POPL 1978, pp. 185–194. ACM NY, USA (1978)

On the Computational Complexity of Problems Related to Distinguishability Sets

Markus Holzer$^{(\boxtimes)}$ and Sebastian Jakobi

Institut für Informatik, Universität Giessen,
Arndtstr. 2, Giessen 35392, Germany
{holzer,sebastian.jakobi}@informatik.uni-giessen.de

Abstract. We study the computational complexity of problems related to distinguishability sets for regular languages. Roughly speaking, the distinguishability set $D(L)$ for a (not necessarily regular) language L consists of all words w that are a common suffix of a word xw in L and of a word yw that does not belong to L. In particular, we investigate the complexity of the representation problem, i.e., deciding for given automata A and B, whether B accepts the distinguishability set of $L(A)$. It is shown that this problem and some of its variants are highly intractable, namely PSPACE-complete. In fact, determining the size of an automaton for $D(L(A))$ is already PSPACE-complete. On the other hand, questions related to the hierarchy induced by iterated application of the D-operator turn out to be much easier. For instance, the question whether for a given automaton A, the accepted language is equal to its own distinguishability set, i.e., whether $L(A) = D(L(A))$ holds, is shown to be NL-complete. As a byproduct of our investigations, we found a compelling characterization of synchronizing automata, namely that a (minimal) automaton A is synchronizing if and only if $D(L(A)) = D^2(L(A))$.

1 Introduction

There is a vast literature documenting the importance of the notion of finite automata and problems thereof as an enormously valuable concept in theoretical computer science and applications. Although the history of finite automata dates back around 60 years, even nowadays this is a vivid area of research. For instance, recently, the language $D(L(A))$ that distinguishes between all non-equivalent states or quotients of a given deterministic finite automaton A was considered in more detail in [1]. There a systematic study of general properties of $D(L(A))$ is conducted from a descriptional and a formal language theoretical point of view. Observe that the idea of distinguishability is not new and has a long and fruitful history, see, e.g., Moore's seminal paper on gedankenexperiments [8]—a brief summary on some developments is given in [1], too. The motivation to study this language and its properties stems from electronic circuit testing. There a property is tested by applying several inputs to the circuit, and checking the produced output. Since circuits and finite automata are closely related by simulating each other, it is natural to ask for the minimality of these models. For

© Springer International Publishing Switzerland 2015
J. Shallit and A. Okhotin (Eds.): DCFS 2015, LNCS 9118, pp. 117–128, 2015.
DOI: 10.1007/978-3-319-19225-3_10

finite automata the minimization problem dates back to the early beginnings of automata theory. This results in testing whether two states are equivalent or not. Thus, it can be described by considering only words from $D(L(A))$, i.e., words that distinguish between non-equivalent states of a given automaton A.

What is missing in the investigation in [1] is the computational complexity of problems related to the language $D(L(A))$, for a deterministic finite automaton A. For instance, from the motivation given in [1] the complexity of the following problem that is related to minimization is relevant: how hard is it to decide for a given word w and an automaton A, whether w belongs to $D(L(A))$? Analogous questions on the representation of the $D(L(A))$ language by automata or on the iterated application of the D-operation, when viewed as an operator $D: 2^{\Sigma^*} \to 2^{\Sigma^*}$ with $L \mapsto D(L)$, can be asked. It turns out that the computational complexity of these problems varies from L- to PSPACE-completeness, which is an enormous span in complexity. Moreover, the results on the PSPACE-completeness are very interesting, since only deterministic finite automata are involved, and normally, standard problems that deal with deterministic devices turn out to be of lower complexity, see, e.g., [3]. In fact, the problems related to representability of distinguishability sets turn out to be highly intractable, namely PSPACE-complete. Even determining the size of an automaton for $D(L(A))$ is already PSPACE-complete. On the other hand, questions related to the hierarchy induced by iterated application of the distinguishability operator turn out to be much easier, namely NL-complete. This significant decrease in complexity goes hand in hand with a very interesting structure of this hierarchy, namely, it collapses to its third level, i.e., $D^2(L) = D^3(L)$, for *every* language L [1]. As a spin-off of our investigations, we found a compelling characterization of synchronizing automata, namely that a (minimal) automaton A is synchronizing if and only if $D(L(A)) = D^2(L(A))$. During the last decade, synchronizing automata and Černý's Conjecture were a very active research area, see, e.g., [10] for a survey on these topics. Our investigation on the computational complexity of the distinguishability operator D can be seen as a first step towards a better understanding of other distinguishability operators as, e.g., described in [1].

The paper is organized as follows: the next section provides basic definitions concerning finite automata and the distinguishability operator D. Moreover, also some important properties of the D-operation are listed. After that, we first discuss the complexity of deciding whether a word belongs to the distinguishability set of a given language. We will see that the complexity of this problem depends on a subtle detail in the problem definition. Then we investigate the complexity of problems related to the representability of distinguishability sets. We close our studies with a summary of the obtained results and give hints for further research. Due to space constraints some proofs are omitted.

2   Preliminaries

We recall some definitions on finite automata as contained in [2]. A *deterministic finite automaton* (DFA) is a quintuple $A = (Q, \Sigma, \delta, q_0, F)$, where Q is the finite set of *states*, Σ is the finite set of *input symbols*, $q_0 \in Q$ is the *initial state*, $F \subseteq Q$ is the set of *accepting states*, and $\delta \colon Q \times \Sigma \to Q$ is the *transition function*. The *language accepted* by the DFA A is defined as

$$L(A) = \{ w \in \Sigma^* \mid \delta(q_0, w) \in F \},$$

where the transition function is recursively extended to $\delta \colon Q \times \Sigma^* \to Q$.

Let $L \subseteq \Sigma^*$ be a language. Then the Myhill-Nerode equivalence relation $\equiv_L$ on L is defined by $x \equiv_L y$ if and only if for every word $w \in \Sigma^*$ we have $xw \in L \iff yw \in L$. If $A = (Q, \Sigma, \delta, q_0, F)$ is a DFA accepting the language L, and $R_p = R_q$, for some states $p, q \in Q$, then we write $p \equiv_A q$. Here $R_q = \{ w \in \Sigma^* \mid \delta(q, w) \in F \}$ refers to the *right-language* of the state q. A DFA is *minimal* if it has no pair of equivalent states and no unreachable states. The *left-quotient*, or *quotient* for short, of a language L by a word w is the language $w^{-1} \cdot L = \{ x \mid wx \in L \}$. A quotient corresponds to an equivalence class of $\equiv_L$. If a language L is regular, then the number of distinct quotients is finite. It is exactly the number of states of the *minimal* DFA accepting L. This number is called the *state complexity* of L, and is denoted by $\mathsf{sc}(L)$. In a minimal DFA, for every state $q \in Q$, the language R_q is exactly a quotient. If some quotient of a regular language L is $\emptyset$, then the minimal DFA of L has a dead state.

Let $L \subseteq \Sigma^*$ be a language. For two words $x, y \in \Sigma^*$ with $x \not\equiv_L y$, there exists at least one word w, such that $xw \in L \iff yw \notin L$. Let $A = (Q, \Sigma, \delta, q_0, F)$ be a DFA with $L = L(A)$. For two states $p, q \in Q$ with $p \not\equiv_A q$, there exists at least one word w such that $\delta(p, w) \in F \iff \delta(q, w) \notin F$. We say that the word w *distinguishes* between the words x and y, in the former case, and the states p and q in the latter case. Next we define the set of words that distinguishes between two words x and y w.r.t. the language L. Let $x, y \in \Sigma^*$, then set

$$\mathsf{D}_L(x, y) = \{ w \in \Sigma^* \mid xw \in L \iff yw \notin L \}.$$

Naturally, we define the distinguishability language of L by

$$\mathsf{D}(L) = \{ w \in \Sigma^* \mid \exists x, y \in \Sigma^* : xw \in L \iff yw \notin L \}.$$

It is easy to see that this definition is equivalent to $\mathsf{D}(L) = \bigcup_{x,y \in \Sigma^*} \mathsf{D}_L(x, y)$. In the same way, for a DFA $A = (Q, \Sigma, \delta, q_0, F)$, we define $\mathsf{D}_A(p, q)$ for $p, q \in Q$ by

$$\mathsf{D}_A(p, q) = \{ w \in \Sigma^* \mid \delta(p, w) \in F \iff \delta(q, w) \notin F \},$$

and

$$\mathsf{D}(A) = \{ w \in \Sigma^* \mid \exists p, q \in Q : \delta(p, w) \in F \iff \delta(q, w) \notin F \}.$$

A DFA is *reduced* if all states are reachable from the initial state (accessible), and all states can reach a final state (useful), except at most one that is a sink state or dead state, i.e., a non-accepting state where all output transitions are self loops. Observe that the following property holds [1].

Lemma 1. *Let L be a language accepted by a* reduced *deterministic finite automaton A. Then $\mathsf{D}(L) = \mathsf{D}(A)$.*

In order to explain these definitions in more detail we give a small example, which we literally take from [1].

Example 2. Consider the language $L = ((0+1)(0+1))^*(\lambda+1)$. Easy calculations show that $D(L) = \lambda + (0+1)^*0$ and $D^2(L) = \lambda$, where $D^2(L)$ is an abbreviation for $D(D(L))$. Note that, since $D^2(L) = \lambda$, all other iterations of the D-operator to $D^2(L)$ give λ as a result, too. The DFAs for L, $D(L)$, and $D^2(L)$ are depicted in Fig. 1 from left to right, respectively.

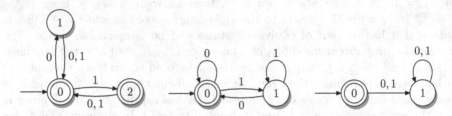

Fig. 1. The DFAs for $L = ((0+1)(0+1))^*(\lambda+1)$, $D(L)$, and $D^2(L)$ from left to right. Here $D^2(L)$ is an abbreviation for $D(D(L))$.

Finally, we list some important properties of D-sets, which can be found in [1]. First of all, the set $D(L)$ is suffix closed, for any language L. An alternative characterization of the distinguishability set of $L \subseteq \Sigma^*$ is

$$D(L) = \mathsf{suff}(L) \cap \mathsf{suff}(\overline{L}),$$

where $\mathsf{suff}(L)$ is the language of all suffixes of L and $\overline{L}$ refers to the complement of L, i.e., $\overline{L} = \Sigma^* \setminus L$. By this characterization it is easy to see that if L is a finite language, then we have $D(L) = \mathsf{suff}(L)$. The iteration of the D-operator is defined by $D^0(L) = L$ and $D^{n+1}(L) = D(D^n(L))$, for $n \geq 0$, and induces a hierarchy

$$D^{n+1}(L) \subseteq D^n(L),$$

for $n \geq 1$. Observe, that $D(L) \subseteq L$ does not hold in general. Nevertheless, if L is suffix closed, then $D(L) \subseteq L$. Concerning the strictness of the hierarchy, surprisingly, it is finite and collapses to its second level, i.e., $D^3(L) = D^2(L)$, for any language L. Moreover, $D(L) = L$ if and only if L is suffix closed with $\emptyset$ as one of its quotients. Finally, $D^2(L) = D(L)$ if and only if $D(L)$ has $\emptyset$ as a quotient.

We classify problems on distinguishability sets w.r.t. their computational complexity. Consider the inclusion chain $L \subseteq NL \subseteq P \subseteq NP \subseteq PSPACE$. Here L (NL, respectively) refers to the set of problems accepted by deterministic (nondeterministic, respectively) logspace bounded Turing machines, P (NP, respectively) is the set of problems accepted by deterministic (nondeterministic, respectively) polynomial time bounded Turing machines, and $PSPACE$ is the set of problems accepted by deterministic or nondeterministic polynomial space bounded Turing machines. Further, for a complexity class C, the set coC is the set of complements of languages from C. Hardness and completeness is always meant w.r.t. deterministic logspace-bounded many-one reducibility, unless otherwise stated.

Many of the hardness results in our paper are obtained by reductions from the following problem:

- DFA-UNION-UNIVERSALITY. Given DFAs $A_1, A_2, \ldots, A_n$ with common input alphabet Σ, decide whether $\bigcup_{i=1}^n L(A_i) = \Sigma^*$.

This problem is well known to be PSPACE-complete, even if Σ is a binary alphabet [5,7]. Another classical decision problem which will be used later is the following:

- GRAPH-REACHABILITY. Given a directed graph $G = (V, E)$ and two of its vertices s and t, decide whether there is a path from s to t.

Most variants of this problem are NL-complete, but if the instances are restricted to graphs of out-degree one, which means that every vertex has exactly one outgoing edge, then the problem becomes L-complete [11].

3 Computational Complexity Results on Distinguishability Sets

In this section we study the computational complexity of several natural problems related to distinguishability sets. Perhaps the first problem that comes to one's mind is the following membership problem: given an automaton and a word, does the word belong to the corresponding distinguishability set? Here we have to be careful with the exact problem definition, because for a given DFA A the two distinguishability sets $D(L(A))$ and $D(A)$ may differ, if A is *not* reduced. Interestingly, this subtle difference influences the computational complexity of the problem. Therefore we define the following two problems:

- DFA-D-MEMBERSHIP. Given a DFA A with input alphabet Σ and a word $w \in \Sigma^*$, decide whether $w \in D(A)$.
- LANGUAGE-D-MEMBERSHIP. Given a DFA A with input alphabet Σ and a word $w \in \Sigma^*$, decide whether $w \in D(L)$, for $L = L(A)$.

The first problem can be shown to be L-complete, where for L-hardness an NC^1 reduction from the reachability problem for out-degree one graphs is applied.

Theorem 3. DFA-D-MEMBERSHIP *is* L-*complete under* NC^1 *many-one reductions.*[1] □

---

[1] An NC^1 many-one reduction from a problem A to B is a Dlogtime-uniform family $C = (C_n)$ of Boolean circuits of polynomial size and logarithmic depth with AND-, OR-, and NOT-gates of bounded fan-in, such that $x \in A$ if and only if $C_{|x|}(x) \in B$.

The LANGUAGE-D-MEMBERSHIP turns out to be NL-complete. For hardness a reduction from the reachability problem for graphs of out-degree two is used.

Theorem 4. LANGUAGE-D-MEMBERSHIP *is* NL-*complete.*   □

The rest of this section is in three parts. First, we consider problems on the description size of distinguishability sets. Then we take a closer look at the problem of deciding whether for two given automata one describes the distinguishability set of the other. Third, and finally, problems related to the finite hierarchy induced by the iteration of the D-operator are investigated.

3.1   Results on the Size of Distinguishability Sets

We study the complexity of the following problem and variants thereof:

– D-SET-SIZE. Given a DFA A and an integer k, decide whether $sc(D(L)) \leq k$, where $L = L(A)$?

In [1] the following result was shown: let A be a n-state DFA accepting the language L. Then $2^n - n$ states are sufficient and necessary in the worst case for a DFA to accept the language $D(L)$.

Our first result shows that D-SET-SIZE problem has already a very high complexity, and is intractable. Using the equality $D(L) = suff(L) \cap suff(\overline{L})$ leads to a construction of an exponentially sized DFA on which an NL-algorithm for testing $sc(D(L)) \leq k$ can be applied. This gives the following result.

Lemma 5. D-SET-SIZE *is contained in* PSPACE.   □

Now we provide the lower bound of the D-SET-SIZE problem.

Lemma 6. D-SET-SIZE *is* PSPACE-*hard w.r.t. deterministic polytime many-one reductions.*

Proof. We give a polynomial time reduction from the DFAs union universality problem. As a problem instance let $A_i = (Q_i, \Sigma, \delta_i, s_i, F_i)$, for $1 \leq i \leq n$, be DFAs over common input alphabet Σ. We may assume that $\bigcup_{i=1}^{n} L(A_i)$ contains λ and Σ—this can be easily checked by the reduction, and if the union does not contain λ and Σ, we have a "no" instance. Therefore we can safely add DFAs A_{n+1} and A_{n+2} for the languages $\{\lambda\}$ and Σ to the instance without changing its membership to the union universality problem. In the next step of the reduction we minimize all DFAs (therefore using the polynomial time and not logspace reduction) and we add a new input symbol $\# \notin \Sigma$ on which every DFA goes back to its initial state. Let us denote the resulting DFAs by $A_i' = (Q_i', \Sigma_\#, \delta_i', s_i', F_i')$, for $1 \leq i \leq n+2$, where $\Sigma_\# = \Sigma \cup \{\#\}$. Notice that the DFAs A_i' are still minimal and that moreover $\bigcup_{i=1}^{n} L(A_i) = \Sigma^*$ if and only if $\bigcup_{i=1}^{n+2} L(A_i') = \Sigma_\#^*$.

Now we construct the DFA $A = (Q, \Gamma, \delta, q_0, F)$ for the instance of the D-SET-SIZE problem as follows: the state set is $Q = \bigcup_{i=1}^{n+2} Q_i' \cup \{q_0, q_f, q_s\}$, final states

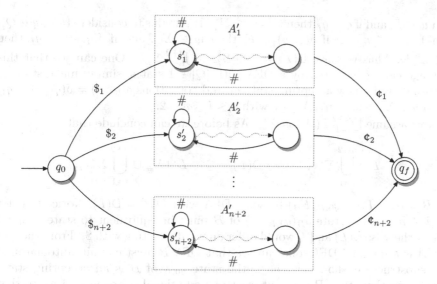

Fig. 2. The DFA A constructed from the instance of the union universality problem. The DFAs $A_1', A_2', \ldots, A_{n+2}'$ are only sketched in the boxes; all these DFAs contain only reachable states. The sink state q_s and transitions leading to it are omitted. The transitions leading from the boxes to state q_f indicate that from every state in the box for DFA A_i' there is a $\text{\textcent}_i$-transition to state q_f. The set of final states of A consists of the final states of the input DFAs A_i', together with q_f.

are $F = \bigcup_{i=1}^{n+2} F_i' \cup \{q_f\}$, the input alphabet is $\Gamma = \Sigma_\# \cup \{\$_i, \text{\textcent}_i \mid 1 \le i \le n+2\}$. The transition function δ is defined as follows—see Fig. 2: for $1 \le i \le n + 2$, $p_i \in Q_i'$ and $a \in \Sigma_\#$ we have

$$\delta(q_0, \$_i) = s_i', \qquad \delta(p_i, a) = \delta_i'(p_i, a), \qquad \delta(p_i, \text{\textcent}_i) = q_f.$$

All undefined transitions lead to the sink state q_s. From the fact that the DFAs A_i are minimal, one can see that also A is a minimal DFA. The integer k for the instance of the D-SET-SIZE problem is $k = |Q| + 1$.

Let $L = L(A)$. We show that $\bigcup_{i=1}^{n+2} L(A_i') = \Sigma_\#^*$ if and only if $\mathrm{sc}(\mathsf{D}(L)) \le k$. First assume $\bigcup_{i=1}^{n+2} L(A_i') = \Sigma_\#^*$. We show that in this case

$$\mathsf{D}(L) = L \cup \Sigma_\#^* \cup \bigcup_{i=1}^{n+2} \Sigma_\#^* \text{\textcent}_i$$

holds. This is seen as follows. We have $L \subseteq \mathsf{D}(L)$ because A has a sink state. Since $\bigcup_{i=1}^{n+2} \$_i \Sigma_\#^* \text{\textcent}_i \subseteq L$ and $\mathsf{D}(L)$ is suffix closed, we obtain $\bigcup_{i=1}^{n+2} \Sigma_\#^* \text{\textcent}_i \subseteq \mathsf{D}(L)$. Similarly, since $\bigcup_{i=1}^{n+2} \$_i L(A_i') \subseteq L$, we have $\bigcup_{i=1}^{n+2} L(A_i') = \Sigma_\#^* \subseteq \mathsf{D}(L)$. For the converse inclusion, notice that $w \in \mathsf{D}(L)$ implies that there is some state $q \in Q$ such that $\delta(q, w) \in F$. Clearly, this state q cannot be the sink state q_s. If $q = q_0$

then $w \in L$, and if $q = q_f$ then $w = \lambda \in \Sigma_{\#}^*$. It remains to consider states $q \in Q_i'$, with $1 \le i \le n+2$: if $\delta(q,w) \in F_i$ then $w \in \Sigma_{\#}^*$, and if $\delta(q,w) = q_f$ then $w \in \Sigma_{\#}^* \mathcal{c}_i$. Thus we have $\mathsf{D}(L) = L \cup \Sigma_{\#}^* \cup \bigcup_{i=1}^{n+2} \Sigma_{\#}^* \mathcal{c}_i$. One can see that this language can be accepted by a DFA with $|Q| + 1$ states: simply make state q_0 accepting, add a new accepting state q_1 and transitions $\delta(q_0, a) = \delta(q_1, a) = q_1$, for all $a \in \Sigma_{\#}$ and $\delta(q_1, \mathcal{c}_i) = q_f$ with $1 \le i \le n+2$.

Now assume $\bigcup_{i=1}^{n+2} L(A_i') \neq \Sigma_{\#}^*$. As before we can conclude that

$$L \cup \bigcup_{i=1}^{n+2} \Sigma_{\#}^* \mathcal{c}_i \quad \subseteq \quad \mathsf{D}(L) \quad \subseteq \quad L \cup \Sigma_{\#}^* \cup \bigcup_{i=1}^{n+2} \Sigma_{\#}^* \mathcal{c}_i.$$

Let $B = (Q_B, \Gamma, \delta_B, q_{0,B}, F_B)$ be some DFA with $L(B) = \mathsf{D}(L)$. Notice that for $1 \le i \le n+2$, the state $\delta_B(q_{0,B}, \$_i)$ of B must be equivalent to state $\delta(q_0, \$_i)$ of A—otherwise $\mathsf{D}(L)$ and L would differ on words starting with $\$_i$. From the fact that A is a minimal DFA, one can see that DFA B must contain automaton A as a substructure—however, the initial state $q_{0,B}$ of B is an accepting state. Further, we show that B needs at least two additional states p_1 and p_2, so that $\mathsf{sc}(\mathsf{D}(L)) > |Q| + 1$.

State p_1 is witnessed as follows. Because $\bigcup_{i=1}^{n+2} L(A_i') \neq \Sigma_{\#}^*$, it must be $\bigcup_{i=1}^{n} L(A_i) \neq \Sigma^*$. So there is a word $w \in \Sigma^*$ with $w \notin \bigcup_{i=1}^{n} L(A_i)$. Notice that also $w \notin \bigcup_{i=1}^{n+2} L(A_i')$. We argue that the word $\#w$ cannot belong to $\mathsf{D}(L)$: if $\#w \in \mathsf{D}(L)$ then there is a state $q \in Q$ such that $\delta(q, \#w) \in F$. Clearly $q \neq q_s$ and $q \neq q_f$. Further, state q cannot be q_0, because the only transitions from q_0 that do not enter the sink state are on symbols $\$_i$. So it can only be $q \in Q_i'$, for some i with $1 \le i \le n+2$. But then $\delta(q, \#w) \in F$ implies $\delta_i'(s_i', w) \in F_i'$, a contradiction to $w \notin \bigcup_{i=1}^{n} L(A_i)$. Thus, $\#w \notin \mathsf{D}(L)$. Let $p_1 = \delta_B(q_{0,B}, \#w)$. Since $\#w$ is not accepted by B, we have p_1 cannot be an accepting state. In particular $p_1 \neq q_{0,B}$ and $p_1 \neq q_f$. But also $p_1 \neq q_s$ because $\#w\mathcal{c}_i$ must be accepted for $1 \le i \le n+2$. Finally, state p_1 cannot belong to any set Q_i', for $1 \le i \le n+2$, because otherwise p_1 would be reachable from the initial state $q_{0,B}$ of by B by some word $\$_i u$. But since $\delta_B(p_1, \mathcal{c}_j)$ is an accepting state for every j with $1 \le j \le n+2$, a word $\$_i u \mathcal{c}_j \notin \mathsf{D}(L)$, with $i \neq j$, would be accepted by B—a contradiction. Therefore p_1 must be a new state.

For the existence of another state p_2 recall the DFAs A_{n+1} and A_{n+2} we added to the union universality instance. These automata ensure that $\Sigma \subseteq \mathsf{D}(L)$. So let $p_2 = \delta_B(q_{0,B}, a)$ for some $a \in \Sigma$. Since p_2 must be an accepting state, we have $p_2 \neq p_1$ and $p_2 \neq q_s$. Moreover $p_2 \neq q_{0,B}$ since otherwise $a\$_1\mathcal{c}_1 \notin \mathsf{D}(L)$ would be accepted by B. Since $a\mathcal{c}_i \in \mathsf{D}(L)$, for $1 \le i \le n+2$, we have $p_2 \neq q_f$. Finally, we have $p_2 \notin Q_i'$, for $1 \le i \le n+2$, because otherwise a word of the form $\$_i u \mathcal{c}_j \notin \mathsf{D}(L)$ would be accepted by B. Hence $\mathsf{sc}(\mathsf{D}(L)) \ge |Q| + 2$, which concludes the proof. □

As an immediate consequence of the previous two lemmata we obtain the next theorem.

Theorem 7. D-SET-SIZE *is* PSPACE-*complete w.r.t. deterministic polytime reductions.*   □

3.2 Results on the Representation of Distinguishability Sets

In this subsection we consider decision problems on the representation of distinguishability sets. The L-VERSUS-D problem is defined as follows:

- L-VERSUS-D. Given two DFAs A and B, decide whether $L' = \mathsf{D}(L)$, for $L = L(A)$ and $L' = L(B)$?

By a similar proof as in the previous subsection, the L-VERSUS-D problem can be classified to be PSPACE-complete

Theorem 8. L-VERSUS-D *is* PSPACE-*complete.* $\square$

In the L-VERSUS-D problem we have to decide whether $L' = \mathsf{D}(L)$. Let us now take a closer look at the complexity of deciding the two inclusions $L' \subseteq \mathsf{D}(L)$ and $L' \supseteq \mathsf{D}(L)$. It turns out that the former inclusion is the hard part, while the latter is easy. For the upper bound in the next lemma, we construct an NFA for the language $\mathsf{D}(L) \cap \overline{L'}$ and test this for emptiness, which can be done in NL [6].

Lemma 9. *Given two deterministic finite automata A and B with $L = L(A)$ and $L' = L(B)$, it is* NL-*complete to decide whether $L' \supseteq \mathsf{D}(L)$.* $\square$

With Theorem 8 and Lemma 9 we can now prove the following result.

Corollary 10. *Given two deterministic finite automata A and B with $L = L(A)$ and $L' = L(B)$, it is* PSPACE-*complete to decide whether $L' \subseteq \mathsf{D}(L)$.* $\square$

3.3 Results on the Finite Hierarchy of Distinguishability Sets

Recall that the iteration of the D-operator induces a finite hierarchy of the form $\mathsf{D}(L) \supseteq \mathsf{D}^2(L) = \mathsf{D}^n(L)$, for $n \geq 3$, regardless on the language L on which we started the D-iteration. In case the language L is suffix closed, we also have $L \supseteq \mathsf{D}(L)$, which is not the case in general. In the forthcoming we consider problems that are related to the levels of this hierarchy. The first problem is

- L-EQUALS-D. Given a deterministic finite automaton A, decide whether $L = \mathsf{D}(L)$, where $L = L(A)$?

Analogously one defines the problems: D-EQUALS-DSQUARE and L-EQUALS-DSQUARE. In contrast to the problems investigated in the previous subsections, it turns out that here the computational complexity of the problems is much easier.

We start with the D-EQUALS-DSQUARE problem. First, we give a characterization of the languages L that satisfy $\mathsf{D}(L) = \mathsf{D}^2(L)$. To this end we introduce a new notion. A language $L \subseteq \Sigma^*$ is *language-synchronizing* if there exists a word $w \in \Sigma^*$ such that $xwz \in L$ if and only if $ywz \in L$, for every $x, y, z \in \Sigma^*$. Any word w with this property is called a *language-reset word* for the language L. Now we come to our characterization result.

Theorem 11. *A language $L \subseteq \Sigma^*$ is language-synchronizing if and only if the equality $\mathsf{D}(L) = \mathsf{D}^2(L)$ holds.*

Proof. Recall that in [1] it was shown that $D(L) = D^2(L)$ holds if and only if $D(L)$ has $\emptyset$ as a quotient. Thus, it suffices to prove that the latter condition is equivalent to the fact that L is language-synchronizing. Assume that L is language-synchronizing and that w is any language-reset word. Then $xwz \in L$ if and only if $ywz \in L$, for every words x, y, and z. Therefore, the words wz do not distinguish between any words x and y. Thus $wz \notin D(L)$, for every word $z \in \Sigma^*$. This in turn implies $w^{-1} \cdot D(L) = \emptyset$. Hence we have shown that if L is language-synchronizing, then $D(L)$ has $\emptyset$ as a quotient. Conversely, we argue as follows. Let $D(L)$ have $\emptyset$ as a quotient, i.e., there is a word $w \in \Sigma^*$ such that $w^{-1} \cdot D(L) = \emptyset$. This means that all continuations of w by any word z do not belong to the set $D(L)$. In other words, the condition $xwz \in L$ if and only if $ywz \in L$, for every $x, y, z \in \Sigma^*$, holds, which is the definition of language-synchronizability. Therefore, we have shown that $D(L)$ having $\emptyset$ as a quotient implies that L is language-synchronizing. □

In fact, language-synchronization is a generalization of synchronizability on DFAs—see, e.g., [10] for a survey on synchronizing finite automata. A DFA $A = (Q, \Sigma, \delta, q_0, F)$ is *synchronizing* if there exists a word $w \in \Sigma^*$ whose action resets A, i.e., it leaves the automaton in one particular state no matter which state in Q one starts—in other words $\delta(p, w) = \delta(q, w)$ for every $p, q \in Q$. Any word w with this property is called a *reset word* for the automaton A. The following alternative characterization of synchronizability is given in [10].

Lemma 12. *A deterministic finite automaton $A = (Q, \Sigma, \delta, q_0, F)$ is synchronizing if and only if for every states $p, q \in Q$ there exists a word $w \in \Sigma^*$ such that $\delta(p, w) = \delta(q, w)$.*

The next theorem shows that language-synchronization and synchronization on DFAs coincide.

Theorem 13. *A regular language L is language-synchronizing if and only if it is accepted by a synchronizing deterministic finite automaton.* □

A close inspection of the proof of Theorem 13 reveals that synchronizability of DFAs is stable under DFA minimization: if a DFA A is synchronizing, the first part of the proof shows that $L(A)$ is language-synchronizing, and then the second part shows that its minimal DFA is synchronizing. In particular, this means that a regular language L is language-synchronizing if and only if its *minimal* DFA is synchronizing. Together with Theorem 11 we immediately obtain the following characterization.

Corollary 14. *Let L be a regular language. Then $D(L) = D^2(L)$ if and only if the minimal deterministic finite automaton for L is synchronizing.* □

We end our small detour on synchronizability with the following computational complexity result. Again, a reduction from the graph reachability problem can be used to prove NL-hardness.

Lemma 15. *The problem of deciding whether a given deterministic finite automaton is synchronizing is* NL-*complete. This even holds if the problem instances are restricted to* minimal *deterministic finite automata.*   □

After this preliminary result on the characterization of (regular) languages that satisfy $D(L) = D^2(L)$ we are now ready to determine the computational complexity of the D-EQUALS-DSQUARE problem.

Theorem 16. D-EQUALS-DSQUARE *is* NL-*complete.*   □

Next we also classify the complexity of the L-EQUALS-D problem.

Theorem 17. L-EQUALS-D *is* NL-*complete.*

Proof. For the containment in NL, we use that a language L satisfies $L = D(L)$ if and only if L is suffix closed and has $\emptyset$ as a quotient [1]. Both properties can be verified in NL since this complexity class is closed under complementation [4,9]. The details are left to the reader.

For proving NL-hardness let $G = (V, E)$ with vertices s and t be an instance of the graph reachability problem, where every vertex $v \in V \setminus \{t\}$ has out-degree two, and the only outgoing edge of t is a self-loop. We construct a DFA A over alphabet $\Sigma = \{a, b\}$, with state set $Q = V \cup \{q_0, f\}$, from which q_0 is the initial state and state t is the only accepting state. The transitions of A in states $v \in V$ resemble the edges of G (the self-loop on t induces loops on both inputs a and b), and the transitions of the additional states are $\delta(q_0, a) = s$ and $\delta(q_0, b) = \delta(f, a) = \delta(f, b) = f$. Clearly, if there is no path from s to t in G, then both languages $L = L(A)$ and $D(L)$ are empty, hence $L = D(L)$. Conversely, if G contains a path from s to t, then we have $D(L) = \Sigma^*$ (because of states s and t). However, since the initial state of A is not accepting we have $L \neq D(L)$.   □

It remains to consider the L-EQUALS-DSQUARE problem, i.e., the problem of deciding for a given DFA A with $L = L(A)$, whether $L = D^2(L)$. In fact, we already solved this problem by Theorem 17 because $L = D^2(L)$ holds if and only if $L = D(L)$, which can be seen as follows: if $L = D(L)$, applying D to both sides yields $D(L) = D^2(L)$, hence $L = D^2(L)$. Conversely, assume $L = D^2(L)$, then applying D yields $D(L) = D^3(L)$. Since we know from [1] that $D^3(L) = D^2(L)$ is always true, we obtain $L = D(L)$. Therefore we obtain the following result.

Corollary 18. L-EQUALS-DSQUARE *is* NL-*complete.*   □

4   Conclusions

We have investigated the computational complexity of problems related to the distinguishability set, for short D-set, of regular languages. The obtained results nicely fit into known ones for finite automata problems. Nevertheless, it is remarkable that some of the problems under consideration turned out to be of very high complexity, namely PSPACE-complete, although *only* deterministic devices were involved.

In the original paper [1], where the properties of the D-language were investigated in detail, also other types of distinguishability sets were mentioned. By definition, the D-sets are related to the suffix operation. Conversely, prefix and infix operation on languages induced certain forms of distinguishability operators, too. Due to their close nature to the original D-operator we feel that the computational complexity of problems related to these forms of operators are probably the same as for the D-operator. On the other hand, there may be operators of interest, the complexity of which may highly differ from ours. One example of this kind is the $\underline{\mathsf{D}}$-operator of [1], which is defined as

$$\underline{\mathsf{D}}_L(x, y) = \min\{\, w \in \Sigma^* \mid w \in \mathsf{D}_L(x, y) \,\}$$

and $\underline{\mathsf{D}}(L) = \{\,\underline{\mathsf{D}}_L(x, y) \mid x \not\equiv_L y\,\}$, where the minimum refers to the smallest element with respect to the lexicographical order. Due to the additional properties of the $\underline{\mathsf{D}}$-operation and the fact that the $\underline{\mathsf{D}}$-sets are always finite for regular languages, we think that this operator is a natural host for further investigations on the computational complexity of related $\underline{\mathsf{D}}$-set problems.

References

1. Câmpeanu, C., Moreira, N., Reis, R.: The distinguishability operation on regular languages. In: Bensch, S., Freund, R., Otto, F. (eds.) Proceedings of the 6th International Workshop on Non-Classical Models of Automata and Applications, number 304 in books@ocg.at, Kassel, Germany, pp. 85–100 (2014). Österreichische Computer Gesellschaft
2. Harrison, M.A.: Introduction to Formal Language Theory. Addison-Wesley, Boston (1978)
3. Holzer, M., Kutrib, M.: Descriptional and computational complexity of finite automata–a survey. Inform. Comput. **209**(3), 456–470 (2011)
4. Immerman, N.: Nondeterministic space is closed under complementation. SIAM J. Comput. **17**(5), 935–938 (1988)
5. Jiang, T., Ravikumar, B.: Minimal NFA problems are hard. SIAM J. Comput. **22**(6), 1117–1141 (1993)
6. Jones, N.: Space-bounded reducibility among combinatorial problems. J. Comput. System Sci. **11**, 68–85 (1975)
7. Kozen, D.: Lower bounds for natural proof systems. In: Proceedings of the 18th Annual Symposium on Foundations of Computer Science, pp. 254–266 (1977)
8. Moore, E.F.: Gedanken experiments on sequential machines. In: Shannon, C.E., McCarthy, J. (eds.) Automata Studies, Annals of Mathematics Studies, pp. 129–153. Princeton University Press (1956)
9. Szelepcsényi, R.: The method of forced enumeration for nondeterministic automata. Acta Inform. **26**(3), 279–284 (1988)
10. Volkov, M.V.: Synchronizing automata and the černý conjecture. In: Martín-Vide, C., Otto, F., Fernau, H. (eds.) LATA 2008. LNCS, vol. 5196, pp. 11–27. Springer, Heidelberg (2008)
11. Wagner, K., Wechsung, G.: Computational Complexity. Mathematics and its applications (East Europeans series). VEB Deutscher Verlag der Wissenschaften, Berlin (1986)

Prefix-Free Subsets of Regular Languages and Descriptional Complexity

Jozef Štefan Jirásek and Juraj Šebej[✉]

Institute of Computer Science, Faculty of Science, P. J. Šafárik University,
Jesenná 5, 040 01 Košice, Slovakia
{jirasekjozef,juraj.sebej}@gmail.com

Abstract. We study maximal prefix-free subsets of regular languages and their descriptional complexity. We start with finite subsets and give the constructions of deterministic finite automata for largest and smallest finite maximal prefix-free subsets. Then we consider infinite maximal prefix-free subsets, and we show that such subsets can be effectively found if they exist. Finally, we prove that if a regular language has a non-regular maximal prefix-free subset, then it has uncountably many maximal prefix-free subsets.

1 Introduction

A language is prefix-free if it does not contain two distinct strings, one of which is a prefix of the other. In prefix codes, such as variable-length Huffman codes or country calling codes, there is no codeword that is a proper prefix of any other codeword. With such a code, a receiver can identify each codeword without any special marker between words. Motivated by prefix codes, the class of prefix-free regular languages has been recently investigated [1–6].

In this paper, we are interested in finding prefix-free subsets, or more precisely, in maximal prefix-free subsets of formal languages. We start with the following problem: assume that we exclude from a given language L the set of strings that are proper prefixes of some other strings in L. What can be said about the resulting language L_1? We prove that if L is accepted by an n-state deterministic finite automaton (DFA), then the language L_1 is accepted by a DFA of at most n states. Moreover, if L is finite, then L_1 is a largest finite maximal prefix-free subset of L. On the other hand, in many cases, the set L_1 is empty, so it is not maximal if L is non-empty.

Then we exclude from a regular language L the set of strings, the proper prefixes of which are in L. We denote the resulting prefix-free set by L_2. We prove that if L_2 is infinite, then L does not have any finite maximal prefix-free subset, and if L_2 is finite, then L_2 is a smallest finite maximal prefix-free subset of L. In the second case, we show that we can effectively find a largest finite maximal subset of L if such a subset exists.

J. Šebej—The author was supported by the Slovak Grant Agency for Science under contract VEGA 1/0142/15.

© Springer International Publishing Switzerland 2015
J. Shallit and A. Okhotin (Eds.): DCFS 2015, LNCS 9118, pp. 129–140, 2015.
DOI: 10.1007/978-3-319-19225-3_11

In the last part of the paper, we study infinite maximal prefix-free subsets of regular languages. We prove that if the sizes of finite prefix-free subsets of L grow without limit, then L has an infinite prefix-free subset. For an infinite regular language, either all the maximal prefix-free subsets are finite, or we can effectively find an infinite maximal prefix-free subset. We always discuss the descriptional complexity of resulting subsets. Finally, we prove that if a regular language has a non-regular maximal prefix-free subset, then it has uncountably many maximal prefix-free subsets.

2   Preliminaries

Let Σ be a finite alphabet and Σ^* the set of all strings over Σ. The empty string is denoted by ε. A *language* is any subset of Σ^*. We denote the size of a set A by $|A|$, and its power set by 2^A. For details, we refer the reader to [8,9].

A *deterministic finite state automaton* (DFA) is a 5-tuple $A = (Q, \Sigma, \cdot, s, F)$, where Q is a finite non-empty set of states; Σ is a finite alphabet; $\cdot : Q \times \Sigma \to Q$ is the transition function which can be naturally extended to the domain $Q \times \Sigma^*$, $s \in Q$ is the initial state; and $F \subseteq Q$ is the set of final states. A non-final state q is a *dead state* if $q \cdot a = q$ for each a in Σ. The language accepted by the DFA A is defined as the set $L(A) = \{w \in \Sigma^* \mid s \cdot w \in F\}$. We let $A = (n, \ell, F)$ denote a unary DFA $A = (\{0, 1, \ldots, n-1\}, \{a\}, \cdot, 0, F)$, where $i \cdot a = i + 1$ for $i = 0, 1, \ldots, n-2$, and $(n-1) \cdot a = \ell$ [7].

The *state complexity* of a regular language L, denoted by $sc(L)$, is the number of states in the minimal DFA accepting the language L.

Let $A = (Q, \Sigma, \cdot, s, F)$ be a DFA. A state $q \in Q$ *accepts a string* w if $q \cdot w \in F$; in such a case we say that q is useful. A DFA $A = (Q, \Sigma, \cdot, s, F)$ is a trim DFA if every state of A is *reachable* from the initial state s, and every state, except for the dead state, is *useful*. A state q is an ε-*state* if it accepts only the empty string.

Let u, v, w be strings over Σ and L be a language over Σ. If $w = uv$, then u is a prefix of w. If, moreover $v \neq \epsilon$, then u is a proper prefix of w. We write $u \leq_p v$ if u is a prefix of v, and $u \lneq_p v$ if u is a proper prefix. A language is prefix-free if it does not contain two strings, one of which is a proper prefix of the other. Next, $[w] = \{u \in \Sigma^* \mid u \leq_p w\} \cup \{u \in \Sigma^* \mid w \leq_p u\}$; that is, $[w]$ is the set of all the strings that are comparable to w. Finally, let $L_w = \{x \mid wx \in L\}$, that is, L_w is the left quotient of L by w. The set $[w]$ is regular for every string w, and if L is regular, then L_w is regular as well.

A set P is a *maximal prefix-free subset* of a language L, if (i) $P \subseteq L$, (ii) P is prefix-free, and (iii) for every string w in $L \setminus P$, the language $P \cup \{w\}$ is not prefix-free. A set P is a *largest finite maximal prefix-free subset* of a language L, if (i) P is a finite maximal prefix-free subset of L, and (ii) for every finite maximal prefix-free subset P' of L, we have $|P'| \leq |P|$. A *smallest finite maximal prefix-free subset* is defined symmetrically. Now for a given regular language L, we present a construction of a DFA $A_{F'}$ which accepts a prefix-free subset of L. Then we state a sufficient condition so that the resulting subset is maximal.

Let $A = (Q, \Sigma, \cdot, s, F)$ be a DFA and $F' \subseteq F$. Construct the DFA

$$A_{F'} = (Q, \Sigma, \cdot', s, F'), \tag{1}$$

where $q \cdot' a$ is undefined (or it goes to an added dead state if we consider complete DFAs) if $q \in F'$, and $q \cdot' a = q \cdot a$ if $q \notin F'$.

Lemma 1. *The set $L(A_{F'})$ is a prefix-free subset of $L(A)$.* □

Lemma 2. *If each state in $F \setminus F'$ is useful in $A_{F'}$, then $L(A_{F'})$ is a maximal prefix-free subset of $L(A)$.* □

3 L' and $L_1 = L \setminus L'$

With the aim of finding prefix-free subsets of a language L, we start by defining the set L' as a set containing such strings of L which are proper prefixes of some other strings in L. Then we define a prefix-free subset L_1 of L by $L_1 = L \setminus L'$. For a regular language L, we study the properties of L' and L_1 in this section.

Definition 3. *For a language L, define languages L' and L_1 as follows:*

$$L' = \{w \in L \mid \text{there is a non-empty string } x \text{ such that } wx \in L\}, \tag{2}$$
$$L_1 = L \setminus L'. \tag{3}$$

It follows directly from the definition that L_1 is prefix-free. We show that if L is regular, then L' is regular as well. Moreover, a DFA for L' can be obtained from a DFA for L by a modification of the set of final states of A.

Construction 4 (The construction of a DFA A' for the language L'). *Let $A = (Q, \Sigma, \cdot, s, F)$ be a minimal DFA for L. Construct the DFA*

$$A' = (Q, \Sigma, \cdot, s, F'),$$

where $F' = \{q \in F \mid \text{there is a non-empty string } w \text{ such that } q \cdot w \in F\}$. □

Thus, the DFA A' is the same as the DFA A, except for the set of final states. All the final states in A that accept only the empty string are non-final in the DFA A'. All the remaining final states in A are final in A'.

Proposition 5. *Let L be a language accepted by a minimal n-state DFA A. Then the language L' given by (2) is accepted by the DFA A' described in Construction 4, and we have $\mathrm{sc}(L') \le n$. Moreover, for every k with $1 \le k \le n$, there is a unary regular language L with $\mathrm{sc}(L) = n$ such that $\mathrm{sc}(L') = k$.* □

The following construction describes a DFA for L_1 of at most n states.

Construction 6 (The construction of a DFA A_1 for the language L_1). *Let $A = (Q, \Sigma, \cdot, s, F)$ be a minimal DFA. Construct the DFA*

$$A_1 = (Q, \Sigma, \cdot, s, F_1),$$

where $F_1 = \{q \in F \mid \text{there is no non-empty string } w \text{ such that } q \cdot w \in F\}$. □

Thus, all the final states of A that accept only the empty string are final in the DFA A_1, and all the remaining final states of A are non-final in A_1.

Proposition 7. *Let L be a language accepted by a minimal n-state DFA A. Then the language L_1 given by (2–3) is accepted by the DFA A_1 described in Construction 6, and we have $\mathrm{sc}(L_1) \leq n$. If L is a unary language, then $\mathrm{sc}(L_1) = 1$ or $\mathrm{sc}(L_1) = n$. Moreover, for every k with $1 \leq k \leq n$, there is a binary language L with $\mathrm{sc}(L) = n$ such that $\mathrm{sc}(L_1) = k$.* □

Theorem 8. *If a language L is finite, then the set L_1 defined by (2–3) is a largest finite maximal prefix-free subset of L.* □

4 L'' and $L_2 = L \setminus L''$

In this section, we define the language L'' as the language containing such strings in L whose proper prefix is in L. Then we define the language L_2 by $L_2 = L \setminus L''$, and study the properties of L'' and L_2.

Definition 9. *For a language L, define languages L'' and L_2 as follows:*

$$L'' = \{w \in L \mid w = uv \text{ for some string } u \text{ in } L \text{ and some } v \neq \varepsilon\}, \quad (4)$$

$$L_2 = L \setminus L''. \quad (5)$$

Construction 10 (The construction of a DFA A'' for the language L'').
Let $A = (Q, \Sigma, \cdot, s, F)$ be a minimal automaton for L. Construct the DFA

$$A'' = (\bigcup_{q \in Q} \{q, q'\}, \Sigma, \odot, s, \{q' \mid q \in F\}),$$

where we add a copy q' of each state q, and $\odot$ is defined as follows:

$$q \odot a = \begin{cases} p, & \text{if } q \notin F, \text{ and } q \cdot a = p; \\ p', & \text{if } q \in F, \text{ and } q \cdot a = p; \end{cases}$$
$$q' \odot a = \quad p', \quad \text{if } q \cdot a = p.$$

Thus we make a copy of the DFA A, and redirect every transition going from a final state of A to the corresponding state in the copied automaton.

Proposition 11. *Let L be a language accepted by a minimal n-state DFA A. Then the language L'' given by (4) is accepted by the DFA A'' described in Construction 10, and $\mathrm{sc}(L'') \leq 2n$. Moreover, for every k with $0 \leq k \leq n$, there is a unary language L with $\mathrm{sc}(L) = n$ and $\mathrm{sc}(L'') = n + k$.* □

Construction 12 (The construction of a DFA A_2 for the language L_2).
Let $A = (Q, \Sigma, \cdot, s, F)$ be a DFA. Construct the DFA $A_2 = (Q \cup \{d\}, \Sigma, \odot, s, F)$, where $\odot$ is defined as follows:

$$q \odot a = \begin{cases} q \cdot a, & \text{if } q \notin F; \\ d, & \text{if } q \in F. \end{cases}$$

Thus, the DFA A_2 is the same as the DFA A, except for the transitions in final states.

Proposition 13. *Let L be a language accepted by a minimal n-state DFA. Then the language L_2 given by (4–5) is accepted by the DFA A'' described in Construction 12 and we have $\operatorname{sc}(L'') \leq n+1$. Moreover, for every k with $2 \leq k \leq n+1$, there is a unary language L such that $\operatorname{sc}(L) = n$ and $\operatorname{sc}(L_2) = k$.* □

Proposition 14. *Let L be a regular language. Then the set L_2 given by (4–5) is a maximal prefix-free subset of L. If L_2 is infinite, then L does not have any finite maximal prefix-free subset. If L_2 is finite, then it is a smallest finite maximal prefix-free subset of L.* □

5 Largest Finite Maximal Prefix-Free Subsets

In this section, we consider a regular language L accepted by a trim DFA A, and such that the set L_2 defined by (4–5) is finite. Recall that L_2 is accepted by the DFA A_2, obtained from the DFA A for L by turning every final state of A to an ε-state, that is, we redirected all the out-transitions from every final state to the dead state d. By Proposition 14, the set L_2 is a smallest finite maximal prefix-free subset of L. The aim of this section is to find a largest finite maximal prefix-free subset of L if such a set exists.

For a final state q of a DFA $A = (Q, \Sigma, \cdot, s, F)$, we define the set of so-called first-accepted strings as follows:

$$P(q) = \{u \mid u \neq \varepsilon, \; q \cdot u \in F, \text{ and } q \cdot u' \notin F \text{ if } \varepsilon \neq u' \lneqq_p u\},$$

that is, the set $P(q)$ contains a non-empty string u if it is accepted from q and if every non-empty proper prefix of u is rejected from q. Let us partition the set of final states of A into four groups:

$$F_0 = \{q \in F \mid |P(q)| = 0\},$$
$$F_\infty = \{q \in F \mid |P(q)| = \infty\},$$
$$F_1 = \{q \in F \mid |P(q)| = 1\},$$
$$F_2 = \{q \in F \mid 2 \leq |P(q)| < \infty\},$$

that is, in F_0, F_∞, F_1, and F_2, we have the final states of A with zero, infinitely many, one, and two but finitely many first-accepted strings. Notice that for a final state q_f, we can effectively determine to which group it belongs: We first add a copy q_f' of the state q_f to the DFA A. We make the state q_f' a new initial and rejecting state. For each symbol a in Σ, we add the transition on a from q_f' to $q_f \cdot a$. Then we change every final state to an ε-state. Finally, we test whether the language accepted by the resulting DFA is empty, finite, or infinite.

Now consider the states in F_1. If $q \in F_1$, then there is exactly one first-accepted string w_1 for q. The string w_1 can move the state q either to a state in F_0, or to a state in F_∞, or to a state in F_2, or again to a state in F_1, and this case can occur several times, and eventually we either get a state in $F_0 \cup F_\infty \cup F_2$, or we get a cycle of states in F_1. Hence, let us partition the set F_1 as follows (here a transition $p \xrightarrow{w} q$ denotes that w is a first-accepted string of the state p and $p \cdot w = q$):

$F_{1,0} = \{q \in F_1 \mid$ there is a sequence of transitions
$\qquad q = q_0 \xrightarrow{w_1} q_1 \xrightarrow{w_2} \cdots \xrightarrow{w_k} q_k$ with $q_i \in F_1$ if $0 \le i < k$ and $q_k \in F_0\}$,
$F_{1,\infty} = \{q \in F_1 \mid$ there is a sequence of transitions
$\qquad q = q_0 \xrightarrow{w_1} q_1 \xrightarrow{w_2} \cdots \xrightarrow{w_k} q_k$ with $q_i \in F_1$ if $0 \le i < k$ and $q_k \in F_\infty\}$,
$F_{1,2} = \{q \in F_1 \mid$ there is a sequence of transitions
$\qquad q = q_0 \xrightarrow{w_1} q_1 \xrightarrow{w_2} \cdots \xrightarrow{w_k} q_k$ with $q_i \in F_1$ if $0 \le i < k$ and $q_k \in F_2\}$,
$F_{1,1} = \{q \in F_1 \mid$ there is a sequence of transitions
$\qquad q = q_0 \xrightarrow{w_1} q_1 \xrightarrow{w_2} \cdots \xrightarrow{w_k} q_k = q_j, 0 \le j \le k, q_i \in F_1$ if $0 \le i \le k\}$.

Next, let $P^*(q) = P(q)$ if $q \notin F_1$, and let $P^*(q) = P(q_k)$ if $q \in F_1$ and q goes (in a unique way) to the state q_k not in F_1. Then we have the following observation.

Proposition 15. *Let $q \in F$ and u be a string such that $s \cdot u = q$. Then the set $\{uv \mid v \in P(q)\}$ is a maximal prefix-free subset of L in the subtree T_u of the tree $(\Sigma^*, \le_p)$.*

Now consider the final states in $F_2 \cup F_{1,2}$. Let us construct a graph

$$G = (F_2 \cup F_{1,2}, E),$$

in which (p, q) is in E iff there is a first-accepted string w in $P(p)$ such that $p \cdot w = q$.

Construction 16 (Construction of the DFA B). *Let $A = (Q, \Sigma, \cdot, s, F)$ be a trim DFA. Construct the DFA B from A as follows:*

If a state q in $F_2 \cup F_{1,2}$ goes to a cycle in G, then we will assign a special mark to the state q which we will use later in the proof. If a state q in $F_2 \cup F_{1,2}$ does not go to any cycle in G, we make the state q non-final. Finally, we turn all the remaining final states of A to ε-states, that is, we remove all the transitions going from all the remaining final states of A.

Lemma 17. *If a marked state is reachable in the DFA B, then there is no largest maximal finite prefix-free subset of L.*

Proof. Recall that a state in $F_2 \cup F_{1,2}$ is marked if it goes to a cycle in the graph G. If a marked state is reachable in the DFA B, then there exists an infinite sequence of strings $u_0, u_1, u_2, \ldots$, and an infinite sequence of states $q_1, q_2, \ldots$ in $F_2 \cup F_{1,2}$ such that

$$s \xrightarrow{u_0} q_1 \xrightarrow{u_1} q_2 \xrightarrow{u_3} q_3 \xrightarrow{u_4} \cdots,$$

where $p \xrightarrow{w} q$ denotes that $w \in P(p)$ and $p \cdot w = q$; notice that u_0 indeed can be taken first-accepted as we can consider the first marked state on the corresponding path from s. Moreover, the states from F_2 occur infinitely many times in the sequence $q_1, q_2, \ldots$ since every state from $F_{1,2}$ eventually, and in a unique way, goes to a state from F_2. Now we define an infinite sequence of sets by

$N_1 = (L(A_F) \setminus \{u_0\}) \cup \{u_0 v \mid v \in P(q_1)\}$, and
$N_{j+1} = (N_j \setminus \{u_0 u_1 \cdots u_j\}) \cup \{u_0 u_1 \cdots u_j v \mid v \in P(q_{j+1})\}$, $j = 1, 2, \ldots$.

Throughout this section we assume that $L(A_F)$ is a maximal finite prefix-free subset of L. It follows that the set N_1, obtained from $L(A_F)$ by replacing the string u_0 with the set $\{u_0 v \mid v \in P(q_1)\}$, is a maximal finite prefix-free subset of L since the set $P(q_1)$ of first-accepted strings of q_1 is finite and maximal prefix-free in the tree T_{u_0}. If $q_1 \in F_2$, then $|N_1| > |L(A_F)|$, and if $q_1 \in F_{1,2}$, then $|N_1| = |L(A_F)|$.

Now assuming that N_j is a maximal finite prefix-free subset of L, by a similar reasoning we get that N_{j+1} is a maximal finite prefix-free with $|N_{j+1}| > |N_j|$ if $q_{j+1} \in F_2$, and with $|N_{j+1}| = |N_j|$ if $q_{j+1} \in F_{1,2}$. Since the states from F_2 occur in the sequence infinitely many times, the sizes of maximal finite prefix-free subsets grow without limit, so there is no largest finite maximal prefix-free subset of L. Our proof is complete. $\square$

Lemma 18. *If no marked state is reachable in the DFA B, then $L(B)$ is a finite maximal prefix-free subset of L.*

Proof. By Lemma 1, the set $L(B)$ is prefix-free, and by Lemma 2, it is maximal. Notice that every final state of B is an ε-state. Therefore all the final states can be merged into a unique final state f. Moreover, we can omit all the unreachable or useless states of B. We can prove that $L(B)$ is finite. $\square$

Lemma 19. *If no marked state is reachable in the DFA B, then there exists a largest maximal finite prefix-free subset of L.*

Proof. Assume, to get a contradiction, that there is no largest finite maximal prefix-free subset of L, that is, the sizes of finite maximal prefix-free subsets grow without limit. Let ℓ be the number of prefixes of the strings in $L(B)$, that is,

$$\ell = |\{u \mid u \leq_p w \text{ for a string } w \text{ in } L(B)\}|.$$

Let M be a finite maximal prefix-free subset of L with $|M| > \ell + |L(B)|$. Let $K = \{v \in M \mid v \text{ is a prefix of some string in } L(B)\}$. Since $L(B)$ is maximal, for every string v in $M \setminus K$, there is a string w in $L(B)$ such that $w \leq_p v$. Since $|M \setminus K| > |L(B)|$, there is a string w in $L(B)$ and there are at least two, but finitely many non-empty strings $v_1, \ldots, v_k$ such that $wv_1, \ldots, wv_k$ are in $M \setminus K$.

The initial state s goes to a final state p of B by w. Next, in A, the final state p goes to some final states $f_1, \ldots, f_k$ by $u_1, \ldots, u_k$, respectively. It follows that p cannot be a state in $F_0 \cup F_{1,0} \cup F_{1,1}$ since $|P^*(p)| \geq 2$. Next, since M is finite, the state p cannot be a state in $F_\infty \cup F_{1,\infty}$ because otherwise M would not be maximal.

It follows that p is a state in $F_2 \cup F_{1,2}$. Since no marked state is reachable in B, the state p must be non-final in B. This is a contradiction with $w \in L(B)$, and the lemma follows. $\square$

Lemma 20. *If no marked state is reachable in the DFA B, then $L(B)$ is a largest finite maximal prefix-free subset of L.*

Proof. Recall that every final state of B is an ε-state. Therefore all the final states can be merged into a unique final state f. Moreover, we can omit all the unreachable or useless states of B. By the previous lemma, there is a largest maximal prefix-free subset M of L. Let us show that there is no string w in $L(B)$ such that a string in M is a proper prefix of w.

Assume, to get a contradiction, that such a string w exists in $L(B)$. Let u be a string in M such that $w = uv$ and $v \neq \varepsilon$. Then in A, the initial state s goes to a final state p by u, while in B, the initial state s goes to the state p by u and then to the state f by v. It follows that the state p is non-final in B, so $p \in F_2 \cup F_{1,2}$. However, then we can replace u in M with $P^*(p)$ and get a larger finite maximal prefix-free subset. This is a contradiction since M is a largest finite maximal prefix-free subset.

Hence each string in $L(B)$ is either in M or it is a proper prefix of a string in M. Let $K = L(B) \cap M$ and $K' = L(B) \setminus K$. Let us show that $|K'| = |M \setminus K|$. Assume for a contradiction that $|K'| < |M \setminus K|$. Since $L(B)$ is maximal, it covers $M \setminus K$. Since $|K'| < |M \setminus K|$, at least two strings in $M \setminus K$ must be covered by the same string in K'. However, in such a case, we get the same contradiction as in the proof of Lemma 19. Hence $|L(B)| = |M|$, which concludes our proof.   □

Let us summarize the results of this section in the following theorem.

Theorem 21. *Let L be a language accepted by a trim DFA A. Let the set L_2 given by (4–5) be finite. Let B be the DFA described in Construction 16. If a state with a mark is reachable in B, then the sizes of finite maximal prefix-free subsets of L grow without limit. Otherwise $L(B)$ is a largest finite maximal prefix-free subset of L.*   □

6   Infinite Prefix-Free Subsets

Now we turn our attention to infinite prefix-free subsets. Let us start with the following result.

Theorem 22. *Let a language L have arbitrarily large finite prefix-free subsets. Then L also has an infinite prefix-free subset.*

Proof. Consider the tree $(\Sigma^*, \leq_p)$ where $u \leq_p v$ iff u is a prefix of v. Then a prefix-free subset of L is an *antichain* in $(\Sigma^*, \leq_P)$.

Let $[w] = \{u \in \Sigma^* \mid u \leq_p w\} \cup \{u \in \Sigma^* \mid w \leq_p u\}$ denote the set of all the strings that are comparable to w.

First, we define an infinite chain $C = \{u_n \mid n \geq 0\}$ inductively as follows: Let $u_0 = \varepsilon$. Assume we have defined u_n such that $[u_n]$ has arbitrarily large finite antichains in L. Consider a finite set $\{u_n a \mid a \in \Sigma\}$ of immediate successors of u_n. There must exist a symbol σ in the finite set Σ such that $[u_n \sigma]$ has arbitrarily large finite antichains in L. Set $u_{n+1} = u_n \sigma$. Then we get an infinite chain $C = \{u_n \mid n \geq 0\}$ such that for every n, the set $[u_n]$ has arbitrarily large finite antichains in L.

Now we are going to define inductively an infinite antichain $\{v_n \mid n \geq 0\}$. Let $k_0 = 0$. For a given n, let A_n be an antichain in $[u_{k_n}]$ such that $|A_n| \geq 2$. Since the intersection of a chain and an antichain has at most one element, there is a string v_n in A_n such that $v_n \notin C$. It follows that there is an integer k_{n+1} with $k_{n+1} > k_n$ such that $v_n \notin [u_{k_{n+1}}]$. For k_{n+1}, we define w_{n+1} as described above. If $n < m$, then $v_n \in [u_{k_n}] \setminus [u_{k_m}]$. Since $u_{k_m} \leq_p v_m$, we have $[v_m] \subseteq [u_{k_m}]$. Since $v_n \notin [u_{k_m}]$, we have $v_n \notin [v_m]$. So the strings v_m and v_n are incomparable. Thus $\{v_n \mid n \geq 0\}$ is an infinite antichain in L, and the theorem follows.   □

The next result shows that for a language with state complexity n, we can effectively find an infinite maximal prefix-free subset of state complexity $2n$ if such a subset exists.

Lemma 23. *Let L be an infinite regular language. Then either all the maximal prefix-free subsets of L are finite, or there exists a regular infinite maximal prefix-free subset of L, and such a subset can be effectively found (in time $O(n+m)$, where n is the number of states and m is the number of transitions in a DFA for L), and it is of state complexity at most $2n$.*

Proof. Let L be accepted by a trim DFA $A = (Q, \Sigma, \cdot, s, F)$. The DFA A can be viewed as a labeled directed graph $D(A) = (V, E)$ with multiple arcs, where $V = Q$, and there is an arc from p to q labeled by a iff $p \cdot a = q$. Since L is infinite, the digraph $D(A)$ must contain a cycle which can be reached from s; here, by a cycle we understand a sequence $p_0, a_1, p_1, a_2, \ldots, a_\ell, p_\ell = p_0$, where (p_{i-1}, a_i, p_i) for $i = 0, \ldots, \ell$ is a labeled arc in G.

(1) Assume that there is a cycle C in $D(A)$, and that there is a vertex p on C and an arc (p, q) labeled by a such that (p, a, q) is not on C. Since A is a trim automaton, the state p is reachable from s, and an accepting state f of A is reached from the state q by a (short enough) string $w = a_1 a_2 \cdots a_k$, where $a_i \in \Sigma$. It follows that there is a sequence $q_0, q_1, \ldots, q_k$ of (pairwise distinct) states in A such that $q_0 = q$, $q_k \in F$, and

$$p \xrightarrow{a} q = q_0 \xrightarrow{a_1} q_1 \xrightarrow{a_2} q_2 \xrightarrow{a_3} \cdots \xrightarrow{a_k} q_k.$$

We make a copy q_j' of each state q_j in this sequence, and add the transitions $p \xrightarrow{a} q_0' \xrightarrow{a_1} q_1' \xrightarrow{a_2} q_2' \xrightarrow{a_3} \cdots \xrightarrow{a_k} q_k'$. Next, we remove the transition (p, a, q), and for each σ with $\sigma \neq a_{i+1}$, we add the transition $(q_i', \sigma, q_i \cdot \sigma)$ for $i = 0, 1, \ldots, k-1$. We make the state q_k' final. Denote by A' the resulting DFA. Then A' is equivalent to the DFA A.

Now we make non-final states all the final states of A' on the cycle C, as well as all the final states on the path from s to p. We turn all the remaining final states in A' into ε-states. Since every state in the cycle C and in the path from s to p reaches the state q_k in the resulting DFA, by Lemma 2, the resulting DFA accepts a maximal prefix-free subset of L. Since p is on the cycle C, and q_k' is outside C, the resulting set is an infinite maximal prefix-free subset of L.

(2) If no cycle in G has a vertex p as in case (1), then G is either a cycle, or it consists of an acyclic part and several disjoint cycles. Since A is a trim DFA,

each cycle must contain a final state. However, it follows that every maximal prefix-free subset of L must be finite since we can have at most one string per cycle in G, and only finitely many strings can be accepted outside the cycles.

We can find a cycle in case (1), or find out that no such cycle exists in time $O(n + m)$, where n is the number of states and m is the number of transitions in A. Since in (1), we make at most one copy for each state in A, the resulting DFA for an infinite maximal prefix-free subset of L has at most $2n$ states.   □

The next observation shows that sometimes $2n$ states are necessary to accept a regular infinite maximal prefix-free subset.

Proposition 24. *Let $n \geq 2$. There exists a language L with $\mathrm{sc}(L) = n$ such that there is a regular infinite maximal prefix-free subset P of L with $\mathrm{sc}(P) = 2n$, and for any other infinite maximal prefix-free subset R of L, we have $\mathrm{sc}(R) \geq 2n$.*   □

Proposition 25. *Let $n \geq 4$. There exist a regular language L with $\mathrm{sc}(L) = n$, such that the state complexities of regular infinite maximal prefix-free subsets of L grow without limit.*

Proof. Let $L = \{a\}^{\geq (n-3)} \cup b^+ a$ be a regular language accepted by a minimal n-state DFA. Then for every k with $k \geq n - 3$, the set $P_k = b^+ a \cup \{a^k\}$ is a regular infinite maximal prefix-free subset of L with $\mathrm{sc}(P_k) = k$.   □

Now we are going to prove that if a regular language L has a non-regular maximal prefix-free subset, then L has uncountably many maximal prefix-free subsets.

Theorem 26. *If a regular language has a non-regular maximal prefix-free subset, then it has uncountably many maximal prefix-free subsets.*

Proof. Let L be a regular language and P be a non-regular maximal prefix-free subset of L. Since $P \subseteq L$, for each string w in P, we have $\varepsilon \in L_w$. Let

$$P^+ = \{w \in P \mid L_w \neq \{\varepsilon\}\}, \tag{6}$$
$$P^- = \{w \in P \mid L_w = \{\varepsilon\}\}, \tag{7}$$

that is a string of P is in P^+ if it is a proper prefix of some string in L; otherwise, it is in P^-. We have $P = P^+ \cup P^-$. Our first aim is to prove that P^+ is infinite.

Assume, to get a contradiction, that P^+ is finite. Let

$$R = L \setminus \bigcup_{w \in P^+} [w],$$

that is, the language R consists of all the strings in L that are incomparable to any string in P^+. Since P is prefix-free, we have $P^- \subseteq R$. Since P^+ is finite and $[w]$ is regular for every w, the language R is regular. Next, let $R^- = \{w \in R \mid R_w = \{\varepsilon\}\}$. Then R^- is regular since to get a DFA for R^- from a DFA A for R, we turn all the final states of A that are not ε-states into non-final states. Let us show that $P^- = R^-$. First let $w \in P^-$. Then $w \in R$ and $L_w = \{\varepsilon\}$. Since $R \subseteq L$, we have $R_w \subseteq L_w = \{\varepsilon\}$. Hence $R_w = \{\varepsilon\}$, so $w \in R^-$.

Now let $w \in R^-$. Therefore $w \notin P^+$. Then $w \in R$, so w is incomparable to any string in P^+. Let $u \in P^-$. Let us show that u and w are incomparable. Since $L_u = \{\varepsilon\}$, the string u cannot be a proper prefix of w. Since $P^- \subseteq R$ and $R_w = \{\varepsilon\}$, the string w cannot be a proper prefix of u. Thus w and u are incomparable. Since P is maximal, the string w must be in P^-.

Hence $P^- = R^-$, which means that P^- is regular. However, then P is also regular since $P = P^+ \cup P^-$, and we assumed that P^+ is finite. This is a contradiction. Thus P^+ must be infinite.

Now our aim is to show that for each subset of the infinite set P^+, we can find a unique maximal prefix-free subset of L. To this aim, for each w in P^+, fix a non-empty string u_w such that $w u_w \in L$; such a string must exist by (6).

Let $S \subseteq P^+$. Define a set S' by $S' = S \cup \{w u_w \mid w \in P^+ \setminus S\}$. Then S' is a prefix-free subset of L. Let S'' be a maximal prefix-free subset of L such that $S' \subseteq S''$. Let us show that if S and T are two distinct subsets of P^+, then $S'' \neq T''$. Without loss of generality, we may assume that there is a string w with $w \in S$ and $w \notin T$. Then $w \in S''$ and $w u_w \in T''$. Since T'' is prefix-free, we have $w \notin T''$. Thus $S'' \neq T''$.

Since P^+ is infinite, the language L has uncountably many maximal prefix-free subsets. Our proof is complete. □

Example 27. Let $L = a^* b^*$. Then $P = \{a^n b^n \mid n \geq 1\}$ is a maximal non-regular prefix-free subset of L. By the lemma above, the language L has uncountably many maximal prefix-free subsets.

Notice that using the notation in the previous proof, we have $P^+ = P$. For each w in P^+, let $u_w = b$. For a subset S of P^+, let $S' = S \cup \{wb \mid w \in P^+ \setminus S\}$. Then S' is a maximal prefix-free subset of L, and if $S \neq T$, then $S' \neq T'$. □

7 Conclusions

We investigated the prefix-free subsets of regular languages and their descriptional complexities. We showed that for a regular language L of state complexity n, we can find a maximal prefix-free subset of L of state complexity n, and a regular infinite maximal prefix-free subset, if it exists, of state complexity $2n$. Notice, that if a regular language has an infinite maximal prefix-free subset, then it also has a *regular* infinite maximal prefix-free subset.

We also showed that if a regular language L has a finite maximal prefix-free subset, then we can effectively find a largest finite maximal subset if such a subset exists. Next, we proved that if the sizes of finite prefix-free subsets of a language grow without limit, then the language has an infinite prefix-free subset.

Finally, we proved that if a regular language has a non-regular maximal prefix-free subset, then it has uncountably many maximal prefix-free subsets.

Acknowledgment. We would like to thank Peter Eliáš for his help with the proof of Theorem 22.

References

1. Eom, H.-S., Han, Y.-S., Salomaa, K.: State complexity of k-union and k-intersection for prefix-free regular languages. In: Jurgensen, H., Reis, R. (eds.) DCFS 2013. LNCS, vol. 8031, pp. 78–89. Springer, Heidelberg (2013). http://dx.doi.org/10.1007/978-3-642-39310-5_9

2. Eom, H., Han, Y., Salomaa, K., Yu, S.: State complexity of combined operations for prefix-free regular languages. In: Paun, G., Rozenberg, G., Salomaa, A. (eds.) Discrete Mathematics and Computer Science. In: Memoriam Alexandru Mateescu (1952–2005), pp. 137–151. The Publishing House of the Romanian Academy (2014)

3. Han, Y., Salomaa, K., Wood, D.: Nondeterministic state complexity of basic operations for prefix-free regular languages. Fundam. Inform. **90**(1–2), 93–106 (2009). http://dx.doi.org/10.3233/FI-2009-0008

4. Han, Y., Salomaa, K., Wood, D.: Operational state complexity of prefix-free regular languages. In: Ésik, Z., Fülöp, Z. (eds.) Automata, Formal Languages, and Related Topics – Dedicated to Ferenc Gécseg on the occasion of his 70th birthday, pp. 99–115. Institute of Informatics, University of Szeged, Hungary (2009)

5. Jirásek, J., Jirásková, G., Krausová, M., Mlynárčik, P., Šebej, J.: Prefix-free languages: right quotient and reversal. In: Jürgensen, H., Karhumäki, J., Okhotin, A. (eds.) DCFS 2014. LNCS, vol. 8614, pp. 210–221. Springer, Heidelberg (2014). http://dx.doi.org/10.1007/978-3-319-09704-6_19

6. Krausová, M.: Prefix-free regular languages: closure properties, difference, and left quotient. In: Kotásek, Z., Bouda, J., Černá, I., Sekanina, L., Vojnar, T., Antoš, D. (eds.) MEMICS 2011. LNCS, vol. 7119, pp. 114–122. Springer, Heidelberg (2012). http://dx.doi.org/10.1007/978-3-642-25929-6_11

7. Nicaud, C.: Average state complexity of operations on unary automata. In: Kutyłowski, M., Wierzbicki, T.M., Pacholski, L. (eds.) MFCS 1999. LNCS, vol. 1672, pp. 231–240. Springer, Heidelberg (1999). http://dx.doi.org/10.1007/3-540-48340-3_21

8. Sipser, M.: Introduction to the Theory of Computation. PWS Publishing Company, Boston (1997)

9. Yu, S.: Regular languages. In: Rozenberg, G., Salomaa, A. (eds.) Word, Language, Grammar, Handbook of Formal Languages, vol. 1, pp. 41–110. Springer, Heidelberg (1997)

Transducer Descriptions of DNA Code Properties and Undecidability of Antimorphic Problems

Lila Kari[1], Stavros Konstantinidis[2], and Steffen Kopecki[1,2](✉)

[1] The University of Western Ontario, London, ON, Canada
{lila,steffen}@csd.uwo.ca
[2] Saint Mary's University, Halifax, NS, Canada
s.konstantinidis@smu.ca

Abstract. This work concerns formal descriptions of DNA code properties and related (un)decidability questions. This line of research allows us to give a property as input to an algorithm, in addition to any regular language, which can then answer questions about the language and the property. Here we define DNA code properties via transducers and show that this method is strictly more expressive than that of regular trajectories, without sacrificing the efficiency of deciding the satisfaction question. We also show that the maximality question can be undecidable. Our undecidability results hold not only for the fixed DNA involution but also for any fixed antimorphic permutation. Moreover, we also show the undecidability of the antimorphic version of the Post correspondence problem, for any fixed antimorphic permutation.

Keywords: Codes · DNA properties · Trajectories · Transducers · Undecidability

1 Introduction

The study of *formal* methods for describing independent language properties (widely known as code properties) provides tools that allow one to give a property as input to an algorithm and answer questions about this property. Examples of such properties include *classic* ones [4,17,27,28] like the properties of being a prefix code, a bifix codes, or an error-detecting language, as well as DNA code properties [2,10,11,13–15,18,20–22,25] like the property of being a θ-nonoverlapping or a θ-compliant language. A formal description method should be expressive enough to allow one to describe many desirable properties. Examples of formal methods for describing *classic* code properties are the implicational conditions method of [16], the trajectories method of [5], and the transducer methods of [8]. The latter two have been implemented to some extent in the Python package FAdo [9]. A formal method for describing DNA code properties is the method of trajectory DNA code properties [6,22].

Typical questions about properties are the following:

© Springer International Publishing Switzerland 2015
J. Shallit and A. Okhotin (Eds.): DCFS 2015, LNCS 9118, pp. 141–152, 2015.
DOI: 10.1007/978-3-319-19225-3_12

Satisfaction Problem: given the description of a property and the description of a regular language, decide whether the language satisfies the property.

Maximality Problem: given the description of a property and the description of a regular language that satisfies the property, decide whether the language is maximal with respect to the given property.

Construction Problem: given the description of a property and a positive integer n, find a language of n words (if possible) satisfying the given property.

In the above problems regular languages are described via (non-deterministic) finite automata (NFA). Depending on the context, properties are described via trajectory regular expressions or transducer expressions. The satisfaction problem is the most basic one and can be answered usually in efficient polynomial time. The maximality problem as stated above can be decidable, in which case it is normally PSPACE-hard. For existing transducer and trajectory properties, both problems can be answered using the online (formal) language server LaSer [24], which relies on FAdo. For the construction problem a simple statistical algorithm is included in FAdo, but we think that this problem is far from being well-understood.

The *general objective* of this research is to develop methods for formally describing DNA code properties that would allow one to express various combinations of such properties and be able to get answers to questions about these properties in an actual implementation. The contributions of this work are as follows:

1. The definition of a new simple formal method for describing many DNA code properties, called θ-*transducer properties*, some of which *cannot* be described by the existing transducer and trajectory methods for classic code properties; see Sect. 3.
2. The demonstration that the new method of transducer DNA code properties is properly more expressive than the method of trajectories; see Sect. 4.
3. The demonstration that the maximality problem can be decidable for some transducer DNA code properties but undecidable for some others; see Sect. 5.
4. The demonstration that some classic undecidable problems (like PCP) remain undecidable when rephrased in terms of *any fixed* (anti-)morphic permutation θ of the alphabet, with the case $\theta = $ id corresponding to these classic problems, where id is the *(morphic) identity*; see Sect. 6.

Even though, our main motivation is the description of DNA-related properties, we follow the more general approach, which considers properties described by transducers involving a fixed (anti-)morphic permutation θ; again, the classical transducer properties are obtained by letting $\theta = $ id.

This is a condensed conference version which does not contain full proofs of our results. The full version of this paper, containing all proofs, can be accessed on arXiv [19]. It also contains additional examples of DNA properties which can be described by θ-transducer properties: these properties naturally extend the hierarchy of DNA properties that is used in [13,18,20].

2   Basic Notions and Background Information

In this section we provide our notation for formal languages, (anti-)morphic permutations, transducers, and language properties. We assume the reader to be familiar with the fundamental concepts of language theory; see e. g., [12, 26]. Then, in Sect. 2.2 we recall the method of transducers for describing classic code properties, and in Sect. 2.3 we recall the method of trajectories for describing DNA-related properties.

2.1   Formal Languages and (Anti-)morphic Permutations

For an alphabet A and a language L over A we have the notation: $A^+ = A^* \setminus \{\varepsilon\}$, where ε is the empty word; and $L^c = A^* \setminus L$. For an integer $k \geq 2$ we define the *generic alphabet* $A_k = \{0, 1, \ldots, k-1\}$ *of size* k. Throughout this paper we only consider alphabets with at least two letters because our investigations would become trivial over unary alphabets.

Let $w \in A^*$ be a word. Unless confusion arises, by w we also denote the singleton language $\{w\}$, e. g., $L \cup w$ means $L \cup \{w\}$. If $w = xyz$ for some $x, y, z \in A^*$, then x, y, and z are called a *prefix*, an *infix* (or a *factor*), and a *suffix* of w, respectively. For a language $L \subseteq A^*$, the set $\mathrm{Pref}(L) = \{x \in A^* \mid \exists y \in A^* : xy \in L\}$ denotes the language containing all prefixes of words in L. If $w = a_1 a_2 \cdots a_n$ for letters $a_1, a_2, \ldots, a_n \in A$, then $|w| = n$ is the *length of* w; for $b \in A$, $|w|_b = |\{i \mid a_i = b, 1 \leq i \leq n\}|$ is the tally of b occurring in w; the i-th letter of w is $w_{[i]} = a_i$ for $1 \leq i \leq n$; the infix of w from the i-th letter to the j-th letter is $w_{[i;j]} = a_i a_{i+1} \cdots a_j$ for $1 \leq i \leq j \leq n$; and the *reverse* of w is $w^R = a_n a_{n-1} \cdots a_1$.

Consider a generic alphabet A_k with $k \geq 2$. The *identity* function on A_k is denoted by id_k; when the alphabet is clear from the context, the index k is omitted. For a *permutation* (or bijection) $\theta \colon A_k \to A_k$, and for $i \in \mathbb{Z}$, the permutation θ^i is the *i-fold composition* of θ; i. e., $\theta^0 = \mathrm{id}_k$, $\theta^i = \theta \circ \theta^{i-1}$, and $\theta^{-i} = (\theta^i)^{-1} = (\theta^{-1})^i$ for $i > 0$. An *involution* θ is a permutation such that $\theta = \theta^{-1}$.

A permutation θ over A_k can naturally be extended to operate on words in A_k^* as (a) *morphic permutation* $\theta(uv) = \theta(u)\theta(v)$, or (b) *antimorphic permutation* $\theta(uv) = \theta(v)\theta(u)$, for $u, v \in A_k^*$. As before, the inverse θ^{-1} of the (anti-)morphic permutation θ over A_k^* is the (anti-)morphic extension of the permutation θ^{-1} over A_k^*. The identity id_k always denotes the morphic extension of id_k while the antimorphic extension of id_k, called the *mirror image* or reverse, is usually denoted by the exponent ^{R}.

Example 1. The *DNA involution*, denoted by δ, is an antimorphic involution on $\Delta = \{\mathtt{A}, \mathtt{C}, \mathtt{G}, \mathtt{T}\}$ such that $\delta(\mathtt{A}) = \mathtt{T}$ and $\delta(\mathtt{C}) = \mathtt{G}$, which implies $\delta(\mathtt{T}) = \mathtt{A}$ and $\delta(\mathtt{G}) = \mathtt{C}$.

2.2   Describing Classic Code Properties by Transducers

A *(language) property* $\mathcal{P}$ is any set of languages. A language L *satisfies* $\mathcal{P}$, or *has* $\mathcal{P}$, if $L \in \mathcal{P}$. Here by a property $\mathcal{P}$ we mean an *(n-)independence* in the sense of [17]: there exists $n \in \mathbb{N} \cup \{\aleph_0\}$ such that a language L satisfies $\mathcal{P}$ if and only if all nonempty subsets $L' \subseteq L$ of cardinality less than n satisfy $\mathcal{P}$. A language L satisfying $\mathcal{P}$ is *maximal* (with respect to $\mathcal{P}$) if for every word $w \in L^c$ we have $L \cup w$ does not satisfy $\mathcal{P}$—note that, for any independence $\mathcal{P}$, every language in $\mathcal{P}$ is a subset of a maximal language in $\mathcal{P}$ [17]. As we shall see further below the focus of this work is on 3-independence properties that can also be viewed as independent with respect to a binary relation in the sense of [28].

A *transducer* $\mathbf{t}$ is a non-deterministic finite state automaton with output; see e. g., [3,30]. Here we only consider transducers whose input and output alphabets are equal: a transducer is a quintuple $\mathbf{t} = (Q, A, E, I, F)$, where A is the input and output alphabet, Q is a finite set of states, E is a set of directed edges between states from Q which are labelled by word pairs $(u, v) \in A^* \times A^*$, I is a set of initial states, and F a set of final states. If $\mathbf{t}$ realizes (x, y) then we write $y \in \mathbf{t}(x)$. We say that the set $\mathbf{t}(x)$ contains all possible outputs of $\mathbf{t}$ on input x. The transducer $\mathbf{t}^{-1}$ is the inverse of $\mathbf{t}$; that is, $x \in \mathbf{t}^{-1}(y)$ if and only if $y \in \mathbf{t}(x)$ for all words x, y. Let θ be an (anti-)morphic permutation and $\mathbf{t}$ be a transducer which are both defined over the same alphabet A. The transducer $\mathbf{t}$ is called *θ-input-preserving* if for all $w \in A^+$ we have $\theta(w) \in \mathbf{t}(w)$; $\mathbf{t}$ is called *θ-input-altering* if for all $w \in A^+$ we have $\theta(w) \notin \mathbf{t}(w)$. We use the simpler terms *input-preserving* and *input-altering* $\mathbf{t}$, respectively, when $\theta = \mathrm{id}$. Note that $\theta(w) \in \mathbf{t}(w)$ is equivalent to $w \in \theta^{-1}(\mathbf{t}(w))$ as well as $\mathbf{t}^{-1}(\theta(w)) \ni w$.

Definition 1 ([8])**.** *An input-altering transducer* $\mathbf{t}$ *describes the property that consists of all languages* L *such that*

$$\mathbf{t}(L) \cap L = \emptyset. \tag{1}$$

An input-preserving transducer $\mathbf{t}$ *describes the property that consists of all languages* L *such that*

$$w \notin \mathbf{t}(L \setminus w), \quad \text{for all } w \in L. \tag{2}$$

A property is called an input-altering (resp., input-preserving) transducer property, if it is described by an input-altering (resp., input-preserving) transducer.

Note that every input-altering transducer property is also an input-preserving transducer property. Input-altering transducers can be used to describe properties like prefix codes, bifix codes, and hypercodes. Input-preserving transducers are intended for error-detecting properties, where in fact the transducer plays the role of the communication channel.

Many input-altering transducer properties can be described in a simpler manner by *trajectory regular expressions* [5,8], that is, regular expressions over $\{0, 1\}$. For example, the expression 0^*1^* describes prefix codes and the expression $1^*0^*1^*$ describes infix codes. On the other hand, there are natural transducer properties that cannot be described by trajectory expressions [8].

2.3   Describing DNA-Related Properties by Trajectories

In [2,10,11,13–15,18,20–22,25] the authors consider numerous properties of languages inspired by reliability issues in DNA computing. We state three of these properties below. Let θ be an antimorphic permutation over A_k^*. Recall that, in the DNA setting, $\theta = \delta$ is an involution, and therefore, we have $\theta^2 = \mathrm{id}$.

(A) A language L is θ-*nonoverlapping* if $L \cap \theta(L) = \emptyset$.
(B) L is θ-*compliant* if $\forall w \in \theta(L), x, y \in A_k^* : xwy \in L \implies xy = \varepsilon$.
(C) L is *strictly* θ-*compliant* if it is θ-nonoverlapping and θ-compliant.

Many of the existing DNA-related properties can be modelled using the concept of a bond-free property, first defined in [22] and later rephrased in [6] in terms of trajectories. We follow the formulation in [6]. Let $\bar{e}_1$ and $\bar{e}_2$ be two regular trajectory expressions. First, we define the following language operators.

$$\Phi_{\bar{e}_1,\bar{e}_2}(L) = (((L \leadsto_{\bar{e}_1} A^+) \cap A^+) \sqcup_{\bar{e}_2} A^*) \cup (((L \leadsto_{\bar{e}_1} A^*) \cap A^+) \sqcup_{\bar{e}_2} A^+), \quad (3)$$

$$\Phi_{\bar{e}_1,\bar{e}_2}^{\mathsf{s}}(L) = ((L \leadsto_{\bar{e}_1} A^*) \cap A^+) \sqcup_{\bar{e}_2} A^*. \quad (4)$$

The language operations $\sqcup_{\bar{a}}$ and $\leadsto_{\bar{a}}$ are *shuffle* (or scattered insertion) and *scattered deletion*, respectively, over the set of trajectories $\bar{a}$; see [6,23] for details.

Definition 2 ([6]). *Let θ be an involution (or more generally a permutation) and $\bar{e}_1, \bar{e}_2$ be two regular trajectory expressions. The* bond-free property *described by $(\bar{e}_1, \bar{e}_2)$ is*

$$\mathcal{B}_\theta(\bar{e}_1, \bar{e}_2) = \{L \subseteq A^* \mid \theta(L) \cap \Phi_{\bar{e}_1,\bar{e}_2}(L) = \emptyset\}. \quad (5)$$

The strictly bond-free property *described by $(\bar{e}_1, \bar{e}_2)$ is*

$$\mathcal{B}_\theta^{\mathsf{s}}(\bar{e}_1, \bar{e}_2) = \{L \subseteq A^* \mid \theta(L) \cap \Phi_{\bar{e}_1,\bar{e}_2}^{\mathsf{s}}(L) = \emptyset\}. \quad (6)$$

A regular θ-trajectory property is a bond-free property described by $(\bar{e}_1, \bar{e}_2)$, or a strictly bond-free property described by $(\bar{e}_1, \bar{e}_2)$, for some pair $(\bar{e}_1, \bar{e}_2)$.

3   New Transducer-Based DNA-Related Properties

A question that arises from the discussion in Sects. 2.2 and 2.3 is whether existing transducer-based properties include DNA-related properties. It turns out that this is not the case; see Proposition 1. In this section, we define new transducer-based properties that are appropriate for DNA-related applications, we demonstrate Proposition 1, and discuss how existing DNA-related properties can be described with transducers.

Definition 3. *A transducer $\mathbf{t}$ and an (anti-)morphic permutation θ, defined over the same alphabet, describe 3-independent properties in two ways:*

1. *strict θ-transducer property (S-property)*: L satisfies the property $\mathcal{S}_{\theta,t}$ if

$$\theta(L) \cap t(L) = \emptyset \qquad (7)$$

2. *weak θ-transducer property (W-property)*: L satisfies the property $\mathcal{W}_{\theta,t}$ if

$$\forall w \in L : \theta(w) \notin t(L \setminus w) \qquad (8)$$

Any of the properties $\mathcal{S}_{\theta,t}$ or $\mathcal{W}_{\theta,t}$ is called a θ-transducer property.

The difference between $\mathcal{S}$-properties and $\mathcal{W}$-properties is that $\mathcal{S}_{\theta,t}$ includes no language containing a word w such that $\theta(w) \in t(w)$, while this case is allowed for some $L \in \mathcal{W}_{\theta,t}$. For fixed t, θ, and L, Condition (7) implies that for all $w \in L$ we have $\theta(w) \cap t(L \setminus w) = \emptyset$ which is equivalent to Condition (8). In other words, if L satisfies $\mathcal{S}_{\theta,t}$, then L satisfies $\mathcal{W}_{\theta,t}$ as well. If $\theta = \mathrm{id}$ and t is input-altering, or input-preserving, then the above defined properties specialize to the existing ones stated in Definition 1.

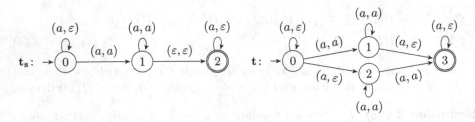

Fig. 1. Together with θ, the left transducer describes the strictly θ-compliant property and the right one describes the θ-compliant property. See Example 2 for explanations.

Example 2. In Fig. 1, an arrow with label (a, a) represents a set of edges with labels (a, a) for all $a \in A$; and similarly for an arrow with label (a, ε). For any word xwy, the left transducer t_s can delete x, then keep w (which has to be non-empty), and then delete y. Thus, $t_s(L) \cap \theta(L) = \emptyset$ if and only if L is strictly θ-compliant. Now let xwy with $xy \neq \varepsilon$ and $w \neq \varepsilon$. If y is nonempty, the right transducer t can delete x, then keep w, and then delete y using the upper path (containing state 1); and if x is nonempty, t can delete x, then keep w, and then delete y using the lower path (containing state 2). Thus, $t(L) \cap \theta(L) = \emptyset$ if and only if L is θ-compliant. Using FAdo [9] format the left transducer can be specified by the following string, assuming alphabet $\{a, b\}$.

```
@Transducer 2 * 0\n0 a @epsilon 0\n0 b @epsilon 0\n0 a a 1\n
0 b b 1\n1 a a 1\n1 b b 1\n1 @epsilon @epsilon 2\n2 a @epsilon 2\n
2 b @epsilon 2\n
```

The next result demonstrates that existing transducer properties are not suitable for describing even simple DNA-related properties.

Proposition 1. *The δ-nonoverlapping property is not describable by any input-preserving transducer.*

4   Expressiveness of Transducer-Based Properties

In this section we examine the descriptive power of the newly defined transducer DNA-related properties, that is, the θ-transducer properties. In Theorem 1 we show that these properties properly include the regular θ-trajectory properties. On the other hand, in Proposition 2 we show that there is an independent DNA-related property that is not a θ-transducer property.

Proposition 2. *The θ-free property (defined below) [13] is not a θ-transducer property.*

(D) *A language $L \subseteq A^*$ is θ-free if and only if $L^2 \cap A^+\theta(L)A^+ = \emptyset$.*

The following DNA language property is considered in Theorem 1

$$\mathcal{H} = \{L \subseteq \Delta^* \mid H(u, \delta(v)) \geq 2, \ \text{for all } u, v \in L\},$$

where $H(\cdot, \cdot)$ is the Hamming distance function with the assumption that its value is ∞ when applied on different length words. Note that $\mathcal{H}$ is described by δ and the transducer shown in Fig. 2.

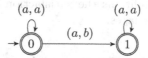

Fig. 2. The transducer **t** describing, together with δ, the $\mathcal{S}$-property $\mathcal{H}$: the displayed transducer **t** realizes (u, v) if and only if $H(u, v) < 2$; therefore, $\delta(L) \cap \mathbf{t}(L) = \emptyset$ if and only if $H(u, \delta(v)) \geq 2$ for all $u, v \in L$.

Example 3. The DNA language $L_1 = \{\text{AGG}, \text{CCA}\}$ does not satisfy $\mathcal{H}$ because $H(\text{CCA}, \delta(\text{AGG})) = 1$. The DNA language $L_2 = \{\text{AAA}, \text{CCT}\}$ satisfies $\mathcal{H}$ because $\delta(\text{AAA}) = \text{TTT}$ and all words $u \in L_2$ contain at most one T.

Theorem 1

1. *Let θ be an antimorphic involution. Every regular θ-trajectory property is a θ-transducer property (in particular an $\mathcal{S}$-property).*
2. *Property $\mathcal{H}$ is a δ-transducer property, but not a (regular) δ-trajectory one.*

5   The Satisfaction and Maximality Problems

For $\theta = \text{id}$ and for input-altering and -preserving transducers the satisfaction and maximality problems are decidable [8]. In particular, for a regular language L given via an automaton **a**, Condition (1) can be decided in time $\mathcal{O}(|\mathbf{t}||\mathbf{a}|^2)$, where the function $|\cdot|$ returns the size of the machine in question (= number of states plus number of edges plus the length of all labels on the edges). Condition (2) can be decided in time $\mathcal{O}(|\mathbf{t}||\mathbf{a}|^2)$, as noted in Remark 1 below. The maximality problem is decidable, but PSPACE-hard, for both input-altering and -preserving transducer properties.

Remark 1. Let $\mathbf{s} = \mathbf{t} \downarrow \mathbf{a} \uparrow \mathbf{a}$ be the transducer obtained by two product constructions: first on the input of $\mathbf{t}$ with $\mathbf{a}$; then, on the output of the resulting transducer with $\mathbf{a}$. In [8] the authors suggest to decide whether or not L satisfies the input-preserving transducer property $\mathcal{W}_{\mathrm{id},\mathbf{t}}$ by testing if the transducer $\mathbf{s}$ is functional. However, deciding $L \in \mathcal{W}_{\mathrm{id},\mathbf{t}}$ can be done by the cheaper test of whether or not $\mathbf{s}$ implements a (partial) identity function. Using the identity test from [1], we obtain that Condition (2) can be decided in time $\mathcal{O}(|\mathbf{t}||\mathbf{a}|^2)$ when the alphabet is considered constant. Also note that the identity test does not require that $\mathbf{t}$ is input-preserving if $\theta = \mathrm{id}$. When θ is antimorphic, however, the identity test does not work anymore and we have to resort to the more expensive functionality test for θ-input-preserving transducers.

In this work we are interested in the case when $\theta \neq \mathrm{id}$ is antimorphic; furthermore, the θ-input-altering or -preserving restrictions on the transducer are not necessarily present in the definition of $\mathcal{W}$-*properties* or $\mathcal{S}$-*properties*. Table 1 summarizes under which conditions the satisfaction and maximality problems are decidable for regular languages.

Table 1. (Un-)decidability of the satisfaction and the maximality problems for a fixed antimorphic permutation θ, a given transducer $\mathbf{t}$, and a regular language L given via an automaton $\mathbf{a}$.

Problem	Property $\mathcal{S}_{\theta,\mathbf{t}}$		Property $\mathcal{W}_{\theta,\mathbf{t}}$	
	No restriction	$\mathbf{t}$ is θ-i.-altering	No restriction	$\mathbf{t}$ is θ-i.-preserving
Satisfaction	Decidable in $\mathcal{O}(|\mathbf{t}||\mathbf{a}|^2)$ as in [8]		Decidable (Theorem 2)	Decidable in $\mathcal{O}(|\mathbf{t}|^2|\mathbf{a}|^4)$ as in [8]
Maximality	Undecidable (Corollary 2)	Decidable (Theorem 3), PSPACE-hard (Corollary 1)		

Remark 2. We note that deciding the satisfaction question for any θ-trajectory property involves testing the emptiness conditions in (5) or (6), which requires time $\mathcal{O}(|\mathbf{a}|^2|\mathbf{a}_1||\mathbf{a}_2|)$, where $\mathbf{a}_1, \mathbf{a}_2$ are automata corresponding to $\bar{e}_1, \bar{e}_2$. Such a property can be expressed as θ-transducer $\mathcal{S}$-property (recall Theorem 1) using a transducer of size $\mathcal{O}(|\mathbf{a}_1||\mathbf{a}_2|)$ and, therefore, the satisfaction question can still be solved within the same asymptotic time complexity.

5.1   The Satisfaction Problem for Non-restricted $\mathcal{W}$-properties

We establish the decidability of non-restricted transducer $\mathcal{W}$-*properties* for regular languages. We are not concerned with the complexity of this algorithm; optimizing the algorithm and analysing its complexity is part of future research.

Let $\mathbf{t}$ be a transducer, θ be an antimorphic permutation, and L be a regular language over the alphabet A. Let $\mathbf{a}_L$ and $\mathbf{a}_{\theta(L)}$ be the NFAs accepting the languages L and $\theta(L)$, respectively. Let $\mathbf{s} = \mathbf{t} \downarrow \mathbf{a}_L \uparrow \mathbf{a}_{\theta(L)}$ be the product transducer such that $y \in \mathbf{s}(x)$ if and only if $y \in \mathbf{t}(x)$, $x \in L$, and $y \in \theta(L)$.

Let $T_\mathbf{s} = \left\{ (x_1, x_2, x_3) \in (A^*)^3 \mid |x_1 x_2 x_3| \leq |\mathbf{s}| \right\}$ be a set of word triples. Note that the length restrictions for the words ensures that $T_\mathbf{s}$ is a finite set. For each triple $t = (x_1, x_2, x_3) \in T_\mathbf{s}$ we define a relation

$$R_t = \left\{ (x_1(x_2)^k x_3, \theta(x_1(x_2)^k x_3)) \mid k \in \mathbb{N} \right\} \subseteq A^* \times A^*.$$

Lemma 1. *The regular language L satisfies $\mathcal{W}_{\theta, \mathbf{t}}$ if and only if the relation realized by $\mathbf{s}$ is included in $\bigcup_{t \in T_\mathbf{s}} R_t$.*

The inclusion in Lemma 1 is decidable by performing the following two tests: (1) verify that $\mathbf{s} \subseteq \bigcup_{(x_1, x_2, x_3) \in T_s} (x_1 x_2^* x_3) \times \theta(x_1 x_2^* x_3)$; and (2) verify that $|x| = |y|$ for all pairs (x, y) that label an accepting path in $\mathbf{s}$. Note that the inclusion test can be performed because the right-hand-side relation is recognizable [3]. The second test follows the same ideas as the algorithm outlined in [1] which decides whether or not a transducer implements a partial identity function.

Theorem 2. *Let L be a regular language given as automaton, $\mathbf{t}$ be a given transducer, and θ be a given antimorphic involution (all defined over A). It is decidable whether L satisfies $\mathcal{W}_{\theta, \mathbf{t}}$ or not.*

5.2   The Maximality Problem

Here we show how to decide maximality of a regular language L with respect to a θ-transducer property; see Theorem 3. This result only holds when we consider $\mathcal{W}$-properties or when we consider $\mathcal{S}$-properties for θ-input-altering transducers. As in the case of existing transducer properties, it turns out that the maximality problem is PSPACE-hard; see Corollary 1. When we consider general $\mathcal{S}$-properties, the maximality problem becomes undecidable; see Corollary 2.

Theorem 3. *For an antimorphic permutation θ, a transducer $\mathbf{t}$, and a regular language L, all defined over A_k^*, such that either $L \in \mathcal{W}_{\theta, \mathbf{t}}$, or $L \in \mathcal{S}_{\theta, \mathbf{t}}$ and $\mathbf{t}$ is θ-input altering, we have that L is maximal with property $\mathcal{W}_{\theta, \mathbf{t}}$ (resp., $\mathcal{S}_{\theta, \mathbf{t}}$) if and only if*

$$L \cup \theta^{-1}(\mathbf{t}(L)) \cup \mathbf{t}^{-1}(\theta(L)) = A_k^*. \tag{9}$$

We note that it is PSPACE-hard to decide whether or not Eq. (9) holds when L is given as an NFA because it is PSPACE-hard to decide universality of a regular language given as an NFA ($L \subseteq A_k^*$ is universal if $L = A_k^*$) [29].

Corollary 1. *For an antimorphic permutation θ, a transducer $\mathbf{t}$, and a regular language L given as NFA, all defined over A_k^*, such that either $L \in \mathcal{W}_{\theta, \mathbf{t}}$, or $L \in \mathcal{S}_{\theta, \mathbf{t}}$ and $\mathbf{t}$ is θ-input altering, we have that it is PSPACE-hard to decide whether or not L is maximal with property $\mathcal{W}_{\theta, \mathbf{t}}$ (resp., $\mathcal{S}_{\theta, \mathbf{t}}$).*

In the rest of this section we show that it is undecidable whether or not a transducer is θ-input-preserving. This question relates directly to the maximality problem of the empty language $\emptyset$ with respect to the property $\mathcal{S}_{\theta,\mathbf{t}}$, as stated in Corollary 2. The following Theorem can be proven using a reduction from the famous, undecidable Post correspondence problem (PCP) to the problem of deciding whether a given transducer is θ-input-preserving or not.

Theorem 4. *For every fixed antimorphic permutation θ over A_k^* with $k \geq 2$ it is undecidable whether or not a given transducer is θ-input-preserving.*

This leads to the undecidability of the maximality problem of a regular language L with respect to a θ-transducer-property $\mathcal{S}_{\theta,\mathbf{t}}$.

Corollary 2. *For every fixed antimorphic permutation θ over A_k^* with $k \geq 2$, it is undecidable whether or not the empty language $\emptyset$ is maximal with respect to the property $\mathcal{S}_{\theta,\mathbf{t}}$, for a given transducer $\mathbf{t}$.*

Note that a singleton language $\{w\}$ satisfies $\mathcal{S}_{\theta,\mathbf{t}}$ if and only if $\theta(w) \notin \mathbf{t}(w)$. Thus, the corollary follows because $\emptyset$ is maximal with property $\mathcal{S}_{\theta,\mathbf{t}}$ if and only if $\mathbf{t}$ is θ-input-preserving.

6   Undecidability of the θ-PCP and the θ-Input-Altering Transducer Problem

In analogy with the undecidable PCP, we introduce the θ version of the PCP and prove that it is undecidable as well; see Theorem 5. Further, we utilize the θ version of the PCP in order to show that it is undecidable whether or not a transducer is θ-input-altering; see Corollary 3.

Definition 4. *For a fixed antimorphic permutation θ over A_k^*, we introduce the θ-Post correspondence problem (θ-PCP): given words $\alpha_0, \alpha_1, \ldots, \alpha_{\ell-1} \in A_k^+$ and $\beta_0, \beta_1, \ldots, \beta_{\ell-1} \in A_k^+$, decide whether or not there exists a non-empty sequence of integers $i_1, \ldots, i_n \in A_\ell = \{0, 1, \ldots, \ell - 1\}$ such that*

$$\alpha_{i_1} \alpha_{i_2} \cdots \alpha_{i_n} = \theta(\beta_{i_1} \beta_{i_2} \cdots \beta_{i_n}).$$

Theorem 5. *For every fixed antimorphic permutation θ over A_k^* with $k \geq 2$ the θ-PCP is undecidable.*

We can utilize the θ-PCP in order to prove that it is undecidable whether or not a transducer is θ-input-altering, even for one-state transducers.

Corollary 3. *For every fixed antimorphic permutation θ over A_k^* with $k \geq 2$ it is undecidable whether or not a given (one-state) transducer is θ-input-altering.*

Corollary 3 follows because the θ-PCP instance $\alpha_0, \ldots, \alpha_{\ell-1}, \beta_0, \ldots, \beta_{\ell-1}$ has a solution if and only if the the one-state transducer $\mathbf{t}$ with edges $q \xrightarrow{(\alpha_i, \theta^2(\beta_i))} q$ for $i = 0, \ldots, \ell - 1$ is not θ-input-altering.

7 Conclusions

We have defined a transducer-based method for describing DNA code properties which is strictly more expressive than the trajectory method. In doing so, the satisfaction question remains efficiently decidable. The maximality question for some types of properties is decidable, but it is undecidable for others. While some versions of the maximality question for trajectory properties are decidable, the case of any given pair of regular trajectories and any given regular language is not addressed in [6], so we consider this to be an interesting problem to solve.

The maximality questions are phrased in terms of any fixed antimorphic permutation. This direction of generalizing decision questions is also applied to the classic Post correspondence problem, where we demonstrate that it remains undecidable. A consequence of this is that the question of whether a given transducer is θ-input-altering is also undecidable. It is interesting to note that if, instead of fixing θ, we fix the transducer $\mathbf{t}$ to be the identity, or the transducer defining the $\mathcal{S}$-property $\mathcal{H}$ (see Fig. 2 in Sect. 4), then the question of whether or not $\theta(L) \cap \mathbf{t}(L) = \emptyset$ is decidable (given any regular language L and antimorphic permutation θ).

The topic of studying description methods for code properties requires further attention. One important aim is the actual implementation of the algorithms, as it is already done for several classic code properties [9,24]. An immediate plan is to incorporate in those implementations what we know about DNA code properties. Another aim is to increase the expressive power of our description methods. The formal method of [16] is quite expressive, using a certain type of first order formulae to describe properties. It could perhaps be further worked out in a way that some of these formulae can be mapped to transducers. We also note that if the defining method is too expressive then even the satisfaction problem could become undecidable; see for example the method of multiple sets of trajectories in [7].

References

1. Allauzen, C., Mohri, M.: Efficient algorithms for testing the twins property. J. Autom. Lang. Comb. **8**(2), 117–144 (2003)
2. Baum, E.: DNA sequences useful for computation. In: 2nd DIMACS Workshop on DNA-based Computers, pp. 122–127. Princeton University (1996)
3. Berstel, J.: Transductions and Context-Free Languages. B.G. Teubner, Stuttgart (1979)
4. Berstel, J., Perrin, D., Reutenauer, C.: Codes and Automata. Cambridge University Press, New York (2009)
5. Domaratzki, M.: Trajectory-based codes. Acta Informatica **40**, 491–527 (2004)
6. Domaratzki, M.: Bond-free DNA language classes. Nat. Comput. **6**, 371–402 (2007)
7. Domaratzki, M., Salomaa, K.: Codes defined by multiple sets of trajectories. Theor. Comput. Sci. **366**, 182–193 (2006)
8. Dudzinski, K., Konstantinidis, S.: Formal descriptions of code properties: decidability, complexity, implementation. IJFCS **23**(1), 67–85 (2012)

9. FAdo: Tools for formal languages manipulation. http://fado.dcc.fc.up.pt/. Accessed February 2015
10. Fan, C.M., Wang, J.T., Huang, C.C.: Some properties of involution binary relations. Acta Informatica (2014). doi:10.1007/s00236-014-0208-8
11. Genova, D., Mahalingam, K.: Generating DNA code words using forbidding and enforcing systems. In: Dediu, A.-H., Martín-Vide, C., Truthe, B. (eds.) TPNC 2012. LNCS, vol. 7505, pp. 147–160. Springer, Heidelberg (2012)
12. Hopcroft, J.E., Ullman, J.D.: Introduction to Automata Theory, Languages, and Computation. Addison-Wesley, Cambridge (1979)
13. Hussini, S., Kari, L., Konstantinidis, S.: Coding properties of DNA languages. Theor. Comput. Sci. **290**, 1557–1579 (2003)
14. Jonoska, N., Kari, L., Mahalingam, K.: Involution solid and join codes. Fundamenta Informaticae **86**, 127–142 (2008)
15. Jonoska, N., Mahalingam, K., Chen, J.: Involution codes: with application to DNA coded languages. Nat. Comput. **4**, 141–162 (2005)
16. Jürgensen, H.: Syntactic monoids of codes. Acta Cybernetica **14**, 117–133 (1999)
17. Jürgensen, H., Konstantinidis, S.: Codes. In: Rozenberg and Salomaa [26], pp. 511–607
18. Kari, L., Kitto, R., Thierrin, G.: Codes, involutions, and DNA encodings. In: Brauer, W., Ehrig, H., Karhumäki, J., Salomaa, A. (eds.) Formal and Natural Computing. LNCS, vol. 2300, pp. 376–393. Springer, Heidelberg (2002)
19. Kari, L., Konstantinidis, S., Kopecki, S.: Transducer descriptions of DNA code properties and undecidability of antimorphic problems (2015). arXiv preprint arXiv:1503.00035
20. Kari, L., Konstantinidis, S., Losseva, E., Wozniak, G.: Sticky-free and overhang-free DNA languages. Acta Informatica **40**, 119–157 (2003)
21. Kari, L., Konstantinidis, S., Sosík, P.: Bond-free languages: formalizations, maximality and construction methods. IJFCS **16**, 1039–1070 (2005)
22. Kari, L., Konstantinidis, S., Sosík, P.: On properties of bond-free DNA languages. Theor. Comput. Sci. **334**, 131–159 (2005)
23. Kari, L., Sosík, P.: Aspects of shuffle and deletion on trajectories. Theor. Comput. Sci. **332**, 47–61 (2005)
24. LaSer: Independent LAnguage SERver. http://laser.cs.smu.ca/independence/. Accessed February 2015
25. Mauri, G., Ferretti, C.: Word design for molecular computing: a survey. In: Chen, Junghuei, Reif, John H. (eds.) DNA 2003. LNCS, vol. 2943, pp. 37–47. Springer, Heidelberg (2004)
26. Rozenberg, G., Salomaa, A. (eds.): Handbook of Formal Languages, vol. I. Springer, Berlin (1997)
27. Shyr, H.: Free Monoids and Languages, 2nd edn. Hon Min Book Company, Taichung (1991)
28. Shyr, H., Thierrin, G.: Codes and binary relations. In: Malliavin, M.P. (ed.) Séminaire d'Algèbre Paul Dubreil, Paris 1975–1976 (29ème Année). Lecture Notes in Mathematics, vol. 586, pp. 180–188. Springer, Berlin (1977)
29. Stockmeyer, L., Meyer, A.: Word problems requiring exponential time (preliminary report). In: Proceedings of the 5th Annual ACM Symposium on Theory of Computing, pp. 1–9. ACM (1973)
30. Yu, S.: Regular languages. In: Rozenberg and Salomaa [26], pp. 41–110

On Simulation Cost of Unary Limited Automata

Martin Kutrib[(✉)] and Matthias Wendlandt

Institut für Informatik, Universität Giessen, Arndtstr. 2, 35392 Giessen, Germany
{kutrib,matthias.wendlandt}@informatik.uni-giessen.de

Abstract. A k-limited automaton is a linear bounded automaton that may rewrite each tape cell only in the first k visits, where $k \geq 0$ is a fixed constant. It is known that these automata accept context-free languages only. We investigate the descriptional complexity of deterministic limited automata accepting unary languages. Since these languages are necessarily regular, we study the cost in the number of states when a k-limited automaton is simulated by finite automata. For the conversion of a $4n$-state 1-limited automaton into one-way or two-way deterministic or nondeterministic finite automata a lower bound of $n \cdot F(n)$ states is shown, where F denotes Landau's function. So, even the ability deterministically to rewrite any cell only once gives an enormous descriptional power. For the simulation cost for removing the ability to rewrite each cell $k \geq 1$ times, that is, the cost for the simulation of (sweeping) k-limited automata by deterministic finite automata, we obtain a lower bound of $n \cdot F(n)^k$. A polynomial upper bound is shown for the simulation by two-way deterministic finite automata, where the degree of the polynomial is quadratic in k. If the k-limited automaton is rotating, the upper bound reduces to $O(n^{k+1})$. A lower bound of $\Omega(n^{k+1})$ is derived even for nondeterministic two-way finite automata. So, for rotating k-limited automata, the trade-off for the simulation is tight in the order of magnitude.

1 Introduction

The cost for the simulation of one formal model by another is one of the main topics of descriptional complexity, where the cost are measured in close connection to the sizes of the models. Such simulations are of particular interest when both formal models capture the same family of languages. A fundamental result is that nondeterministic finite automata can be simulated by deterministic finite automata by paying the cost of exponentially many states (see, for example, [16]). Among the many models characterizing the regular languages, an interesting variant is the linear bounded automata where the rewrite operations are restricted. If any tape cell may be visited only a constant number of times, it is shown in [5] that even linear-time computations cannot accept non-regular languages. This result has been improved to $O(n \log n)$ time in [4]. Recent results [24] show that the upper as well as the lower bound for converting a weight-reducing machine of this type into a deterministic finite automaton is doubly exponential.

© Springer International Publishing Switzerland 2015
J. Shallit and A. Okhotin (Eds.): DCFS 2015, LNCS 9118, pp. 153–164, 2015.
DOI: 10.1007/978-3-319-19225-3_13

A related result in [1] shows that if a two-way finite automaton is allowed to freely place a pebble on the tape, then again no non-regular language can be accepted, even if the time is unlimited. A doubly exponential upper and a lower bound for the simulation by a deterministic finite automaton is derived in [16].

A generalization of the machines studied in [5] is introduced by Hibbard [6]. He investigated linear bounded automata that may rewrite each tape cell only in the first k visits, where k is a fixed constant. However, afterwards the cells can still be visited any number of times (but without rewriting their contents). It is shown in [6] that the nondeterministic variant characterizes the context-free languages provided $k \geq 2$, while there is a tight and strict hierarchy of language classes depending on k for the deterministic variant. One-limited automata, deterministic and nondeterministic, can accept only regular languages. From these results it follows that any unary k-limited automaton accepts regular languages only.

Recently, the study of limited automata from the descriptional complexity point of view has been initiated by Pighizzini and Pisoni [20,21]. In [21] it was shown that the deterministic 2-limited automata characterize the deterministic context-free languages, which complements the result on nondeterministic machines. Furthermore, conversions between 2-limited automata and pushdown automata are investigated. For the deterministic case the upper bound for the conversion from 2-limited automata to pushdown automata is doubly exponential. Conversely, the trade-off is shown to be polynomial. Comparisons between 1-limited automata and finite automata are done in [20]. In particular, a double exponential trade-off between nondeterministic 1-limited automata and one-way deterministic finite automata is shown. For deterministic 1-limited automata the conversion costs a single exponential increase in size. These results imply an exponential trade-off between nondeterministic and deterministic 1-limited automata, and they show that 1-limited automata can have less states than equivalent two-way nondeterministic finite automata.

For a restricted variant of limited automata, so-called strongly limited automata, it is shown that context-free grammars as well as pushdown automata can be transformed in strongly limited automata and vice versa with polynomial cost [19].

Here, we consider deterministic k-limited automata accepting unary languages. The descriptional complexity of unary regular languages has been studied in many ways. On one hand, many automata models such as one-way finite automata, two-way finite automata, pushdown automata, or context-free grammars for unary languages are investigated and compared to each other with respect to simulation results and the size of the simulation (see, for example, [3,15,18,23]). On the other hand, many results concerning the state complexity of operations on unary languages have been obtained (see, for example, [7,10,14,22]).

The results on the expressive power of limited automata imply that any unary language accepted by some k-limited automaton is regular. So, it is of interest to investigate the descriptional complexity in comparison with the models mentioned above. We establish upper and lower bounds for the conversion of unary deterministic k-limited automata to one-way and two-way finite automata.

2   Preliminaries

We write Σ^* for the set of all words over the finite alphabet Σ. The empty word is denoted by λ, the reversal of a word w by w^R, and for the length of w we write $|w|$. We use $\subseteq$ for inclusions and $\subset$ for strict inclusions.

Let $k \geq 0$ be an integer. A deterministic k-limited automaton is a restricted linear bounded automaton. It consists of a finite state control and a read-write tape whose initial contents is the input word in between two endmarkers. At the outset of a computation, the automaton is in the designated initial state and the head of the tape scans the left endmarker. Depending on the current state and the currently scanned symbol on the tape, the automaton changes its state, rewrites the current symbol on the tape, and moves the head one cell to the left or one cell to the right. However, the rewriting is restricted so that the machine may rewrite each tape cell only in the first k visits. Subsequently, the cell can still be scanned but the content cannot be changed any longer. So, a deterministic 0-limited automaton is a two-way deterministic finite automaton. An input is accepted if the machine reaches an accepting state and halts.

The original definition of such devices in [6] is based on string rewriting systems whose sentential forms are seen as configurations of automata. Let $u_1 u_2 \cdots u_{i-1} s u_i u_{i+1} \cdots u_n$ be a sentential form that represents the tape contents $u_1 u_2 \cdots u_n$ and the current state s. Basically, in [6] rewriting rules are provided of the form $s u_i \to u_i' s'$ which means that the state changes from s to s', the tape cell to the right of s is scanned and rewritten from u_i to u_i', and $u_{i-1} s \to s u_{i-1}'$, which means that the state changes from s to s', the tape cell to the left of s is scanned and rewritten from u_{i-1} to u_{i-1}'. In this context, an automaton that changes its head direction on a cell scans the cell twice. In [20, 21] and below, limited automata are defined in a way that reflects this behavior.

Formally, a *deterministic k-limited automaton* (k-DLA, for short) is a system $M = \langle S, \Sigma, \Gamma, \delta, \triangleright, \triangleleft, s_0, F \rangle$, where S is the finite, nonempty set of *internal states*, Σ is the finite set of *input symbols*, Γ is the finite set of *tape symbols* partitioned into $\Gamma_k \cup \Gamma_{k-1} \cup \cdots \cup \Gamma_0$ where $\Gamma_0 = \Sigma$, $\triangleright \notin \Gamma$ is the *left* and $\triangleleft \notin \Gamma$ is the *right endmarker*, $s_0 \in S$ is the *initial state*, $F \subseteq S$ is the set of *accepting states*, and $\delta : S \times (\Gamma \cup \{\triangleright, \triangleleft\}) \to S \times (\Gamma \cup \{\triangleright, \triangleleft\}) \times \{-1, 1\}$ is the partial transition function, where -1 means to move the head one cell to the left, 1 means to move it one cell to the right, and whenever $(s', y, d) = \delta(s, \triangleright)$ is defined then $y = \triangleright$, $d = 1$ and whenever $(s', y, d) = \delta(s, \triangleleft)$ is defined then $y = \triangleleft$, $d = -1$.

In order to implement the limited number of rewrite operations, δ is required to satisfy the following condition. For each $(s', y, d) = \delta(s, x)$ with $x \in \Gamma_i$, (1) if $i = k$ then $x = y$, (2) if $i < k$ and $d = 1$ then $y \in \Gamma_j$ with $j = \min\{\lceil \frac{i}{2} \rceil \cdot 2 + 1, k\}$, and (3) if $i < k$ and $d = -1$ then $y \in \Gamma_j$ with $j = \min\{\lceil \frac{i+1}{2} \rceil \cdot 2, k\}$.

It is worth mentioning that these conditions make the a priori global condition of a head turn on some cell local. The clever transformation of the original definition to the automata world used in [20, 21] gives that, if a cell content is from Γ_i then the head position is always to the right of that cell if i is odd, and it is to the left of the cell if i is even, as long as $i < k$.

A *configuration* of the k-DLA M is a triple (s, v, h), where $s \in S$ is the current state, $v \in \triangleright \Gamma^* \triangleleft$ is the current tape contents, and $h \in \{0, 1, \ldots, |w| + 1\}$ gives the current head position. If h is 0, the head scans the symbol $\triangleright$, if it satisfies $1 \leq i \leq |w|$, then the head scans the ith letter of w, and if it is $|w| + 1$, then the head scans the symbol $\triangleleft$. The *initial configuration* for input w is set to $(s_0, \triangleright w \triangleleft, 0)$. During the course of its computation, M runs through a sequence of configurations. One step from a configuration to its successor configuration is denoted by $\vdash$. Let $a_0 = \triangleright$ and $a_{n+1} = \triangleleft$, for $n \geq 0$, then we set $(s, \triangleright a_1 a_2 \cdots a_h \cdots a_n \triangleleft, h) \vdash (s', \triangleright a_1 a_2 \cdots a_h' \cdots a_n \triangleleft, h + d)$ if and only if $(s', a_h', d) = \delta(s, a_h)$.

A k-DLA *halts*, if the transition function is undefined for the current configuration. An input is *accepted* if the automaton halts at some time in an accepting state, otherwise it is rejected. The *language* $L(M)$ *accepted* by M is the set of all accepted inputs.

A k-DLA is said to be *sweeping* if the direction of the head movement changes only on the endmarkers.

In order to clarify the notions we continue with an example that is later used for lower bounds.

Example 1. Let $k \geq 1$ and $n \geq 2$. The finite unary language $L = \{a^{n^{k+1}}\}$, that consists of one word only, is accepted by a sweeping k-limited automaton $M = \langle S, \{a\}, \Gamma, \delta, \triangleright, \triangleleft, s_0, F \rangle$ with $n + 2$ states and $2k + 1$ tape symbols.

The principal idea of the construction is to sweep k times across the tape, where in each sweep $n - 1$ out of every n non-marked symbols are marked. In this way, in the mth sweep it is checked whether the number of non-marked symbols is a multiple of n^m. In the final $(k + 1)$st sweep it is checked whether exactly n non-marked symbols exist. If yes, the length ℓ of the input is $\frac{\ell}{n^k} = n$ which implies $\ell = n^{k+1}$.

Formally, we set $S = \{s_0, s_1, \ldots, s_{n-1}, s_+, s_-\}$, $F = \{s_+\}$, and the tape alphabet to be $\Gamma = \{a, a_1, a_1', a_2, a_2', \ldots, a_k, a_k'\}$.

Whenever the head reaches an endmarker M has to be in state s_0. Otherwise the computation halts rejecting. In this way it is verified whether the input length is a multiple of n^m, respectively.

1. $\delta(s_0, \triangleright) = (s_0, \triangleright, 1)$
2. $\delta(s_0, \triangleleft) = (s_0, \triangleleft, -1)$

The first sweep is realized by Transitions 3 and 4.

3. $\delta(s_i, a) = (s_{i+1}, a_1', 1)$,   for $0 \leq i \leq n - 2$
4. $\delta(s_{n-1}, a) = (s_0, a_1, 1)$

For sweeps $2 \leq m \leq k$, Transitions 5 to 7 are used. Let $d = 1$ if m is odd and $d = -1$ if m is even.

5. $\delta(s_i, a_{m-1}) = (s_{i+1}, a_m', d)$,   for $0 \leq i \leq n - 2$
6. $\delta(s_{n-1}, a_{m-1}) = (s_0, a_m, d)$
7. $\delta(s_i, a_{m-1}') = (s_i, a_m', d)$,   for $0 \leq i \leq n - 1$

Finally, in the $(k+1)$st sweep the states are reused in the same way to count up to n unmarked symbols. But after the first cycle state s_+ is entered instead of state s_0. If M reaches another unmarked symbol in state s_+ it rejects. Otherwise the computation halts accepting on the endmarker in state s_+.

8. $\delta(s_i, a_k) = (s_{i+1}, a_k, d)$,   for $0 \leq i \leq n-2$
9. $\delta(s_{n-1}, a_k) = (s_+, a_k, d)$
10. $\delta(s_i, a'_k) = (s_i, a'_k, d)$,   for $0 \leq i \leq n-1$
11. $\delta(s_+, a'_k) = (s_+, a'_k, d)$
12. $\delta(s_+, a_k) = (s_-, a_k, d)$ ∎

As is often the case in connection with unary languages, Landau's function

$$F(n) = \max\{\mathrm{lcm}(c_1, c_2, \ldots, c_l) \mid l \geq 1, c_1, c_2, \ldots, c_l \geq 1 \text{ and } c_1 + c_2 + \cdots + c_l = n\}$$

plays a crucial role, where lcm denotes the *least common multiple*. It is well known that the c_i always can be chosen to be relatively prime. Moreover, an easy consequence of the definition is that the c_i can always be chosen so that $c_1, c_2, \ldots, c_l \geq 2$, $c_1 + c_2 + \cdots + c_l \leq n$, and $\mathrm{lcm}(c_1, c_2, \ldots, c_l) = F(n)$ (cf., for example, [17]). Since F depends on the irregular distribution of the prime numbers, we cannot expect to express $F(n)$ explicitly by n. The function itself was investigated by Landau [12,13] who proved the asymptotic growth rate $\lim_{n \to \infty} \frac{\ln(F(n))}{\sqrt{n \cdot \ln(n)}} = 1$. The upper and lower bounds $F(n) \in e^{\sqrt{n \cdot \ln(n)}(1+o(1))}$ and $F(n) \in \Omega\left(e^{\sqrt{n \cdot \ln(n)}}\right)$ have been derived in [2,25].

3 Simulation Cost of 1-DLA

We start with simulations of unary 1-DLA by finite automata. Upper bounds for general regular languages have been obtained in [20] as follows. Any n-state 1-DLA can be simulated by a one-way deterministic finite automaton (1DFA) with no more than $n \cdot (n+1)^n$ states. The currently best lower bound for the simulations of *unary* 1-DLA by two-way nondeterministic finite automata (2NFA) was also obtained in [20], where it is shown that for infinitely many integers n there is a unary regular language recognized by an n-state, 3-tape-symbol 1-DLA such that each equivalent 2NFA requires a number of states which is quadratic in n. The next theorem improves this lower bound. It is worth mentioning that, to this end, the number of tape symbols is set to $n+1$.

Theorem 2. *Let $n \geq 2$ be a prime number. Then there is a unary $4n$-state and $n+1$ tape symbol 1-DLA M, such that $n \cdot F(n)$ states are necessary for any 2NFA to accept the language $L(M)$.*

Proof. As mentioned above, there are positive integers $c_1, c_2, \ldots, c_l \geq 2$ such that $c_1 + c_2 + \cdots + c_l \leq n$ and $\mathrm{lcm}(c_1, c_2, \ldots, c_l) = F(n)$. The witness language is the singleton $L = \{a^{n \cdot F(n)}\}$. We construct a 1-DLA $M = \langle S, \{a\}, \Gamma, \delta, \triangleright, \triangleleft, s_0, \{p_+\} \rangle$ accepting L with at most $4n$ states and $n+1$ tape symbols, where

$$\Gamma = \{a\} \cup \{t_{i,j} \mid 1 \le i \le l, 0 \le j \le c_i - 1\}, \text{ and}$$
$$S = \{s_0, p_+, p_-, r_0, \ldots, r_{n-2}, r'_0, \ldots, r'_{n-3}, q_{l+1}, \ldots, q_n\}$$
$$\cup \{s_{i,j} \mid 1 \le i \le l, 0 \le j \le c_i - 1\}.$$

A tape symbol $t_{i,j}$ occurring on some cell m means that $m \bmod c_i = j$. An input is accepted only if its length is a multiple of n. We consider the input to be partitioned into blocks of length n. The 1-DLA rewrites the input symbols in each block $xn + 1, xn + 2, \ldots, xn + n$ by $t_{1,j_1}, t_{2,j_2}, \ldots, t_{l,j_l}, \sqcup, \ldots, \sqcup$, where $j_1 = (xn+1) \bmod c_1, \ldots, j_l = (xn+l) \bmod c_l$. For each block these rewritings are successively from left to right. To this end, the states $s_{i,j}$ are used, where state $s_{i,j}$ with the head on cell m means $(m-1) \bmod c_i = j$. The idea of this part of the construction is as follows. When one of the first $l - 1$ cells, say cell ℓ, of the new block has been rewritten (Transition 3), states r_i and r'_i are used to move the head back to cell $\ell + 1$ of the previous block. Basically, state r_i or r'_i means to move the head back for another i cells. The content of cell $\ell + 1$ of the previous block is then used to continue the counting modulo $\ell + 1$ (by states $s_{\ell+1,j}$, Transition 2 and further transitions below) until the next still-unwritten cell is reached (Transition 3). This is cell $\ell + 1$ of the new block. After rewriting cell l of a block the states q_i are used to write the symbol $\sqcup$ to the remaining cells of the block (Transitions 4 and 5). Afterwards, states r_i and r'_i are again used to start the rewriting of the next block (Transition 6). Further roles played by the states r_i and r'_i are explained below. Let $y \in \Gamma \setminus \{\triangleleft, a\}$.

1. $\delta(s_0, \triangleright) = (s_{1,0}, \triangleright, 1)$
2. $\delta(s_{i,j}, y) = (s_{i,(j+1) \bmod c_i}, y, 1),$ for $1 \le i \le l$ and $0 \le j \le c_i - 1$
3. $\delta(s_{i,j}, a) = (r_{n-2}, t_{i,(j+1) \bmod c_i}, -1),$ for $1 \le i \le l - 1$ and $0 \le j \le c_i - 1$
4. $\delta(s_{l,j}, a) = (q_{l+1}, t_{l,(j+1) \bmod c_l}, 1),$ for $0 \le j \le c_l - 1$
5. $\delta(q_i, a) = (q_{i+1}, \sqcup, 1),$ for $l + 1 \le i \le n - 1$
6. $\delta(q_n, a) = (r_{n-2}, \sqcup, -1),$ for $l + 1 \le i \le n - 1$

The very first block is treated differently, since there is no predecessor block. However, whenever the head is moved back to the left endmarker, the index of states r_i and r'_i says how to continue the counting (Transitions 1, 7, and 8).

7. $\delta(r_{n-i}, \triangleright) = (s_{i,0}, \triangleright, 1),$ for $2 \le i \le l$
8. $\delta(r'_{n-i}, \triangleright) = (s_{i,0}, \triangleright, 1),$ for $2 \le i \le l$

Now we turn to the end of the computation and the roles played by the states r_i and r'_i. Let w be the input. Its length $|w|$ is a multiple of n if and only if there is no partial block at the end. It is a multiple of $F(n)$ if and only if $|w|$ is divisible by all the c_i, that is, $|w| \bmod c_i = 0$. In order to test whether the length of the input up to and including the current block is a multiple of $F(n)$, it is sufficient to inspect the first l cells of the block. The test is positive if the contents $t_{\ell,j}$ of all cells $1 \le \ell \le l$ is so that $(j + n - \ell) \bmod c_\ell = 0$. This test is performed while M is in states r_i and r'_i, which move the head to the left. Moreover, since only the first input that meets the criteria may be accepted, M remembers a negative test result by changing from some state r_i to a primed version r'_{i-1} (Transitions 11, 12). Once in a primed state the head is moved back without further tests (Transitions 11, 12).

9. $\delta(r_i, \sqcup) = (r_{i-1}, \sqcup, -1)$,   for $1 \leq i \leq n-3$
10. $\delta(r'_i, \sqcup) = (r'_{i-1}, \sqcup, -1)$,   for $1 \leq i \leq n-3$
11. $\delta(r_i, t_{i',j'}) = (r_{i-1}, t_{i',j'}, -1)$,
    if $(j' + n - i') \mod c_{i'} = 0$, for $1 \leq i \leq n-2, 1 \leq i' \leq l, 0 \leq j' \leq c_{i'} - 1$
12. $\delta(r_i, t_{i',j'}) = (r'_{i-1}, t_{i',j'}, -1)$,
    if $(j' + n - i') \mod c_{i'} \neq 0$, for $1 \leq i \leq n-2, 1 \leq i' \leq l, 0 \leq j' \leq c_{i'} - 1$
13. $\delta(r'_i, t_{i',j'}) = (r'_{i-1}, t_{i',j'}, -1)$, for $1 \leq i \leq n-3, 1 \leq i' \leq l, 0 \leq j' \leq c_{i'} - 1$

When the head reaches its destination, M is in state r_0 or r'_0. If the destination is not the first cell of the block or the test was negative, M takes the cell contents to continue the counting (Transition 14 and 15). If the destination is the first cell of the block and the test was positive the first cell is tested as well. Dependent on the result, either the rewriting of the next block is started (Transition 16) or state p_+ is entered (Transition 17).

14. $\delta(r_0, t_{i',j'}) = (s_{i',j'}, t_{i',j'}, 1)$,   for $2 \leq i' \leq l, 0 \leq j' \leq c_{i'} - 1$
15. $\delta(r'_0, t_{i',j'}) = (s_{i',j'}, t_{i',j'}, 1)$,   for $1 \leq i' \leq l, 0 \leq j' \leq c_{i'} - 1$
16. $\delta(r_0, t_{1,j'}) = (s_{1,j'}, t_{1,j'}, 1)$,   if $(j' + n - 1) \mod c_1 \neq 0$, for $0 \leq j' \leq c_1 - 1$
17. $\delta(r_0, t_{1,j'}) = (p_+, t_{1,j'}, 1)$,   if $(j' + n - 1) \mod c_1 = 0$, for $0 \leq j' \leq c_1 - 1$

Once in state p_+ it is known that the input length, up to and including the current block, is the least multiple of $F(n)$. So, it remains to be tested that there is no further input symbol a at the right of the block. By Transition 18 the head is moved to the right as long as there appears neither the input symbol a nor the right endmarker. If the right endmarker appears the computation halts in the accepting state p_+. If there is a further a to the right of the current block, the rejecting state p_- is entered and the computation halts rejecting (Transition 19). Let $y \in \Gamma \setminus \{\triangleleft, a\}$.

18. $\delta(p_+, y) = (p_+, y, 1)$
19. $\delta(p_+, a) = (p_-, a, 1)$

From the construction follows that M accepts the shortest input that is a multiple of n and a multiple of $F(n)$. Since n is prime, and the c_i are relatively prime and less than n, all c_i and n are relatively prime as well. So, M accepts $a^{n \cdot F(n)}$. The numbers of states and tape symbols claimed follow also from the construction. So, it remains to be verified that no further inputs are accepted by M.

The only possibility to accept is in state p_+ on the right endmarker. Since state p_+ is entered only when the head is on the first cell of a block after the test was positive, we derive that the input is a multiple of n and $F(n)$. Since there is no transition leading from state p_+ to any other state except for p_-, it follows that the input is the shortest word which is a multiple of n and $F(n)$ and, thus, $L(M) = \{a^{n \cdot F(n)}\}$.

Finally, any two-way nondeterministic finite automaton that accepts a unary singleton language needs as least as many states as the length of the sole word in the language.   □

Since even a 2NFA needs at least n states to accept the unary singleton language $\{a^n\}$, the proof of Theorem 2 reveals the same lower bound for one-way and two-way deterministic and nondeterministic finite automata.

Corollary 3. *Let $n \geq 2$ be a prime number. Then there is a unary 4n-state and $n+1$-tape-symbol 1-DLA M, such that $n \cdot F(n)$ states are necessary for any 2DFA, 1DFA, or 1NFA to accept the language $L(M)$.*

4 Simulation Cost of k-DLA

This section is devoted to deriving bounds on the cost for removing the ability to rewrite each cell $k \geq 1$ times. That is, the cost for the simulation of k-DLA by deterministic finite automata. We start with a lower bound for the simulation by 1DFA. Interestingly, this lower bound is greater than the lower bound $n \cdot F(n)^{k-1}$ known for the simulation of unary one-way k-head finite automata [11]. Both types of devices accept only regular unary languages, but only trivial bounds are currently known for the cost of their mutual simulations.

Theorem 4. *Let $k, n \geq 2$ be integers so that n is prime. Then there is a unary sweeping $(n+1)$-state, $2k$-tape-symbol k-DLA M, so that $n \cdot F(n)^k$ states are necessary for any 1DFA to accept the language $L(M)$.*

Proof. For any constants $k \geq 2$ and prime $n \geq 2$, we construct a unary sweeping k-DLA $M = \langle S, \{a\}, \Gamma, \delta, \rhd, \lhd, s_0, \{s_+\} \rangle$, where $S = \{s_0, s_1, \ldots, s_{n-1}, s_+\}$, $\Gamma_0 = \{a\}$, $\Gamma_1 = \{a_1\}$, and $\Gamma_i = \{\sqcup_i, a_i\}$, for $2 \leq i \leq k$.

There are integers $c_1, c_2, \ldots, c_l \geq 2$ so that $c_1 + c_2 + \cdots + c_l \leq n$ and $\text{lcm}(c_1, c_2, \ldots, c_l) = F(n)$. We set $p(1) = 0$, $q(1) = c_1 - 1$, $p(i) = q(i-1) + 1$, $q(i) = p(i) + c_i - 1$, for $2 \leq i \leq l$. So, we obtain in particular $q(l) \leq n - 1$.

Let w be an input. In its first sweep, M rewrites any input cell with the symbol a_1. The purpose of the first sweep is to determine the value $|w| \bmod n$ (Transitions 1 and 2). If the value does not belong to the set $\{p(1), p(2), \ldots, p(l)\}$ the computation halts rejecting (Transitions 3).

1. $\delta(s_0, \rhd) = (s_0, \rhd, 1)$
2. $\delta(s_i, a) = (s_{(i+1) \bmod n}, a_1, 1)$,   for $0 \leq i \leq n-1$
3. $\delta(s_{p(j)}, \lhd) = (s_{p(j)}, \lhd, -1)$,   for $1 \leq j \leq l$

The principal idea of the further computation is as follows. In the first sweep a value $p(j)$ is determined. Now M fixes the j and uses k further sweeps to test whether the length of the input is a multiple of c_j^k. A detailed analysis of the language accepted follows after the construction. In the next $k-1$ sweeps only the states $s_{p(j)}$ to $s_{q(j)}$ are used. During a sweep every c_jth non-blank symbol is kept non-blank, while all the others are rewritten by a blank (Transitions 4, 5, and 6). If the number of non-blank symbols found during the sweep is not a multiple of c_j, that is, M reaches the opposite endmarker not in state $s_{p(j)}$ the computation halts rejecting (Transitions 7 and 8). The following transitions are used for the mth sweep, $2 \leq m \leq k$, where $d = 1$ if m is odd, and $d = -1$ if m is even.

4. $\delta(s_i, a_{m-1}) = (s_{(i+1)}, \sqcup_m, d)$,   for $p(j) \leq i \leq q(j) - 1$
5. $\delta(s_{q(j)}, a_{m-1}) = (s_{p(j)}, a_m, d)$
6. $\delta(s_i, \sqcup_{m-1}) = (s_i, \sqcup_m, d)$,   for $p(j) \leq i \leq q(j)$
7. $\delta(s_{p(j)}, \triangleright) = (s_{p(j)}, \triangleright, 1)$
8. $\delta(s_{p(j)}, \triangleleft) = (s_{p(j)}, \triangleleft, -1)$

After the kth sweep no further rewritings are possible. However, M continues with one more sweep for which the states $s_{p(j)}$ to $s_{q(j)}$ and s_+ are used, where s_+ just replaces $s_{p(j)}$ after the first cycle.

9. $\delta(s_i, a_k) = (s_{(i+1)}, a_k, d)$,   for $p(j) \leq i \leq q(j) - 1$
10. $\delta(s_{q(j)}, a_k) = (s_+, a_k, d)$
11. $\delta(s_i, \sqcup_k) = (s_i, \sqcup_k, d)$,   for $p(j) \leq i \leq q(j)$
12. $\delta(s_+, a_k) = (s_{p(j)+1}, a_k, d)$
13. $\delta(s_+, \sqcup_k) = (s_+, \sqcup_k, d)$

Finally, if M reaches the endmarker with state s_+, the input is accepted since the transition function is undefined for s_+ on endmarkers and $s_+ \in F$.

Now we turn to determining the language $L(M)$. Let $\ell = |w|$ be the length of the input. The first sweep is used to count ℓ modulo n. If the head arrives at the right endmarker in any state not in $\{s_{p(1)}, s_{p(2)}, \ldots, s_{p(l)}\}$, the computation halts and rejects. Let us assume the state is $s_{p(j)}$, for $1 \leq j \leq l$. Then we know $\ell = x_1 \cdot n + p(j)$, for some $x_1 \geq 0$.

For sweep $2 \leq m \leq k$, if the head arrives at the endmarker in any state not equal to $s_{p(j)}$, the computation halts and rejects. Otherwise, the number of non-blank cells have been divided by c_j and the number of non-blank cells found during the sweep is a multiple of c_j. So, we have $\ell = x_2 \cdot c_j^{m-1}$, for some $x_2 \geq 0$. If M accepts after a further sweep, it has checked once more whether the number of non-blank cells found during the sweep is a multiple of c_j. Therefore, we derive $\ell = x_3 \cdot c_j^k$, for some $x_3 \geq 0$. The further reasoning is as for k-head finite automata shown in [11]. We recall it for the sake of completeness.

Together we have that the length of the input has to meet the two properties $\ell = x_1 \cdot n + p(j)$ and $\ell = x_3 \cdot c_j^k$. Since n is prime and c_j is less than n, the numbers n and c_j are relatively prime. We conclude that n and c_j^k are relatively prime as well. So, there is a smallest x' so that $x' c_j^k$ is congruent 1 modulo n. We derive that there is a y' so that $x' c_j^k = y'n + 1$. This implies $p(j)x'c_j^k = p(j)y'n + p(j)$ and, thus, there is an ℓ having the properties mentioned above at all. By extending the length of the input by multiples of nc_j^k an infinite set of input lengths ℓ meeting the properties are derived. More precisely, given such an ℓ, the difference to the next input length longer than ℓ meeting the properties has to be a multiple of n and a multiple of c_j^k. Since both numbers are relatively prime, it has to be a multiple of nc_j^k. The language L_j consisting of all input lengths having these two properties is regular, and every 1DFA accepting unary L_j has a cycle of at least nc_j^k states.

The language $L(M)$ is the union of the languages L_j, $1 \le j \le l$. Since all c_j and n are pairwise relatively prime, all c_j^k and n are pairwise relatively prime. So, an immediate generalization of the proof of the state complexity for the union of two unary 1DFA languages [26] shows that every 1DFA accepting $L(M)$ has a cycle of at least $\operatorname{lcm}\{\, nc_j^k \mid 1 \le j \le l \,\} = n(c_1 c_2 \cdots c_l)^k = n \cdot F(n)^k$ states.   □

Now we turn to an upper bound that shows that removing the ability to rewrite each cell $k \ge 1$ times, but keeping the two-way head movement, costs only a polynomially number of states. From the resulting unary 2DFA an upper bound for one-way devices can be derived by the known bounds for removing the two-way head movement.

Theorem 5. *Let $k, n \ge 1$ be integers and M be a unary n-state sweeping k-DLA. Then $O(n^{\frac{k^2+3k+2}{2}})$ states are sufficient for a 2DFA to accept the language $L(M)$. The 2DFA can effectively be constructed from M.*

Example 1 provides the witness language $L = \{a^{n^{k+1}}\}$ for a lower bound. Since L is a unary singleton, every 2NFA, 2DFA, 1NFA, or 1DFA needs at least n^{k+1} states to accept it. Since the example shows that L is accepted by some $(n+2)$-state k-DLA the following lower bound follows.

Theorem 6. *Let $k \ge 1$ and $n \ge 2$ be integers. Then there is a unary sweeping $(n+2)$-state, $(2k+1)$-tape-symbol k-DLA M, so that n^{k+1} states are necessary for any 2NFA, 2DFA, 1NFA, or 1DFA to accept the language $L(M)$.*

The quadratic degree of the polynomial for the upper bound shown in Theorem 5 is essentially due to the fact that the non-unary tape contents after the first sweep cannot be recomputed. Instead, the computation has to be simulated by states that reflect the contents. The problem with the recomputation is caused by the alternating directions of the sweeps. So, a recomputation would require reversibility of the single sweeps. But in general these sweeps have an irreversible nature. Further restrictions of sweeping two-way automata studied in the literature are so-called rotating automata [8]. A *rotating* k-DLA is a sweeping k-DLA whose head is reset to the left endmarker every time the right endmarker is reached. So, the computation of a rotating machine can be seen as on a circular input with a marker between the last and first symbol. While every unary 2DFA can be made sweeping by adding one more state [9], and unary sweeping 2DFA can be made rotating without increasing the number of states, for unary 2DFA all these modes are almost the same. However, this is not true for limited automata. The next theorem shows that the simulation of rotating k-DLA by 2DFA is cheaper. Moreover, it will turn out that the upper and lower bounds are tight in the order of magnitude. The degree of the polynomials is the same.

Theorem 7. *Let $k, n \geq 1$ be integers and M be a unary n-state rotating k-DLA. Then $O(n^{k+1})$ states are sufficient for a (sweeping) 2DFA to accept the language $L(M)$. The 2DFA can effectively be constructed from M.*

The construction of the sweeping k-DLA accepting the singleton language $L = \{a^{n^{k+1}}\}$ given in Example 1 can easily be modified to the construction of an equivalent rotating k-DLA. So, we have the following lower bound that matches the upper bound in the order of magnitude.

Theorem 8. *Let $k \geq 1$ and $n \geq 2$ be integers. Then there is a unary rotating $(n+2)$-state, $(2k+1)$-tape-symbol k-DLA M, so that n^{k+1} states are necessary for any 2NFA, 2DFA, 1NFA, or 1DFA to accept the language $L(M)$.*

References

1. Blum, M., Hewitt, C.: Automata on a 2-dimensional tape. In: Symposium on Switching and Automata Theory (SWAT 1967), pp. 155–160. IEEE (1967)
2. Ellul, K.: Descriptional Complexity Measures of Regular Languages. Master's thesis. University of Waterloo, Ontario, Canada (2004)
3. Geffert, V., Mereghetti, C., Pighizzini, G.: Converting two-way nondeterministic unary automata into simpler automata. Theoret. Comput. Sci. **295**, 189–203 (2003)
4. Hartmanis, J.: Computational complexity of one-tape Turing machine computations. J. ACM **15**, 325–339 (1968)
5. Hennie, F.C.: One-tape, off-line Turing machine computations. Inform. Control **8**, 553–578 (1965)
6. Hibbard, T.N.: A generalization of context-free determinism. Inform. Control **11**, 196–238 (1967)
7. Holzer, M., Kutrib, M.: Unary language operations and their nondeterministic state complexity. In: Ito, M., Toyama, M. (eds.) DLT 2002. LNCS, vol. 2450, pp. 162–172. Springer, Heidelberg (2003)
8. Kapoutsis, C.A., Královic, R., Mömke, T.: Size complexity of rotating and sweeping automata. J. Comput. System Sci. **78**, 537–558 (2012)
9. Kunc, M., Okhotin, A.: On deterministic two-way finite automata over a unary alphabet. Technical report. 950, Turku Centre for Computer Science (2011)
10. Kunc, M., Okhotin, A.: State complexity of operations on two-way finite automata over a unary alphabet. Theoret. Comput. Sci. **449**, 106–118 (2012)
11. Kutrib, M., Malcher, A., Wendlandt, M.: Simulations of unary one-way multi-head finite automata. Int. J. Found. Comput. Sci. **25**, 877–896 (2014)
12. Landau, E.: Über die Maximalordnung der Permutationen gegebenen Grades. Archiv der Math. und Phys. **3**, 92–103 (1903)
13. Landau, E.: Handbuch der Lehre von der Verteilung der Primzahlen. Teubner, Leipzig (1909)
14. Mera, F., Pighizzini, G.: Complementing unary nondeterministic automata. Theoret. Comput. Sci. **330**, 349–360 (2005)
15. Mereghetti, C., Pighizzini, G.: Optimal simulations between unary automata. SIAM J. Comput. **30**, 1976–1992 (2001)
16. Meyer, A.R., Fischer, M.J.: Economy of description by automata, grammars, and formal systems. In: Symposium on Switching and Automata Theory (SWAT 1971), pp. 188–191. IEEE (1971)

17. Nicolas, J.L.: Sur l'ordre maximum d'un élément dans le groupe S_n des permutations. Acta Arith. **14**, 315–332 (1968)
18. Pighizzini, G.: Deterministic pushdown automata and unary languages. Int. J. Found. Comput. Sci. **20**, 629–645 (2009)
19. Pighizzini, G.: Strongly limited automata. In: Non-Classical Models of Automata and Applications (NCMA 2014). books@ocg.at, vol. 304, pp. 191–206. Austrian Computer Society, Vienna (2014)
20. Pighizzini, G., Pisoni, A.: Limited automata and regular languages. Int. J. Found. Comput. Sci. **25**, 897–916 (2014)
21. Pighizzini, G., Pisoni, A.: Limited automata and context-free languages. Fund. Inform. **136**, 157–176 (2015)
22. Pighizzini, G., Shallit, J.: Unary language operations, state complexity and Jacobsthal's function. Int. J. Found. Comput. Sci. **13**, 145–159 (2002)
23. Pighizzini, G., Shallit, J., Wang, M.W.: Unary context-free grammars and pushdown automata, descriptional complexity and auxiliary space lower bounds. J. Comput. System Sci. **65**, 393–414 (2002)
24. Průša, D.: Weight-reducing hennie machines and their descriptional complexity. In: Dediu, A.-H., Martín-Vide, C., Sierra-Rodríguez, J.-L., Truthe, B. (eds.) LATA 2014. LNCS, vol. 8370, pp. 553–564. Springer, Heidelberg (2014)
25. Szalay, M.: On the maximal order in S_n and S_n^*. Acta Arithm. **37**, 321–331 (1980)
26. Yu, S.: State complexity of regular languages. J. Autom., Lang. Comb. **6**, 221–234 (2001)

On Some Decision Problems for Stateless Deterministic Ordered Restarting Automata

Kent Kwee and Friedrich Otto[✉]

Fachbereich Elektrotechnik/Informatik, Universität Kassel, 34109 Kassel, Germany
{kwee,otto}@theory.informatik.uni-kassel.de

Abstract. The stateless deterministic ordered restarting automata accept exactly the regular languages, and it is known that the trade-off for turning a stateless deterministic ordered restarting automaton into an equivalent DFA is at least double exponential. Here we show that the trade-off for turning a stateless deterministic ordered restarting automaton into an equivalent unambiguous NFA is exponential, which yields an upper bound of $2^{2^{O(n)}}$ for the conversion into an equivalent DFA, thus meeting the lower bound up to a constant. Based on the new transformation we then show that many decision problems, such as emptiness, finiteness, inclusion, and equivalence, are PSPACE-complete for stateless deterministic ordered restarting automata.

Keywords: Restarting automaton · Ordered rewriting · Descriptional complexity · Decision problem

1 Introduction

The *deterministic ordered restarting automaton* (or *det-ORWW-automaton*) was introduced in [9] in the setting of picture languages. While the nondeterministic variant of this type of automaton even accepts some languages that are not context-free, it has been shown in [9] that the deterministic variant accepts exactly the regular languages.

In [10] an investigation of the descriptional complexity of the det-ORWW-automaton was initiated. It was shown that each det-ORWW-automaton can be simulated by an automaton of the same type that has only a single state, which means that for these automata, states are actually not needed. Accordingly, such an automaton is called a *stateless* det-ORWW-automaton (stl-det-ORWW-automaton). For these automata, the size of their working alphabets can be taken as a measure for their descriptional complexity, and it has been shown that these automata are polynomially related in size to the weight-reducing Hennie machines studied by Průša in [12]. Actually, for $n \geq 1$, there exists a regular language that is accepted by a stl-det-ORWW-automaton of size $O(n)$ such that each DFA for this language has size at least 2^{2^n}. On the other hand, each stl-det-ORWW-automaton of size n can be simulated by a DFA of size $2^{2^{O(n^2 \cdot \log n)}}$. Thus, there is a huge gap between the upper and lower bounds.

© Springer International Publishing Switzerland 2015
J. Shallit and A. Okhotin (Eds.): DCFS 2015, LNCS 9118, pp. 165–176, 2015.
DOI: 10.1007/978-3-319-19225-3_14

Here we present a new construction that, for a stl-det-ORWW-automaton of size n, yields an equivalent unambiguous NFA of size $2^{O(n)}$, which implies that there is an equivalent DFA of size $2^{2^{O(n)}}$. Actually, we will show that these bounds are sharp (up to the O-notation). We then exploit our construction to establish that many basic decision problems, like emptiness, universality, finiteness, inclusion, and equivalence, are PSPACE-complete for stl-det-ORWW-automata. In addition, we consider the problem of deciding, given a stl-det-ORWW-automaton, whether the language accepted belongs to a certain subclass of the regular languages. For the subclasses of strictly locally k-testable languages ($k \geq 1$), nilpotent languages, combinatorial languages, and some others, we obtain that the corresponding decision problems are PSPACE-complete, too.

This paper is structured as follows. In Sect. 2, we introduce the stl-det-ORWW-automata, and we restate the main results on them from [10]. Then, in Sect. 3, we present the announced construction of an NFA from a given stl-det-ORWW-automaton, and in Sect. 4 we consider the decision problems mentioned above. The paper closes with Sect. 5, which summarizes our results briefly and states a number of open problems for future work.

2   Stateless Deterministic Ordered Restarting Automata

A *stateless deterministic ordered restarting automaton* (stl-det-ORWW-automaton) is a one-tape machine that is described by a 6-tuple $M = (\Sigma, \Gamma, \rhd, \lhd, \delta, >)$, where Σ is a finite input alphabet, Γ is a finite tape alphabet such that $\Sigma \subseteq \Gamma$, the symbols $\rhd, \lhd \notin \Gamma$ serve as markers for the left and right border of the work space, respectively,

$$\delta : (((\Gamma \cup \{\rhd\}) \cdot \Gamma \cdot (\Gamma \cup \{\lhd\})) \cup \{\rhd\lhd\}) \dashrightarrow \{\mathsf{MVR}\} \cup \Gamma \cup \{\mathsf{Accept}\}$$

is the (partial) *transition function*, and $>$ is a *partial ordering* on Γ. The transition function describes three different types of transition steps:

(1) A *move-right step* has the form $\delta(a_1 a_2 a_3) = \mathsf{MVR}$, where $a_1 \in \Gamma \cup \{\rhd\}$ and $a_2, a_3 \in \Gamma$. It causes M to shift the window one position to the right. Observe that no move-right step is possible, if the window contains the symbol $\lhd$.

(2) A *rewrite/restart step* has the form $\delta(a_1 a_2 a_3) = b$, where $a_1 \in \Gamma \cup \{\rhd\}$, $a_2, b \in \Gamma$, and $a_3 \in \Gamma \cup \{\lhd\}$ such that $a_2 > b$ holds. It causes M to replace the symbol a_2 in the middle of its window by the symbol b and to restart.

(3) An *accept step* has the form $\delta(a_1 a_2 a_3) = \mathsf{Accept}$, where $a_1 \in \Gamma \cup \{\rhd\}$, $a_2 \in \Gamma$, and $a_3 \in \Gamma \cup \{\lhd\}$. It causes M to halt and accept. In addition, we allow an accept step of the form $\delta(\rhd\lhd) = \mathsf{Accept}$.

If $\delta(u)$ is undefined for some word u, then M necessarily halts, when it sees u in its window, and we say that M *rejects* in this situation. Further, the letters in $\Gamma \smallsetminus \Sigma$ are called *auxiliary symbols*.

A *configuration* of a stl-det-ORWW-automaton M is a pair of words (α, β), where $|\beta| \geq 3$, and either $\alpha = \lambda$ (the empty word) and $\beta \in \{\rhd\} \cdot \Gamma^+ \cdot \{\lhd\}$ or

$\alpha \in \{\triangleright\} \cdot \Gamma^*$ and $\beta \in \Gamma \cdot \Gamma^+ \cdot \{\triangleleft\}$; here $\alpha\beta$ is the current content of the tape, and it is understood that the window contains the first three symbols of β. In addition, we admit the configuration $(\lambda, \triangleright\triangleleft)$. A *restarting configuration* has the form $(\lambda, \triangleright w \triangleleft)$; if $w \in \Sigma^*$, then $(\lambda, \triangleright w \triangleleft)$ is also called an *initial configuration*. Furthermore, we use **Accept** to denote the *accepting configurations*, which are those configurations that M reaches by an accept step. We let $\vdash_M$ denote the *single-step computation relation* that M induces on the set of configurations, and the *computation relation* $\vdash_M^*$ of M is the reflexive and transitive closure of $\vdash_M$.

Any computation of a stl-det-ORWW-automaton M consists of certain phases. A phase, called a *cycle*, starts in a restarting configuration, the head is moved along the tape by MVR steps until a rewrite/restart step is performed and thus, a new restarting configuration is reached. If no further rewrite operation is performed, any computation necessarily finishes in a halting configuration – such a phase is called a *tail*. By $\vdash_M^c$ we denote the execution of a complete cycle, and $\vdash_M^{c*}$ is the reflexive transitive closure of this relation. It can be seen as the rewrite relation that M induces on its set of restarting configurations.

An input $w \in \Sigma^*$ is accepted by M if the computation of M which starts with the initial configuration $(\lambda, \triangleright w \triangleleft)$ ends with an accept step. The language consisting of all input words that are accepted by M is denoted by $L(M)$.

As each cycle ends with a rewrite operation, which replaces a symbol a by a symbol b that is strictly smaller than a with respect to the given ordering $>$, we see that each computation of M on an input of length n consists of at most $(|\Gamma| - 1) \cdot n$ cycles and a tail. Thus, M can be simulated by a deterministic single-tape Turing machine in time $O(n^2)$. The following example illustrates the way in which a stl-det-ORWW-automaton works.

Example 1. Let $n \geq 2$ be a fixed integer, and let $M = (\Sigma, \Gamma, \triangleright, \triangleleft, \delta, >)$ be defined by taking $\Sigma = \{a, b\}$ and $\Gamma = \Sigma \cup \{a_i, b_i, x_i \mid 1 \leq i \leq n - 1\}$, by choosing the ordering $>$ such that $a > a_i > x_j$ and $b > b_i > x_j$ hold for all $1 \leq i, j \leq n - 1$, and by defining the transition function δ in such a way that M proceeds as follows: on input $w = w_1 w_2 \cdots w_m$, $w_1, \ldots, w_m \in \Sigma$, M numbers the first $n - 1$ letters of w from left to right, by replacing $w_i = a$ (b) by a_i (b_i) for $i = 1, \ldots, n - 1$. If $w_n \neq a$, then the computation fails, but if $w_n = a$, then M continues by replacing the last $n - 1$ letters of w from right to left using the letters x_1 to x_{n-1}. If the n-th last letter is b or some b_i, then M accepts, otherwise the computation fails again.

Then $L(M) = \{ w \in \{a, b\}^m \mid m > n, w_n = a, \text{ and } w_{m+1-n} = b \}$. As shown in [6], every det-RR(1)-automaton for $L(M)$ has at least $O(2^n)$ states. Here a det-RR(1)-automaton is another type of deterministic restarting automaton that characterizes the regular languages (see [7]).

While nondeterministic ORWW-automata are quite expressive, the deterministic variants are fairly weak. Taking the size of the tape alphabet as the measure for the descriptional complexity of a stl-det-ORWW-automaton, the following results are shown in [10].

Theorem 2

(a) *For each DFA $A = (Q, \Sigma, q_0, F, \varphi)$, there is a stl-det-ORWW-automaton $M = (\Sigma, \Gamma, \triangleright, \triangleleft, \delta, >)$ such that $L(M) = L(A)$ and $|\Gamma| = |Q| + |\Sigma|$.*

(b) *For each stl-det-ORWW-automaton M with an alphabet of size n, there exists a DFA A of size $2^{2^{O(n^2 \log n)}}$ such that $L(A) = L(M)$ holds.*

(c) *For each $n \geq 1$, there exists a regular language $B_n \subseteq \{0, 1, \$\}^*$ such that B_n is accepted by a stl-det-ORWW-automaton over an alphabet of size $O(n)$, but each DFA for accepting B_n has at least 2^{2^n} states.*

Thus, there is a double exponential trade-off for converting a stl-det-ORWW-automaton into a DFA. Observe, however, that the gap between the lower and upper bounds is still huge.

3 Simulating a stl-det-ORWW-automaton by an NFA

Here we present our main result, which consists in the construction of an unambiguous NFA A of size $2^{O(n)}$ from a stl-det-ORWW-automaton M of size n such that A accepts the same language as M. In order to simplify this construction, we require that M only accepts on reaching the right sentinel $\triangleleft$. This is not a restriction, as shown by the following lemma.

Lemma 3. *From a stl-det-ORWW-automaton $M = (\Sigma, \Gamma, \triangleright, \triangleleft, \delta, >)$, one can construct a stl-det-ORWW-automaton $M' = (\Sigma, \Delta, \triangleright, \triangleleft, \delta', >)$ such that $L(M') = L(M)$, $|\Delta| \leq |\Gamma| + 1$, and M' only accepts when its window contains the right sentinel $\triangleleft$.*

To motivate our main construction we consider an example.

Example 4. Let M be a stl-det-ORWW-automaton on the input alphabet $\Sigma = \{a_1, a_2, a_3, a_4, a_5\}$ and the working alphabet $\Gamma = \Sigma \cup \{b_1, b_2, b_3, b_4, c_1, c_2, c_3, c_4\}$ with the ordering $a_i > b_i > c_i$ for all $1 \leq i \leq 4$, and let the transition function be given by the following table:

$$\delta(\triangleright a_1 a_2) = b_1, \quad \delta(\triangleright b_1 a_2) = \text{MVR}, \quad \delta(b_1 a_2 a_3) = \text{MVR}, \quad \delta(a_2 a_3 a_4) = b_3,$$
$$\delta(b_1 a_2 b_3) = b_2, \quad \delta(c_2 c_3 a_4) = \text{MVR}, \quad \delta(\triangleright c_1 b_2) = \text{MVR}, \quad \delta(c_1 b_2 b_3) = \text{MVR},$$
$$\delta(b_2 b_3 a_4) = c_3, \quad \delta(c_2 c_3 c_4) = \text{MVR}, \quad \delta(\triangleright c_1 c_2) = \text{MVR}, \quad \delta(c_1 c_2 c_3) = \text{MVR},$$
$$\delta(c_1 b_2 c_3) = c_2, \quad \delta(c_3 a_4 a_5) = b_4, \quad \delta(c_2 c_3 b_4) = \text{MVR}, \quad \delta(c_3 b_4 a_5) = c_4,$$
$$\delta(\triangleright b_1 b_2) = c_1, \quad \delta(c_3 c_4 a_5) = \text{MVR}, \quad \delta(c_4 a_5 \triangleleft) = \text{Accept.}$$

Given the word $w = a_1 a_2 a_3 a_4 a_5$ as input, M executes the following accepting computation, where the rewritten letters are underlined:

$$(\lambda, \triangleright \underline{a_1} a_2 a_3 a_4 a_5 \triangleleft) \vdash_M^c (\lambda, \triangleright b_1 a_2 \underline{a_3} a_4 a_5 \triangleleft) \vdash_M^c (\lambda, \triangleright b_1 a_2 \underline{b_3} a_4 a_5 \triangleleft) \vdash_M^c$$
$$(\lambda, \triangleright \underline{b_1} b_2 b_3 a_4 a_5 \triangleleft) \vdash_M^c (\lambda, \triangleright c_1 b_2 \underline{b_3} a_4 a_5 \triangleleft) \vdash_M^c (\lambda, \triangleright c_1 \underline{b_2} c_3 a_4 a_5 \triangleleft) \vdash_M^c$$
$$(\lambda, \triangleright c_1 c_2 c_3 \underline{a_4} a_5 \triangleleft) \vdash_M^c (\lambda, \triangleright c_1 c_2 c_3 \underline{b_4} a_5 \triangleleft) \vdash_M^c (\lambda, \triangleright c_1 c_2 c_3 c_4 a_5 \triangleleft) \vdash_M^* \text{Accept.}$$

To encode this computation in a compact way, we introduce a 3-tuple of vectors $T = (L, W, R)$ for each position on the tape of M, where

- W is a sequence of letters $W = (x_1, x_2, \ldots, x_r)$ over Γ such that $x_1 > x_2 > \cdots > x_r$ using the ordering on Γ defined by M,
- L is a sequence of indices $L = (i_1, \ldots, i_{r-1})$ such that $i_1 \leq \cdots \leq i_{r-1} \leq |\Gamma|$,
- R is a sequence of indices $R = (j_1, \ldots, j_{r-1})$ such that $j_1 \leq \cdots \leq j_{r-1} \leq |\Gamma|$.

The idea is that W encodes the sequence of letters that are produced by M in an accepting computation for a particular field, and L and R encode the information on the neighbouring letters to the left and to the right that are used to perform the corresponding rewrite operations. For the computation above we obtain the following sequence of triples, where Λ denotes an empty sequence:

L_0 W_0 R_0	L_1 W_1 R_1	L_2 W_2 R_2	L_3 W_3 R_3	L_4 W_4 R_4	L_5 W_5 R_5	L_6 W_6 R_6
Λ $\triangleright$ Λ	1 a_1 1	2 a_2 2	1 a_3 1	3 a_4 1	Λ a_5 Λ	Λ $\triangleleft$ Λ
	1 b_1 2	3 b_2 3	2 b_3 1	3 b_4 1		
	c_1	c_2	c_3	c_4		

For example, the triple $(2, b_3, 1) \in (L_3, W_3, R_3)$ means that b_3 is rewritten into c_3, while the left neighbouring field contains the second letter of its sequence W_2, and the right neighbouring field contains the first letter of its sequence W_4.

If a letter is not rewritten at all, like a_5, then the corresponding sequences L and R are empty. In fact, there is a consistency condition that must be met by the sequences R_{i-1} and L_i for each index i, as the rewrites at positions $i-1$ and i are executed in some order, and this order is encoded in these sequences. For example, $L_3 = (1, 2)$, which means that a_3 is rewritten into b_3, while tape field 2 still contains the original letter a_2, and b_3 is rewritten into c_3, while tape field 2 contains the next letter b_2. Thus, before the second rewrite at position 3 can occur, the letter a_2 at position 2 has been rewritten into b_2, which is expressed by the fact that $R_2 = (2, 3)$ starts with the number 2. Finally, the second number in R_2 states that b_2 is rewritten into c_2 only after the second rewrite at position 3 has been performed. Hence, $R_2 = (2, 3)$ and $L_3 = (1, 2)$ lead to the sequence of rewrite steps $(1 : a_3 \to b_3), (2 : a_2 \to b_2), (3 : b_3 \to c_3), (4 : b_2 \to c_2)$. $\square$

To formalize the notion of *compatibility* of two finite non-decreasing sequences of integers $R = (r_1, \ldots, r_k)$ and $L = (\ell_1, \ldots, \ell_s)$, where $k, s \geq 0$, we define a multiset $\mathrm{order}(R, L)$ as follows:

$$\mathrm{order}(R, L) = \{\, r_i + i - 1 \mid i = 1, \ldots, k \,\} \cup \{\, \ell_j + j - 1 \mid j = 1, \ldots, s \,\}.$$

Now the pair of sequences (R, L) is called *consistent*, if $\mathrm{order}(R, L) = \{1, 2, \ldots, k+s\}$, that is, it is the integer interval $[1, k+s]$. In the example above we obtain $\mathrm{order}(R_2, L_3) = \mathrm{order}((2, 3), (1, 2)) = \{2, 4, 1, 3\} = \{1, 2, 3, 4\}$, thus we assign a number between 1 and $4 = |R_2| + |L_3|$ to each of the rewrites at positions $i-1$ and i, in this way specifying the order in which these rewrites must be executed.

Based on the above ideas, we will now establish the following general result.

Theorem 5. Let $M = (\Sigma, \Gamma, \triangleright, \triangleleft, \delta_M, >)$ be a stl-det-ORWW-automaton. Then an unambiguous NFA $A = (Q, \Sigma, \Delta_A, q_0, F)$ can be constructed from M such that $L(A) = L(M)$ and $|Q| \in 2^{O(|\Gamma|)}$.

Proof. Let $M = (\Sigma, \Gamma, \rhd, \lhd, \delta_M, >)$ be a stl-det-ORWW-automaton. At the extra cost of at most one additional tape symbol, we can assume by Lemma 3 that M executes an accept step only when its window contains the right sentinel $\lhd$. Let $n = |\Gamma|$. As a first step we construct an NFA B for the *characteristic language* $L_C(M) = \{ w \in \Gamma^* \mid (\lambda, \rhd w \lhd) \vdash_M^* \text{Accept} \}$ of M, which consists of all words over Γ that M accepts.

The NFA $B = (Q, \Gamma, \Delta_B, q_0, F)$ is constructed as follows:

- The set Q contains the initial state q_0, a designated final state q_F, and all pairs of triples of the form $((L_1, W_1, R_1), (L_2, W_2, R_2))$, where, for $i = 1, 2$,

  - W_i is a sequence of letters $W_i = (w_{i,1}, \ldots, w_{i,k_i})$ from Γ of length $1 \leq k_i \leq n$ such that $w_{i,1} > w_{i,2} > \cdots > w_{i,k_i}$, or $W_i = (\rhd)$ and $k_i = 1$,
  - L_i is a sequence of positive integers $L_i = (l_{i,1}, \ldots, l_{i,k_i-1})$ of length $k_i - 1$ such that $l_{i,1} \leq l_{i,2} \leq \cdots \leq l_{i,k_i-1} \leq n$,
  - R_i is a sequence of positive integers $R_i = (r_{i,1}, \ldots, r_{i,k_i-1})$ of length $k_i - 1$ such that $r_{i,1} \leq r_{i,2} \leq \cdots \leq r_{i,k_i-1} \leq n$,
  - the sequences R_1 and L_2 are consistent, that is, $\text{order}(R_1, L_2) = \{1, 2, \ldots, k_1 + k_2 - 2\}$.

The transition relation Δ_B is given through the following rules, where $x \in \Gamma$ and $((L_{i-1}, W_{i-1}, R_{i-1}), (L_i, W_i, R_i))$, $i = 2, 3$, are states from Q:

- $\Delta_B(q_0, \lambda) \ni q_F$, if $\delta_M(\rhd \lhd) = \text{Accept}$.

- $\Delta_B(q_0, x) \ni ((\Lambda, (\rhd), \Lambda), (L_1, W_1, R_1))$, if $x = w_{1,1}$.

- $\Delta_B(((L_1, W_1, R_1), (L_2, W_2, R_2)), x) \ni ((L_2, W_2, R_2), (L_3, W_3, R_3))$, if
  1. $x = w_{3,1}$,
  2. $\forall 1 \leq j \leq k_2 - 1 : \delta_M\left(w_{1,l_{2,j}} w_{2,j} w_{3,r_{2,j}}\right) = w_{2,j+1}$,
  3. $\forall 1 \leq j \leq k_3 - 1 : \delta_M\left(w_{1,l_{3,j}} w_{2,l_{3,j}} w_{3,j}\right) = \text{MVR}$, where $l_{2,k_2} = k_1$ is taken, and
  4. $\delta_M\left(w_{1,k_1} w_{2,k_2} w_{3,k_3}\right) = \text{MVR}$.

- $\Delta_B(((L_1, W_1, R_1), (L_2, W_2, R_2)), \lambda) \ni q_F$, if
  1. R_2 is a sequence of 1's of length $k_2 - 1$,
  2. $\delta_M(w_{1,k_1} w_{2,k_2} \lhd) = \text{Accept}$, and
  3. $\forall 1 \leq j \leq k_2 - 1 : \delta_M\left(w_{1,l_{2,j}} w_{2,j} \lhd\right) = w_{2,j+1}$.

We will prove that $L(B) = L_C(M)$ holds.

Claim 1. $L_C(M) \subseteq L(B)$.

Proof. Let $w \in \Gamma^*$ be a word that belongs to the language $L_C(M)$. Thus, the computation of M that starts with the restarting configuration $(\lambda, \rhd w \lhd)$ is accepting. If $w = \lambda$, then $\delta_M(\rhd \lhd) = \text{Accept}$, which implies that $q_F \in \Delta_B(q_0, \lambda)$. It follows that $w \in L(B)$ holds in this case.

Now assume that $w = w_1 w_2 \cdots w_n$ for some $n \geq 1$ and letters $w_1, \ldots, w_n \in \Gamma$. As $w \in L_C(M)$, we can now use the accepting computation of M for w to construct a representation as in the example above. This representation translates into a sequence of states of B, and it can be shown that this sequence of states yields an accepting computation of B for the input w.   □

Claim 2. $L(B) \subseteq L_C(M)$.

Proof. We have to check that we can deduct a valid computation of M from an accepting computation of B. So let $w \in \Gamma^*$ be any word in $L(B)$, and let $n = |w|$. If $w = \lambda$, then $q_F \in \delta_B(q_0, \lambda)$, which implies that $\delta_M(\triangleright \triangleleft) = \mathsf{Accept}$ holds, which in turn means that $w \in L_C(M)$.

If $w = w_1 \in \Gamma$, then there exist sequences $W_1 = (w_{1,1}, \ldots, w_{1,k_1})$ over Γ and $L_1 = (l_{1,1}, \ldots, l_{1,k_1-1})$ and $R_1 = (r_{1,1}, \ldots, r_{1,k_1-1})$ over $\mathbb{N}$ such that

- $w_{1,1} = w_1$,
- $((\Lambda, (\triangleright), \Lambda), (L_1, W_1, R_1)) \in \Delta_B(q_0, w_1)$, and
- $q_F \in \Delta_B(((\Lambda, (\triangleright), \Lambda), (L_1, W_1, R_1)), \lambda)$.

From the definition of Δ_B it follows that either $k_1 = 1$, and then $\mathsf{Accept} \in \delta_M(\triangleright w_1 \triangleleft)$, or $k_1 > 1$, and then $l_{1,j} = 1 = r_{1,j}$ for all $j = 1, \ldots, k_1 - 1$, $w_{1,j+1} \in \delta_M(\triangleright w_{1,j} \triangleleft)$ for all $j = 1, \ldots, k_1 - 1$, and $\mathsf{Accept} \in \delta_M(\triangleright w_{1,k_1} \triangleleft)$. Hence, we see that the computation of M that begins with the restarting configuration $(\lambda, \triangleright w \triangleleft)$ accepts, that is, $w = w_1 \in L_C(M)$.

Now assume that $w = w_1 \cdots w_n$ for some $n \geq 2$ and letters $w_1, \ldots, w_n \in \Gamma$. As B accepts on input w, there exist sequences $W_i = (w_{i,1}, \ldots, w_{i,k_i})$ over Γ and sequences of integers $L_i = (l_{i,1}, \ldots, l_{i,k_i-1})$ and $R_i = (r_{i,1}, \ldots, r_{i,k_i-1})$, $i = 1, \ldots, n$, such that all of the following conditions are met:

1. $((\Lambda, (\triangleright), \Lambda), (L_1, W_1, R_1)) \in \Delta_B(q_0, w_1)$,
2. $((L_{i-1}, W_{i-1}, R_{i-1}), (L_i, W_i, R_i)) \in$
   $\Delta_B((L_{i-2}, W_{i-2}, R_{i-2}), (L_{i-1}, W_{i-1}, R_{i-1})), w_i)$ for all $i = 2, \ldots, n$,
3. $q_F \in \Delta_B((L_{n-1}, W_{n-1}, R_{n-1}), (L_n, W_n, R_n)), \lambda)$.

From the definition of Δ_B we see that, for all $i = 1, \ldots, n$, $k_i \geq 1$ and $w_{i,1} = w_i$. Now let $N = N(R_1, \ldots, R_n) = \sum_{i=1}^n |R_i| = \sum_{i=1}^n (k_i - 1)$. By induction on N we will prove the following technical statement.

Claim 2.1. The computation of M that begins with the restarting configuration $(\lambda, \triangleright w \triangleleft)$ consists of N cycles and an accepting tail, that is, it has the form

$$(\lambda, \triangleright w \triangleleft) \vdash_M^c (\lambda, \triangleright z^{(1)} \triangleleft) \vdash_M^c \cdots \vdash_M^c (\lambda, \triangleright z^{(N)} \triangleleft) \vdash_M^* (\triangleright u, v \triangleleft) \vdash_M \mathsf{Accept},$$

where $z^{(N)} = uv$ and $|v| = 2$.

Proof. If $N = 0$, then $k_i = 1$ for all $i = 1, \ldots, n$, and hence, $W_i = (w_i)$ and $L_i = R_i = \Lambda$ for all $i = 1, \ldots, n$. From the definition of Δ_B it follows that $\delta_M(w_{i-2} w_{i-1} w_i) = \mathsf{MVR}$ for all $i = 2, \ldots, n$, where $w_0 = \triangleright$ is taken, and $\delta_M(w_{n-1} w_n \triangleleft) = \mathsf{Accept}$. Thus, the computation of M that begins with the restarting configuration $(\lambda, \triangleright w \triangleleft)$ is simply an accepting tail computation.

Now assume that $N \geq 1$. Then $k_i > 1$ for some indices $i \in \{1, \ldots, n\}$, and accordingly, the corresponding sequences L_i and R_i are non-empty. Because of the consistency of the pairs (R_{i-1}, L_i), $i = 1, \ldots, n$, there exists an index $j \in \{1, \ldots, n\}$ such that $l_{j,1} = 1 = r_{j,1}$. Let $s \in \{1, \ldots, n\}$ be the minimal index such that $l_{s,1} = 1 = r_{s,1}$ holds. It follows that $k_s > 1$ and that $W_s = (w_{s,1}, w_{s,2}, \ldots, w_{s,k_s})$, where $w_s = w_{s,1} > w_{s,2}$. Let $\hat{w}$ denote the word $\hat{w} = w_1 \cdots w_{s-1} w_{s,2} w_{s+1} \cdots w_n \in \Gamma^n$. For this word the following result can be shown.

Claim 2.1.1. $(\lambda, \triangleright w \triangleleft) \vdash^c_M (\lambda, \triangleright \hat{w} \triangleleft)$.

We continue with the proof of Claim 2.1 by establishing the following claim, which will allow us to perform the intended inductive step.

Claim 2.1.2. The word $\hat{w}$ is accepted by the NFA B.

Proof. For all $i = 1, \ldots, n$, we define sequences $\hat{W}_i$ over Γ and sequences of integers $\hat{L}_i$ and $\hat{R}_i$ as follows:

$$\hat{W}_i = \begin{cases} (w_{i,2}, \ldots, w_{i,k_i}), & \text{if } i = s, \\ W_i, & \text{otherwise}; \end{cases}$$

$$\hat{L}_i = \begin{cases} (l_{i,2}, \ldots, l_{i,k_i-1}), & \text{if } i = s, \\ (l_{i,1} - 1, \ldots, l_{i,k_i-1} - 1), & \text{if } i = s + 1, \\ L_i, & \text{otherwise}; \end{cases}$$

$$\hat{R}_i = \begin{cases} (r_{i,2}, \ldots, r_{i,k_i-1}), & \text{if } i = s, \\ (r_{i,1} - 1, \ldots, r_{i,k_i-1} - 1), & \text{if } i = s - 1, \\ R_i, & \text{otherwise}, \end{cases}$$

and we take $\hat{k}_i$ to denote the length of the sequence $\hat{W}_i$, $i = 1, \ldots, n$. Then $\hat{k}_s = k_s - 1$, and $\hat{k}_i = k_i$ for all $i \neq s$. In order to unify the notation we write $\hat{w} = w_1 \cdots w_{s-1} w_{s,2} w_{s+1} \cdots w_n$ as $\hat{w} = \hat{w}_1 \cdots \hat{w}_n$. Also we write $\hat{W}_i$ as $\hat{W}_i = (\hat{w}_{i,1}, \ldots, \hat{w}_{i,\hat{k}_i})$, and $\hat{L}_i$ and $\hat{R}_i$ as $\hat{L}_i = (\hat{l}_{i,1}, \ldots, \hat{l}_{i,\hat{k}_i-1})$ and $\hat{R}_i = (\hat{r}_{i,1}, \ldots, \hat{r}_{i,\hat{k}_i-1})$, $i = 1, \ldots, n$. It can now be shown that the above sequences satisfy all of the following conditions::

1. $((\Lambda, (\triangleright), \Lambda), (\hat{L}_1, \hat{W}_1, \hat{R}_1)) \in \Delta_B(q_0, \hat{w}_1)$,
2. $((\hat{L}_{i-1}, \hat{W}_{i-1}, \hat{R}_{i-1}), (\hat{L}_i, \hat{W}_i, \hat{R}_i)) \in$
   $\Delta_B((\hat{L}_{i-2}, \hat{W}_{i-2}, \hat{R}_{i-2}), (\hat{L}_{i-1}, \hat{W}_{i-1}, \hat{R}_{i-1})), \hat{w}_i)$ for all $i = 2, \ldots, n$,
3. $q_F \in \Delta_B((\hat{L}_{n-1}, \hat{W}_{n-1}, \hat{R}_{n-1}), (\hat{L}_n, \hat{W}_n, \hat{R}_n)), \lambda)$.

It follows that the word $\hat{w}$ is accepted by B using the sequence of states defined above. As

$$N(\hat{R}_1, \ldots, \hat{R}_n) = \sum_{i=1}^{n} (\hat{k}_i - 1) = \sum_{i=1}^{n} (k_i - 1) - 1 = N(R_1, \ldots, R_n) - 1 = N - 1,$$

we can apply our induction hypothesis, which implies that the computation of M that begins with the restarting configuration $(\lambda, \triangleright \hat{w} \triangleleft)$ consists of $N - 1$ cycles and an accepting tail. Together with Claim 2.1.1 this says that the computation of M that begins with the restarting configuration $(\lambda, \triangleright w \triangleleft)$ consists of N cycles and an accepting tail, which completes the proof of Claim 2.1.   □

From the claims above we obtain that $L(B) = L_C(M)$ holds. As M is deterministic, there is only a single accepting computation of B for each word $w \in L_C(M)$. It follows that B is unambiguous.   □

Claim 3. $|Q| \in 2^{O(|\Gamma|)}$.

Proof. The set Q of states of B contains the two designated states q_0 and q_F and certain states that consist of pairs of triples of the form (L, W, R), where W is a sequence of letters $W = (a_1, \ldots, a_m)$ from Γ such that $a_1 > \cdots > a_m$, and L and R are sequences of integers $L = (l_1, \ldots, l_{m-1})$ and $R = (r_1, \ldots, r_{m-1})$ such that $1 \le l_1 \le \cdots \le l_{m-1}$ and $1 \le r_1 \le \cdots \le r_{m-1}$. From upper bounds for the number of these sequences we will obtain an upper bound for the size of Q.

From the condition on the sequence W we see that $m \le n = |\Gamma|$, and also $l_{m-1} \le n$ and $r_{m-1} \le n$. The sequence W defines the subset $\{w_1, \ldots, w_m\}$ of Γ, and different sequences W and W' yield different subsets. Hence, the number $2^n - 1$ of non-empty subsets of Γ is an upper bound for the number of different subsequences W.

The sequence L can be interpreted as a multiset over the set of integers $\{1, \ldots, n\}$, because it can contain repetitions. This multiset is of size at most $n-1$ (counting elements with their multiplicities). There are $\binom{n+r-1}{r}$ such multisets of size r (see, e.g., [14]), and hence, the number of possible sequences L is bounded from above by the expression

$$\sum_{r=0}^{n-1} \binom{n+r-1}{r} \le \sum_{r=0}^{n-1} \binom{2n}{r} \le \sum_{r=0}^{2n} \binom{2n}{r} = 2^{2n},$$

and the same is true for the number of possible sequences R. Hence, there are at most $2^{2n} \cdot 2^n \cdot 2^{2n} = 2^{5n}$ different triples of the form (L, W, R), and so the number of states of B is bounded from above by the number 2^{10n}.   □

It follows that B is of size $2^{O(n)}$. From B we now obtain an NFA A for the language $L(M) = L_C(M) \cap \Sigma^*$ by simply deleting all transitions from Δ_B that read a letter $x \in (\Gamma \smallsetminus \Sigma)$. Then it is immediate that A is an unambiguous NFA of size $2^{O(n)}$ that accepts the language $L(A) = L(B) \cap \Sigma^* = L(M)$.   □

For all $n \ge 3$, the language $U_n = \{a^{2^n}\}$ can be shown to be accepted by a stl-det-ORWW-automaton with an alphabet of $3n - 1$ letters, while each NFA for U_n needs at least $2^n + 1$ states. Hence, the bound given in Theorem 5 is sharp up to the O-notation. In addition, we have the following consequence, which is a clear improvement over the upper bound given in Theorem 2 (b).

Corollary 6. *For each stl-det-ORWW-automaton M with alphabet of size n, there exists a DFA C of size $2^{2^{O(n)}}$ such that $L(C) = L(M)$ holds.*

4 Decision Problems for stl-det-ORWW-automata

The *emptiness problem* for an NFA $A = (Q, \Sigma, \delta, q_0, F)$ of size $|Q| = m$ is decidable nondeterministically in space $O(\log m)$ (see, e.g., [5]), and so, by Savitch's Theorem [13] it follows that NFA-Emptiness $\in$ DSPACE$((\log |Q|)^2)$. Based on this observation we can use Theorem 5 to derive the following result.

Theorem 7. *The emptiness problem for stl-det-ORWW-automata is PSPACE-complete.*

Proof. Let $M = (\Sigma, \Gamma, \rhd, \lhd, \delta, >)$ be a stl-det-ORWW-automaton such that $|\Gamma| = n$. By Theorem 5, there exists an NFA A of size $2^{O(n)}$ such that $L(A) = L(M)$. Now we can check emptiness of $L(A)$ deterministically using space $(\log(2^{O(n)}))^2 = O(n^2)$. Thus, we see that stl-det-ORWW-Emptiness $\in$ PSPACE.

Now let $A_1, \ldots, A_t$ be $t \geq 2$ DFAs over a common input alphabet Σ of size k such that A_i has n_i states, $1 \leq i \leq t$. From these DFAs we can construct a stl-det-ORWW-automaton M with a tape aphabet of size $k \cdot (1 + n_1 + \cdots + n_{t-1}) + n_t$ such that $L(M) = \bigcap_{i_1}^t L(A_i)$ [10]. Hence, M has at most $O((k \cdot \sum_{i=1}^t n_i)^3)$ transitions, and so it can be computed from $A_1, \ldots, A_t$ in polynomial time. Now $L(M) \neq \emptyset$ iff $L(A_1) \cap \cdots \cap L(A_t) \neq \emptyset$, which shows that the above construction yields a polynomial-time reduction from the DFA-Intersection-Emptiness Problem to stl-det-ORWW-Emptiness. As the former is PSPACE-complete (see, e.g., [4]), we see that the latter is also PSPACE-hard. Together with the membership in PSPACE shown above, PSPACE-completeness follows.   □

From this theorem we also get the following completeness results.

Corollary 8. *For stl-det-ORWW-automata, universality, finiteness, inclusion, and equivalence are PSPACE-complete.*

Proof. **Universality:** Let M be a stl-det-ORWW-automaton with input alphabet Σ. In polynomial time we can construct a stl-det-ORWW-automaton M^c for the language $L(M^c) = (L(M))^c = \Sigma^* \smallsetminus L(M)$ from M such that M^c uses the same tape alphabet as M [10]. The automaton M is universal, that is, $L(M) = \Sigma^*$, iff $L(M^c) = \emptyset$. PSPACE-completeness of the universality problem now follows from PSPACE-completeness for the emptiness problem.

Inclusion and Equivalence: Let M_1 and M_2 be stl-det-ORWW-automata with alphabets of sizes n_1 and n_2, respectively. In polynomial time we can construct a stl-det-ORWW-automaton M with an alphabet of size $O(n_1 \cdot n_2)$ from M_1 and M_2 such that $L(M) = L(M_1) \cap L(M_2)^c$ [10]. Now $L(M_1) \subseteq L(M_2)$ iff $L(M_1) \cap L(M_2)^c = \emptyset$ iff $L(M) = \emptyset$. It follows that the inclusion problem is in PSPACE, which in turn implies immediately that the equivalence problem is in PSPACE.

On the other hand, let M' be a stl-det-ORWW-automaton that accepts the empty set. Then $L(M) = L(M')$ iff $L(M) \subseteq L(M')$ iff $L(M) = \emptyset$. Thus, PSPACE-completeness of the inclusion and the equivalence problems follows from PSPACE-completeness for the emptiness problem.

Finiteness: Let $M = (\Sigma, \Gamma, \triangleright, \triangleleft, \delta, >)$ be a stl-det-ORWW-automaton. We take a new symbol $\square$, that is, $\square \notin \Gamma$, and define a stl-det-ORWW-automaton $M' = (\Sigma', \Gamma', \triangleright, \triangleleft, \delta', >)$ as follows:

- $\Sigma' = \Sigma \cup \{\square\}$ and $\Gamma' = \Gamma \cup \{\square\}$,
- the transition function δ' is obtained from δ by simply interpreting an occurrence of the symbol $\square$ as an occurrence of the right delimiter $\triangleleft$.

Then $L(M') = L(M) \cup (L(M) \cdot \square \cdot \Sigma'^*)$, which means that $L(M')$ is finite iff $L(M) = \emptyset$. PSPACE-hardness of finiteness now follows from PSPACE-hardness of the emptiness problem.

On the other hand, from a stl-det-ORWW-automaton M with an alphabet of size n we can construct an NFA A of size $2^{O(n)}$ such that $L(M) = L(A)$. Just like emptiness, also infiniteness is decidable for A nondeterministically in space $\log(2^{O(n)}) \in O(n)$, and hence, it is decidable deterministically in space $O(n^2)$. Thus, finiteness for stl-det-ORWW-automata is indeed PSPACE-complete. $\square$

In the literature many subfamilies of the regular languages have been studied (see, e.g., [1,3,11]). Here we only consider some of them, beginning with the *strictly locally testable languages* of [8,15], but the corresponding problem can be stated for any subclass of REG.

A language $L \subseteq \Sigma^*$ is strictly k-testable for some $k \geq 1$ if $L \cap \Sigma^k \cdot \Sigma^* = (A \cdot \Sigma^* \cap \Sigma^* \cdot B) \smallsetminus \Sigma^+ \cdot (\Sigma^k \smallsetminus C) \cdot \Sigma^+$ for some finite sets $A, B, C \subseteq \Sigma^k$. For example, the language $(a + b)^*$ is strictly 1-testable, and the language $a(baa)^+$ is strictly 3-testable, but the language $(aa)^*$ is not strictly locally testable.

For each $k \geq 1$, if a language L is given through a DFA, then it is decidable in polynomial time whether or not L is strictly locally k-testable. Also it is decidable in polynomial time whether L is strictly locally testable [2]. We are interested in the corresponding variant of these problems in which the language considered is given through a stl-det-ORWW-automaton. Here we have the following result.

Theorem 9. *The following problem is PSPACE-complete for each $k \geq 1$:*

INSTANCE: A stl-det-ORWW-automaton M.
QUESTION: Is the language $L(M)$ strictly locally k-testable?

The construction in the proof shows that the problem of deciding strictly locally testability is at least PSPACE-hard for stl-det-ORWW-automata, but it remains open whether this problem is in PSPACE.

Using the same kind of reasoning it can be shown that, for a stl-det-ORWW-automaton, also the problems of deciding whether the accepted language is *nilpotent*, *combinatorial*, *circular*, *suffix-closed*, *prefix-closed*, *suffix-free*, or *prefix-free* (see, e.g., [1,3] for the definitions of these notions) are PSPACE-complete.

5   Concluding Remarks

We have shown that stl-det-ORWW-automata, although being deterministic devices, can provide exponentially more succinct representations for regular languages than NFAs. In addition, we have shown that many decision problems of interest are PSPACE-complete for stl-det-ORWW-automata. However, some open problems remain, for example:

- Can the given upper bounds be further improved by providing small constants in the exponents?
- Is the problem of deciding whether the language $L(M)$ that is accepted by a given stl-det-ORWW-automaton M is strictly locally testable decidable in polynomial space?

References

1. Bordihn, H., Holzer, M., Kutrib, M.: Determination of finite automata accepting subregular languages. Theor. Comp. Sci. **410**, 3209–3222 (2009)
2. Caron, P.: Families of locally testable languages. Theor. Comp. Sci. **242**, 361–376 (2000)
3. Dassow, J.: Subregular restrictions for some language generating devices. In: Freund, R., Holzer, M., Truthe, B., Ultes-Nitsche, U. (eds.) Proceedings of the Fourth Workshop on Non-Classical Models for Automata and Applications (NCMA 2012). books@ocg.at, Band, vol. 290, pp. 11–26. Oesterreichische Computer Gesellschaft, Wien (2012)
4. Garey, M.R., Johnson, D.S.: Computers and Intractability. A Guide to the Theory of NP-Completeness. Freeman, San Francisco (1979)
5. Holzer, M., Kutrib, M.: Descriptional and computational complexity of finite automata - a survey. Inform. Comp. **209**, 456–470 (2011)
6. Hundeshagen, N., Otto, F.: Characterizing the regular languages by nonforgetting restarting automata. In: Mauri, G., Leporati, A. (eds.) DLT 2011. LNCS, vol. 6795, pp. 288–299. Springer, Heidelberg (2011)
7. Kutrib, M., Reimann, J.: Succinct description of regular languages by weak restarting automata. Inform. Comp. **206**, 1152–1160 (2008)
8. McNaughton, R.: Algebraic decision procedures for local testability. Math. Syst. Theor. **8**, 60–76 (1974)
9. Mráz, F., Otto, F.: Ordered restarting automata for picture languages. In: Geffert, V., Preneel, B., Rovan, B., Štuller, J., Tjoa, A.M. (eds.) SOFSEM 2014. LNCS, vol. 8327, pp. 431–442. Springer, Heidelberg (2014)
10. Otto, F.: On the descriptional complexity of deterministic ordered restarting automata. In: Jürgensen, H., Karhumäki, J., Okhotin, A. (eds.) DCFS 2014. LNCS, vol. 8614, pp. 318–329. Springer, Heidelberg (2014)
11. Pin, J.-E.: Syntactic semigroups. In: Rozenberg, G., Salomaa, A. (eds.) Handbook of Formal Languages, vol. 1, pp. 679–746. Springer, Berlin (1997)
12. Průša, D.: Weight-reducing Hennie machines and their descriptional complexity. In: Dediu, A.-H., Martín-Vide, C., Sierra-Rodríguez, J.-L., Truthe, B. (eds.) LATA 2014. LNCS, vol. 8370, pp. 553–564. Springer, Heidelberg (2014)
13. Savitch, J.E.: Relationships between nondeterministic and deterministic tape complexities. J. Comp. Syst. Sci. **4**, 177–192 (1970)
14. Stanley, R.P.: Enumerative Combinatorics, vol. 1, 2nd edn. Cambridge University Press, Cambridge (2012)
15. Zalcstein, Y.: Locally testable languages. J. Comp. Syst. Sci. **6**, 151–167 (1972)

Quantum Queries on Permutations

Taisia Mischenko-Slatenkova[1], Alina Vasilieva[2], Ilja Kucevalovs[2],
and Rūsiņš Freivalds[1,2]([✉])

[1] Institute of Mathematics and Computer Science, University of Latvia,
Raiņa bulvāris 29, Riga 1459, Latvia
Taisia.Mischenko@gmail.com, Rusins.Freivalds@mii.lu.lv
[2] Faculty of Computing, University of Latvia, Raiņa bulvāris 19, Riga 1586, Latvia
Alina.Vasilieva@lu.lv, Ilja.Kucevalovs@intellisoft.lv

Abstract. K. Iwama and R. Freivalds considered query algorithms where
the black box contains a permutation. Since then several authors have
compared quantum and deterministic query algorithms for permutations.
It turns out that the case of n-permutations where n is an odd number is
difficult. There was no example of a permutation problem where quantization can save half of the queries for $(2m + 1)$-permutations if $m \geq 2$. Even
for $(2m)$-permutations with $m \geq 2$, the best proved advantage of quantum query algorithms is the result by Iwama/Freivalds where the quantum
query complexity is m but the deterministic query complexity is $(2m - 1)$.
We present a group of 5-permutations such that the deterministic query
complexity is 4 and the quantum query complexity is 2.

1 Introduction

Many papers on query algorithms consider computation of Boolean functions.
The input of the query algorithm is a black box oracle containing the values of
the variables $x_1 = a_1, x_2 = a_2, \ldots, x_n = a_n$ for an explicitly known Boolean
function $f(x_1, \ldots, x_n)$. The result of the query algorithm is to be the value
$f(a_1, \ldots, a_n)$. The query algorithm can ask for the values of the variables. The
queries are asked individually, and the result of any query influences the next
query to be asked or the result to be output.

The complexity of the query algorithm is defined as the number of queries
asked of the black box oracle. Deterministic query algorithms prescribe the next
query uniquely, depending only on the previously received answers from the black
box oracle. Probabilistic query algorithms allow randomization of the process of
computation. They sometimes allow reduction of the complexity of the algorithm
dramatically [1].

Quantum query algorithms (see a formal definition in [6]) consist of a finite
number of states, where each of them can query the black box oracle and determine how to change states. In fact they alternate *query operations* and *unitary
transformations*. In the steps called *query operations* the states of the algorithm

The research was supported by the project ERAF Nr.2DP/2.1.1.1/13/APIA/
VIAA/027.

J. Shallit and A. Okhotin (Eds.): DCFS 2015, LNCS 9118, pp. 177–184, 2015.
DOI: 10.1007/978-3-319-19225-3_15

are divided into subsets corresponding to the allowed quantum-parallel queries. If each of states $q_{i_1}, \ldots, q_{i_m}$ asks a query "$x_i = ?$" then for every possible answer "$x_i = j$" a unitary operation over the states $q_{i_1}, \ldots, q_{i_m}$ is pre-programmed. In the steps called *unitary transformations* the amplitudes of all states are transformed according to a unitary matrix. Every sequence of steps is ended in a special operation called *measurement*, in which the amplitudes (being complex numbers) for all the states are substituted by real numbers called *probabilities* by the following rule. The complex number $a + bi$ is substituted by the real number $a^2 + b^2$. It follows from the unitarity of all the operations that the total of the probabilities of all states equals 1. Some states are defined to be accepting, and the other states are defined to be rejecting. This distinction is not seen before the measurement. After the measurement the probabilities of the accepting states are summed up and the result is called the accepting probability. We say that the quantum query algorithm is *exact* if the accepting probability is always either 1 or 0.

The notion of *promise* for quantum algorithm was introduced by Deutsch and Jozsa [10], and Simon [13]. In quantum query algorithms for problems under promise, the domain of correctness of the algorithm is explicitly restricted. We are not interested in behavior of the algorithm outside this restriction. For instance, in this paper all the query algorithms are considered under a promise that the target function describes a permutation (in a way precisely stated below).

Recently there have been many papers studying query algorithms computing Boolean functions. Powerful methods to prove lower bounds for quantum query complexity were developed by A. Ambainis [2,3]. A good reference is the survey by Buhrman and de Wolf [6].

We consider in this paper a more general class of functions $f(x_1, \cdots, x_n)$, namely, functions $\{0, 1, 2, \ldots, n-1\}^n \to \{0, 1\}$. The domain $\{0, 1, 2, \ldots, n-1\}^n$ includes a particularly interesting case — permutations. For instance,

$$x_1 = 4, x_2 = 3, x_3 = 2, x_4 = 1, x_5 = 0$$

can be considered as a permutation of 5 symbols $\{0,1,2,3,4\}$ usually described as 43210. Under such a restriction the functions $f : \{0, 1, 2, \ldots, n-1\}^n \to \{0, 1\}$ can be considered as properties of permutations. For instance, the function

$$f(0, 1, 2) = 1, f(1, 2, 0) = 1, f(2, 0, 1) = 1, f(0, 2, 1) = 0, f(1, 0, 2) = 0, f(2, 1, 0) = 0$$

describes the property of 3-permutations to be *even* (as opposed to the property to be *odd*).

The property of a permutation to be even or odd can be defined in many equivalent ways. One of the most popular definitions used below is as follows. A permutation $x_1 = a_1, x_2 = a_2, \ldots, x_n = a_n$ is called even (odd) if it can be obtained from the identical permutation $x_1 = 0, x_2 = 1, \ldots, x_n = n-1$ by an even (odd) number of transpositions, i.e., mutual changes of exactly two elements of the permutation: substituting $x_i = a_i$ and $x_j = a_j$ by $x_i = a_j$ and $x_j = a_i$. It is a well-known fact that the property of being even or odd does not depend on the particular sequence of transpositions. Deciding whether a given permutation is even or odd is called deciding the parity of this permutation.

Our Contribution. It is easy to see that for every n-permutation $(n-1)$ queries uniquely determine the permutation. Hence for arbitrary permutation problems the deterministic query complexity never exceeds $(n-1)$. Theorem 1 below gives us a hope that sometimes half of the number of queries can be eliminated by quantization. The proof of Theorem 1 suggests that Fourier transforms might be used again in counterparts of this theorem for larger values of n. However, it is far from obvious how to organize the pairs of the values of results of the queries into a linear string which is needed to apply the Fourier transform.

The paper [12] attempted to show that quantum query algorithms can use only about a half of the number of queries needed for deterministic query algorithms for deciding parity of permutations. Unfortunately, this attempt was only partially successful. It was proved that for $2m$-permutations it suffices to have m quantum queries but for $2m+1$-permutations it suffices to have $m+1$ quantum queries. In both cases this is more than half of the deterministic query complexity.

In this paper we were not able to construct an effective quantum query algorithm to decide parity of 5-permutations. Instead we propose another permutation problem for 5-permutations where a quantum algorithm is indeed exactly twice as efficient as the best deterministic algorithm.

As proved in [4,5,7–9,11], for every problem where a quantum algorithm is more efficient than any deterministic algorithm, it is crucially important to have a rich structure of symmetries. This is why we use a group of 5-permutations such that it has interesting automorphisms. We believe that our algorithm admits a generalization for larger values of n, but up to now we have been able to use it only for 5-permutations.

2   First Example

Theorem 1. [12] *There is an exact quantum query algorithm deciding the parity of 3-permutations with one query.*

Proof. By way of quantum parallelism, in the state q_1 we ask the query x_1 with an amplitude $\frac{1}{\sqrt{3}}$, in the state q_2 we ask the query x_2 with an amplitude $\frac{1}{\sqrt{3}}$, and in the state q_3 we ask the query x_3 with an amplitude $\frac{1}{\sqrt{3}}$.

If the answer from the black box to the query x_1 is 0, we do not change the amplitude of the state q_1. If the answer is 1, we multiply the existing amplitude by $e^{i\frac{2\pi}{3}}$. If the answer is 2, we multiply the existing amplitude by $e^{i\frac{4\pi}{3}}$.

If the answer from the black box to the query x_2 is 0, we multiply the amplitude of the state q_2 by $e^{i\frac{4\pi}{3}}$. If the answer is 1, we do not change the amplitude. If the answer is 2, we multiply the existing amplitude by $e^{i\frac{2\pi}{3}}$.

If the answer from the black box to the query x_3 is 0, we multiply the amplitude of the state q_3 by $e^{i\frac{2\pi}{3}}$. If the answer is 1, we multiply the existing amplitude by $e^{i\frac{4\pi}{3}}$. If the answer is 2, we do not change the amplitude.

We process the obtained amplitudes of the states q_1, q_2, q_3 by a unitary transformation corresponding to the matrix

$$\begin{pmatrix} (\frac{1}{\sqrt{3}}) & (\frac{1}{\sqrt{3}}) & (\frac{1}{\sqrt{3}}) \\ (\frac{1}{\sqrt{3}}) & (\frac{1}{\sqrt{3}})e^{i\frac{2\pi}{3}} & (\frac{1}{\sqrt{3}})e^{i\frac{4\pi}{3}} \\ (\frac{1}{\sqrt{3}}) & (\frac{1}{\sqrt{3}})e^{i\frac{4\pi}{3}} & (\frac{1}{\sqrt{3}})e^{i\frac{2\pi}{3}} \end{pmatrix}$$

This transformation is a particular case of a Fourier transform. If we are computing $f(0, 1, 2)$, $f(1, 2, 0)$ or $f(2, 0, 1)$ (these are all even 3-permutations) the amplitude of the state q_1 becomes, correspondingly, 1, $e^{i\frac{2\pi}{3}}$ or $e^{i\frac{4\pi}{3}}$. After measuring this state we get the probability 1. If we are computing $f(0, 2, 1)$ (which is an odd permutation) the amplitude of the state q_1 becomes 0 but the amplitude of the state q_2 becomes 1. If we are computing $f(1, 0, 2)$ or $f(2, 1, 0)$ (which are odd permutations) the amplitude of the state q_1 becomes 0, but the amplitude of the state q_3 becomes $e^{i\frac{4\pi}{3}}$ or $e^{i\frac{2\pi}{3}}$, correspondingly.   □

3   Further Results

We define a group GR of 5-permutations consisting of 20 permutations. These permutations are:

$$\begin{array}{l} 01234 \ 12340 \ 23401 \ 34012 \ 40123 \\ 02413 \ 13024 \ 24130 \ 30241 \ 41302 \\ 03142 \ 14203 \ 20314 \ 31420 \ 42031 \\ 04321 \ 10432 \ 21043 \ 32104 \ 43210 \end{array}$$

First of all, we note that the 5-permutations in GR can be represented as linear functions modulo 5.

$$\begin{array}{l} x \quad x+1 \quad x+2 \quad x+3 \quad x+4 \\ 2x \ 2x+1 \ 2x+2 \ 2x+3 \ 2x+4 \\ 3x \ 3x+1 \ 3x+2 \ 3x+3 \ 3x+4 \\ 4x \ 4x+1 \ 4x+2 \ 4x+3 \ 4x+4 \end{array}$$

These permutations can be considered as group with a 2-argument algebraic operation "multiplication of permutations". The properties of this group GR have been well-known for a long time. For instance, it is known that GR is not a commutative group and it is not a cyclic group.

In this section we prove that there is an exact quantum query algorithm deciding the membership in the group GR of 5-permutations with 2 queries. Obviously, 4 deterministic queries are needed for this problem because if a permutation

$$P = \{x_0 = a_0, x_1 = a_1, x_2 = a_2, x_3 = a_3, x_4 = a_4\}$$

is in GR and only 3 queries are asked, there remain two possibilities, namely, either $x_i = a_i, x_j = a_j$ and the permutation is in GR, or $x_i = a_j, x_j = a_i$ and the permutation is not in GR.

We will prove below that there is an exact quantum query algorithm deciding the membership in the group GR of 5-permutations with two queries.

We need a numbering of 20 values of all possible pairs (a_i, a_j) where a_i and a_j are values from the black box (i.e., the elements of the permutation in question). We might arrange these 20 values in accordance with the above-presented table but we prefer a slightly different layout.

$$
\begin{array}{lllll}
01 & 12 & 23 & 34 & 40 \\
02 & 24 & 41 & 13 & 30 \\
04 & 43 & 32 & 21 & 10 \\
03 & 31 & 14 & 42 & 20
\end{array}
$$

These values correspond to the linear functions

$$
\begin{array}{lllll}
x & x+1 & x+2 & x+3 & x+4 \\
2x & 2x+1 & 2x+2 & 2x+3 & 2x+4 \\
4x & 4x+1 & 4x+2 & 4x+3 & 4x+4 \\
3x & 3x+1 & 3x+2 & 3x+3 & 3x+4
\end{array}
$$

There is a certain regularity in this layout. Each row can be obtained from the preceding row multiplying it by 2 modulo 5. Each column can be obtained from the preceding column adding 1 modulo 5.

Now we introduce two distances among the values of these pairs. The distance $D_r[(u, v), (a, b)]$ between the pair (u, v) and the pair (a, b) is the value (the number of the row of (a, b)) — (the number of the row of (u, v)) (mod 4). The distance $D_c[(u, v), (a, b)]$ between the pair (u, v) and the pair (a, b) is the value (the number of the column of (a, b) — (the number of the column of (u, v)) (mod 5).

Our quantum query algorithm (in a way of quantum parallelism) enters (with equal amplitudes $\frac{1}{\sqrt{20}}$) 20 states. In each of these states the algorithm asks one of the 20 possible queries (x_i, x_j) where $x_i \in \{0, 1, 2, 3, 4\}$, $x_j \in \{0, 1, 2, 3, 4\}$, and $i \neq j$. In the result of these queries the amplitude is multiplied either by (-1) or by $(+1)$. The following is a description of the value of these multipliers.

There are 20 values of all possible pairs (x_i, x_j) where $x_i \in \{0, 1, 2, 3, 4\}$, $x_j \in \{0, 1, 2, 3, 4\}$, and $x_i \neq x_j$. For every pair of queries (x_i, x_j) where $x_i \neq x_j$, the answer-pair (a_i, a_j) corresponds to the answer-pair (a_j, a_i) of the query (x_j, x_i). However, for the sake of symmetry, we have considered a quantum query algorithm with 20 possible pairs of queries.

Since the black box contains a permutation, the results of the query are also such that $a_i \in \{0, 1, 2, 3, 4\}$, $a_j \in \{0, 1, 2, 3, 4\}$, and $a_i \neq a_j$. Hence there are exactly 20 values of all possible answer-pairs (a_i, a_j).

For each of these 20 pairs (x_i, x_j) we get the corresponding pair of answers (a_i, a_j) and again $a_i \in \{0, 1, 2, 3, 4\}$, $a_j \in \{0, 1, 2, 3, 4\}$, and $a_i \neq a_j$. We can consider 20 distances $D_r[(x_i, x_j), (a_i, a_j)]$ and 20 distances $D_c[(x_i, x_j), (a_i, a_j)]$.

In the table below each of these pairs of distances is considered as a pair (w, z) where $w = D_r[(x_i, x_j), (a_i, a_j)]$ and $z = D_c[(x_i, x_j), (a_i, a_j)]$. Please observe that while asking the query (x_i, x_j) we automatically get an answer to the query

(x_j, x_i) as well but for this pair the distances may be different. Our quantum query algorithm uses this distinction essentially.

Suppose that the permutation in the black box is 03241 and we consider the pair of queries (x_2, x_4). Then the pair of answers (a_2, a_4) equals $(2, 1)$ and the pair of answers for (x_4, x_2) denoted as (a_4, a_2) equals $(1, 2)$. Then $D_r[(x_2, x_4), (a_2, a_4)]$ is equal to $D_r[(2, 4), (2, 1)] = 1$, $D_c[(2, 4), (2, 1)] = 2$, $D_r[(4, 2), (1, 2)] = 1$, and $D_c[(4, 2), (1, 2)] = 3$.

The following table describes in which cases the multiplier is $+1$ and in which cases it is -1. The first column corresponds to the pair

$$(w_1, z_1) = (D_r[(x_i, x_j), (a_i, a_j)], D_c[(x_i, x_j), (a_i, a_j)]),$$

the second column corresponds to

$$(w_2, z_2) = (D_r[(x_j, x_i), (a_j, a_i)], D_c[(x_j, x_i), (a_j, a_i)]),$$

the last column corresponds to the multiplier $(+1)$ or (-1). For instance, the above-mentioned example is described in the table below as

$$(1, 2)\ (1, 3) \rightarrow +1$$

The table is as follows.

$$(0, 0)\ (0, 0) \rightarrow +1$$
$$(0, 1)\ (0, 4) \rightarrow +1$$
$$(0, 2)\ (0, 3) \rightarrow +1$$
$$(0, 3)\ (0, 2) \rightarrow +1$$
$$(0, 4)\ (0, 1) \rightarrow +1$$
$$(1, 0)\ (1, 0) \rightarrow +1$$
$$(1, 1)\ (1, 4) \rightarrow +1$$
$$(1, 2)\ (1, 3) \rightarrow +1$$
$$(1, 3)\ (1, 2) \rightarrow +1$$
$$(1, 4)\ (1, 1) \rightarrow +1$$
$$(2, 0)\ (2, 0) \rightarrow -1$$
$$(2, 1)\ (2, 4) \rightarrow -1$$
$$(2, 2)\ (2, 3) \rightarrow -1$$
$$(2, 3)\ (2, 2) \rightarrow -1$$
$$(2, 4)\ (2, 1) \rightarrow -1$$
$$(3, 0)\ (3, 0) \rightarrow -1$$
$$(3, 1)\ (3, 4) \rightarrow -1$$
$$(3, 2)\ (3, 3) \rightarrow -1$$
$$(3, 3)\ (3, 2) \rightarrow -1$$
$$(3, 4)\ (3, 1) \rightarrow -1$$

Lemma 1. *If the permutation in the black box is from the group GR then all the 20 multipliers are equal.*

Proof. If the permutation corresponds to the function $ax + b$ where $a = 3$ or $a = 4$ then the multiplier equals (-1).   □

Lemma 2. *If the permutation in the black box is one of the following*

$$01243, 01342, 01423, 01324, 01432$$

then exactly 10 multipliers equal (-1) *and exactly 10 multipliers equal* $(+1)$.

Proof. By explicit counting. □

Lemma 3. *If the permutation in the black box* $f(x)$ *can be obtained from a permutation* $g(x)$ *in the set*

$$\{ 01243, 01342, 01423, 01324, 01432 \}$$

as $f(x) \equiv ag(x) + b(mod\ 5)$ *then exactly 10 multipliers equal* (-1) *and exactly 10 multipliers equal* $(+1)$.

Proof. The definition of the values of multipliers depend only on the distances D_r but not on the distances D_c. Application of a linear function $at + b$ does not change the distance D_r. □

Lemma 4. *If the permutation in the black box is not from the group* GR *then exactly 10 multipliers equal* (-1) *and exactly 10 multipliers equal* $(+1)$.

Proof. The group G_5 of all 5-permutations consists of 120 elements. GR is a subgroup of G_5 consisting of 20 elements. Lagrange's theorem on finite groups shows that G_5 is subdivided into 6 cosets of equal size, one of the cosets being GR. The other 5 cosets $GC_1, GC_2, GC_3, GC_4, GC_5$ can be described as the set of all permutations $f(x)$ such that $f(x) \equiv ag(x) + b(mod\ 5)$ and $g(x) \in GC_i$. It follows from Lemma 3 that exactly 10 multipliers equal (-1) and exactly 10 multipliers equal $(+1)$. □

Theorem 2. *There is an exact quantum query algorithm deciding the membership in the group* GR *of 5-permutations with two queries.*

Proof. Our quantum query algorithm (in a way of quantum parallelism) enters (with equal amplitudes $\frac{1}{\sqrt{20}}$) 20 states. In each of these states the algorithm asks one of the 20 possible queries (x_i, x_j) where $x_i \in \{0, 1, 2, 3, 4\}, x_j \in \{0, 1, 2, 3, 4\}$, and $i \neq j$. In the result of these queries the amplitude is multiplied either by (-1) or by $(+1)$ as described in the table above. Lemmas 1 and 4 ensure that our quantum algorithm accepts all permutations in GR and rejects all permutations not in GR with probability 1. □

References

1. Ablayev, F.M., Freivalds, R.: Why sometimes probabilistic algorithms can be more effective. In: Gruska, J., Rovan, B., Wiedermann, J. (eds.) MPCS 1986. LNCS, vol. 233, pp. 1–14. Springer, Heidelberg (1986)
2. Ambainis, A.: Quantum lower bounds by quantum arguments. J. Comput. Syst. Sci. **64**(4), 750–767 (2002)

3. Ambainis, A.: Polynomial degree vs. quantum query complexity. In: Proceedings of FOCS 1998, pp. 230–240 (1998)
4. Ambainis, A., Freivalds, R.: 1-way quantum finite automata: strengths, weaknesses and generalizations. In: Proceedings of FOCS 1998, pp. 332– 341. Also quant-ph/9802062
5. Ambainis, A., de Wolf, R.: Average-case quantum query complexity. In: Reichel, H., Tison, S. (eds.) STACS 2000. LNCS, vol. 1770, pp. 133–144. Springer, Heidelberg (2000)
6. Buhrman, H., de Wolf, R.: Complexity measures and decision tree complexity: a survey. Theoret. Comput. Sci. **288**(1), 21–43 (2002)
7. Beals, R., Buhrman, H., Cleve, R., Mosca, M., de Wolf, R.: Quantum lower bounds by polynomials. J. ACM **48**(4), 778–797 (2001)
8. Buhrman, H., Cleve, R., de Wolf, R., Zalka, C.: Bounds for small-error and zero-error quantum algorithms. In: Proceedings of FOCS 1999, pp. 358–368 (1999)
9. Cleve, R., Ekert, A., Macchiavello, C., Mosca, M.: Quantum algorithms revisited. Proc. R. Soc. Lond. **A 454**, 339–354 (1998)
10. Deutsch, D., Jozsa, R.: Rapid solutions of problems by quantum computation. Proc. R. Soc. Lond. **A 439**, 553 (1992)
11. Freivalds, R.: Languages recognizable by quantum finite automata. In: Farré, J., Litovsky, I., Schmitz, S. (eds.) CIAA 2005. LNCS, vol. 3845, pp. 1–14. Springer, Heidelberg (2006)
12. Freivalds, R., Iwama, K.: Quantum queries on permutations with a promise. In: Maneth, S. (ed.) CIAA 2009. LNCS, vol. 5642, pp. 208–216. Springer, Heidelberg (2009)
13. Simon, I.: String matching algorithms and automata. In: Bundy, A. (ed.) CADE 1994. LNCS, vol. 814, pp. 386–395. Springer, Heidelberg (1994)

Complement on Free and Ideal Languages

Peter Mlynárčik[✉]

Mathematical Institute, Slovak Academy of Sciences,
Grešaková 6, 040 01 Košice, Slovakia
mlynarcik1972@gmail.com

Abstract. We study nondeterministic state complexity of the complement operation on the classes of prefix-free, suffix-free, factor-free and
subword-free languages and on the class of ideal languages. For the cases
prefix-free and suffix-free we improve the lower bound, and improve the
upper bound for suffix-free languages in the binary case. In all other
cases, we find tight bounds for sufficient alphabet sizes.

1 Introduction

The complement of a formal language L over an alphabet Σ is the language
$L^c = \Sigma^* \setminus L$, where Σ^* is the set of all strings over an alphabet Σ. The complementation is an easy operation on regular languages represented by deterministic
finite automata (DFAs) since to get a DFA for the complement of a regular language, it is enough to interchange the final and non-final states in a DFA for
this language.

On the other hand, complementation on regular languages represented by
nondeterministic finite automata (NFAs) is an expensive task. First, we have
to apply the subset construction to a given NFA, and only after that, we may
interchange the final and non-final states. This gives an upper bound 2^n.

Sakoda and Sipser [9] gave an example of languages over a growing alphabet size meeting this upper bound on the nondeterministic state complexity
of complementation. Birget claimed the result for a three-letter alphabet [1],
but later corrected this to a four-letter alphabet. Holzer and Kutrib [4] proved
the lower bound 2^{n-2} for a binary n-state NFA language. Finally, a binary n-
state NFA language meeting the upper bound 2^n was described by Jirásková
in [5]. In the unary case, the complexity of complementation is known to be in
$e^{\Theta(\sqrt{n\ln(n)})}$ [4,5].

Jirásková and Mlynárčik [7] gave tight bounds in case of prefix- and suffix-
free languages over a ternary alphabet and for binary languages gave the lower
bound $F(n-2) + 1$, where $F(n)$ is the Landau function, and $F(n)$ is in the
class $e^{\Theta(\sqrt{n\ln(n)})}$. The upper bound for binary alphabet was improved to $2^{n-1} -
2^{n-3} + 1$, but only in the prefix-free case. The suffix-free case remained open.

In this paper, we investigate the complementation operation on prefix-free
and suffix-free binary languages, where we give significantly better lower bounds

---

P. Mlynárčik—Research supported by VEGA grant 2/0084/15.

J. Shallit and A. Okhotin (Eds.): DCFS 2015, LNCS 9118, pp. 185–196, 2015.
DOI: 10.1007/978-3-319-19225-3_16

$2^{\lfloor \frac{n}{2} \rfloor - 1}$ and give an improved upper bound for the suffix-free language $2^{n-1} - 2^{n-3} + 2$. We also deal with factor-free and subword-free languages, where we give tight bounds for proper alphabet. For the factor-free case over binary alphabets we get a result similar as to that mentioned above. In the second part of the paper we deal with complementation ideal languages, including right ideals, left ideals, two-sided ideals, and all-sided ideals. In the first three cases we give a tight bound in binary case, and in the last case the tight bound is for an exponentially-growing alphabet.

2   Preliminaries

Recall that a language is prefix-free if it does not contain two distinct strings, one of which is a prefix of the other. The suffix-free languages are defined in a similar way.

To prove the minimality of nondeterministic finite automata, we use a fooling set lower-bound technique [1,8].

Definition 1. *A set of pairs of strings* $\{(x_1, y_1), (x_2, y_2), \ldots, (x_n, y_n)\}$ *is called a fooling set for a language* L *if for all* i, j *in* $\{1, 2, \ldots, n\}$,
(F1) $x_i y_i \in L$, *and*
(F2) *if* $i \neq j$, *then* $x_i y_j \notin L$ *or* $x_j y_i \notin L$.

Lemma 2 ([1,8]). *Let* $\mathcal{F}$ *be a fooling set for a language* L. *Then every NFA (with multiple initial states) for the language* L *has at least* $|\mathcal{F}|$ *states.*   □

Although the difference between the size of fooling set and the size of minimal NFA can be large, we successfully use the fooling set technique throughout the paper [6].

Landau's function is frequently needed, and is defined as follows:

Let n be a positive integer. Then $F(n) = \max\{\text{lcm}(x_1, x_2, \ldots, x_k) | x_1 + x_2 + \cdots + x_k = n\}$. The function $F(n)$ is in the class $e^{\theta(\sqrt{n \ln(n)})}$ (Landau, 1903).

3   Free Languages

Let G be the language accepted by the NFA over $\{a, b\}$ shown in Fig. 1 with $n - 1$ states. Let $L = cG$. The language L is a suffix-free language over $\{a, b, c\}$ recognized by an n-state NFA A, shown in Fig. 2 and $\text{nsc}(L^c) \geq 2^{n-1}$ [7]. Now let us define a homomorphism h as follows: $h(c) = 00$, $h(a) = 10$, $h(b) = 11$ (used in [2]). After applying h on the language L, we have a binary language $K = h(L)$ over $\{0, 1\}$.

Lemma 3. *The language* K *is a suffix-free language.*

Proof. Every string in L contains exactly one symbol c at the beginning, so every string in K begins with the string 00 and this substring does not appear later in string. If there is a string $w = uv$ and $u \neq \varepsilon$, then v does not contain 00 and therefore $v \notin K$. So K is suffix-free.   □

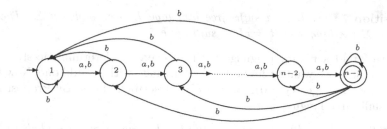

Fig. 1. An NFA of a binary regular language G with $\mathrm{nsc}(G) = 2^{n-1}$

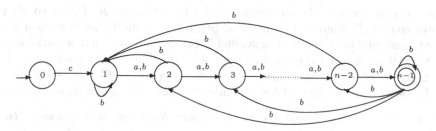

Fig. 2. An NFA of a ternary suffix-free regular language L with $\mathrm{nsc}(G) = 2^{n-1}$

Now let us define NFA A' for the language K. We use the description of automaton A for original language L. Let $A = (Q, \{a, b, c\}, \delta, 0, \{n-1\})$ (NFA shown in Fig. 2). The idea is replace every transition $q \xrightarrow{a} q_a$ by adding a new state q' and two transitions $q \xrightarrow{1} q' \xrightarrow{0} q_a$, and similarly for the symbol b $q \xrightarrow{1} q' \xrightarrow{1} q_b$ and transition $0 \xrightarrow{c} 1$ we replace by adding $0'$ and two transitions $0 \xrightarrow{0} 0' \xrightarrow{0} 1$.

Lemma 4. *The NFA A' defined above recognizes the language K.*   □

Lemma 5. *The NFA A' is a minimal NFA for the language K.*   □

Lemma 6. *Let $n \geq 3$ and K be the binary language defined above. Then $\mathrm{nsc}(K^c) \geq 2^{n-1}$.*

Proof. As shown in [7, Lemma 5], the set $\mathcal{F} = \{(c x_S, y_S) \mid S \subseteq \{1, 2, \ldots, n-1\}\}$ is a fooling set for L^c. Let us define $\mathcal{F}' = \{(h(c x_S), h(y_S)) \mid S \subseteq \{1, 2, \ldots, n-1\}\}$. Let us show that the $\mathcal{F}'$ is a fooling set for K^c.

(F1) For every pair $(h(c x_S), h(y_S))$, we have $c x_S y_S \in L^c$, so $c x_S y_S \notin L$ and since homomorphism h is a bijection, we have $h(c x_S y_S) \notin K$ so $(h(c x_S), h(y_S)) \in K^c$.

(F2) Let $(h(c x_S), h(y_S))$, $(h(c x_T), h(y_T))$ be two distinct pairs. Without loss of generality, let $c x_S y_T \notin L^c$. So $c x_S y_T \in L$, then $h(c x_S y_T) \in K$, so $h(c x_S y_T) \notin K^c$.

Hence $\mathcal{F}'$ is a fooling set for K^c. Since $|\mathcal{F}'| = 2^{n-1}$, $\mathrm{nsc}(K^c) \geq 2^{n-1}$.   □

Proposition 7. *Let L be a suffix-free language L over alphabet Σ. Then for every $x \in \Sigma$ the language $R = xL$ is suffix-free.*   □

Above we found a binary language with an even nondeterministic state complexity, and now we want to find a binary language with an odd one. Now let us consider the language $K_1 = 0K$, where K is described above. By Proposition 7, K_1 is a suffix-free language.

Lemma 8. *Let K and K_1 be binary suffix-free languages mentioned above. Then $\mathrm{nsc}(K_1) = 2n + 1$.*

Proof. Let us consider the automaton A' for the language K. Let us construct an automaton A'' from A' by simply adding a new state $0'$ and transition from $0'$ to the original initial state 0 on symbol 0. State $0'$ becomes a new initial state.

Now let us consider the minimality of A''. Let $\mathcal{F}$ be a fooling set for K. Let us construct $\mathcal{F}'$ from $\mathcal{F}$ as follows: $\mathcal{F}' = \{(0u, v) \mid (u, v) \in \mathcal{F}\} \cup \{\varepsilon, 000(10)^{n-2}\}$.

The set $\mathcal{F}'$ is a fooling set for K_1 and $|\mathcal{F}'| = 2n + 1$, so $\mathrm{nsc}(K_1) = 2n + 1$. □

Lemma 9. *Let $n \geq 3$ and K_1 be the binary language defined above. Then $\mathrm{nsc}(K_1^c) \geq 2^{n-1}$.*

Proof. Let $\mathcal{F}$ be a fooling set for language K^c (see Lemma 6). Let us construct the set $\mathcal{F}' = \{(0u, v) \mid (u, v) \in \mathcal{F}\}$. Let us show that $\mathcal{F}'$ is a fooling set for K_1^c.

(F1) If $uv \in K^c$, then $uv \notin K$, then also $0uv \notin K_1$, so $0uv \in K_1^c$.
(F2) If $(u, v), (x, y) \in \mathcal{F}$ and without loss of generality, $uy \notin K^c$, so $uy \in K$. Then $0uy \in K_1$ and $0uy \notin K_1^c$.

Hence $\mathcal{F}'$ is a fooling set for K_1^c. Since $|\mathcal{F}'| = 2^{n-1}$, $\mathrm{nsc}(K_1^c) \geq 2^{n-1}$.   □

We summarize our results in the following theorem.

Theorem 10. *Let $n \geq 6$. There is a binary suffix-free language L such that $\mathrm{nsc}(L) = n$ and $\mathrm{nsc}(L^c) \geq 2^{\lfloor \frac{n}{2} \rfloor - 1}$.*

Now we consider an upper bound. Let us recall the following result.

Lemma 11. *Let $n \geq 12$. Let L be a binary prefix-free language with $\mathrm{nsc}(L) = n$. Then $\mathrm{nsc}(L^c) \leq 2^{n-1} - 2^{n-3} + 1$. [7, Lemma 9]*

Notice that the proof at [7, Lemma 9] also works for NFAs with multiple initial states. We are also going to use it for suffix-free languages.

Theorem 12. *Let $n \geq 12$. Let L be a binary suffix-free language with $\mathrm{nsc}(L) = n$. Then $\mathrm{nsc}(L^c) \leq 2^{n-1} - 2^{n-3} + 2$.*

Proof. After reversing an NFA for L, we obtain an n-state NFA (possibly with multiple initial states) for a prefix-free language L^R. By Lemma 11, $\mathrm{nsc}((L^R)^c) \leq 2^{n-1} - 2^{n-3} + 1$. Since $(L^R)^c = (L^c)^R$, we have $\mathrm{nsc}((L^c)^R) \leq 2^{n-1} - 2^{n-3} + 1$. It follows that $(L^c)^R$ is accepted by an NFA N which has at most $2^{n-1} - 2^{n-3} + 1$ states. Now we reverse the NFA N, and get a NFA N^R, possibly with multiple initial states. By adding one more state, we get an NFA for L^c with at most $2^{n-1} - 2^{n-3} + 2$ states and with a unique initial state.   □

Similarly as in the case of suffix-free language, we can apply the same homomorphism h on the ternary prefix-free language L from [7, Lemma 3]. We only have to be careful with the proof of the prefix-free property of the language $h(L)$. Now every string in $h(L)$ ends with 00. The only proper prefix of a string in $h(L)$ which ends with 00 has an odd length. But such a string does not belong to $h(L)$. Therefore $h(L)$ is prefix-free.

We can construct NFA A for $h(L)$ with $2n$ states similarly as in the suffix-free case. The main difference between the automaton for the case of a binary suffix-free language, and for a binary prefix-free language is the final state. Similarly as in suffix-free case we can prove that A is minimal and therefore $nsc(h(L)) = 2n$. Finally, we use a similar approach to find a binary prefix-free language with an odd number of states, such that we add a new state n' and the transition from original final state n to n' on symbol 0. State n' become a new final state. Such a language is still prefix-free.

Hence we get the following result for binary prefix-free languages.

When we use the result from Lemma 11 we can state the following result.

Theorem 13 (Complement on Binary prefix-free, suffix-free languages). *Let $n \geq 12$. Let L be a binary prefix-free or suffix free language with $nsc(L) = n$. Then $nsc(L^c) \leq 2^{n-1} - 2^{n-3} + 2$. The lower bound is $2^{\lfloor \frac{n}{2} \rfloor - 1}$.*

In the paper [7, Lemma 8] we presented a binary suffix-free and prefix free language L with $nsc(L) \leq n$, such that every NFA for its complement requires at least $F(n-2)+1$ states, where $F(n)$ is the Landau function. The function $F(n)$ is in $2^{\Theta(\sqrt{n \log(n)})}$; therefore $\lim_{n \to \infty} F(n-2) + 1/2^{\lfloor \frac{n}{2} \rfloor - 1} = 0$. So the lower bound in our Theorem 13 is significantly higher.

After investigation of prefix and suffix free languages we will investigate other free classes of languages: factor-free and subword-free languages. First we present a lemma which we use in our next considerations.

Lemma 14. *Let L be a language such that $\varepsilon \in L$. Let u and v be strings, and let $u \notin L$. Let $\mathcal{A}$ be a set of pairs of strings such that the sets $\mathcal{A} \cup \{(\varepsilon, v)\}$ and $\mathcal{A} \cup \{(u, v)\}$ are fooling sets for L. Then $nsc(L) \geq |\mathcal{A}| + 2$.* □

Let w be a string. We say that a string v is a *factor* of the w iff there are strings x, v, y, such that $w = xvy$. Moreover, if $xy \neq \varepsilon$, we say that v is a *proper factor*. We say a language L is *factor-free* iff there are no two strings u, v in L, such that u is a proper factor of v.

Theorem 15. *Let $n \geq 3$. Let L be a factor-free language over an alphabet Σ such that $nsc(L) = n$. Then $nsc(L^c) \leq 2^{n-2} + 1$, and the bound is tight if $|\Sigma| \geq 3$.*

Proof. We first prove the upper bound. Let A be an n-state NFA for L. Since L is factor-free, it is suffix-free and also prefix-free. It follows that no transition goes to the initial state of A, and all the final states in the subset automaton are equivalent. Hence the subset automaton has at most $2^{n-2} + 2$ reachable and pairwise distinguishable states. After exchanging the final and non-final states, we get a DFA for L^c of at most $2^{n-2}+2$ states. In the same way as for prefix-free languages in [7, Lemma 2], we can use a nondeterminism to save one state. This gives the upper bound $2^{n-2}+1$.

To prove tightness, consider the binary language G accepted by the $(n-2)$-state NFA N shown in Fig. 1. Let $L = c \cdot G \cdot c$. Then L is accepted by an n-state NFA A shown in Fig. 3.

Let $\mathcal{F} = \{(x_S, y_S) \mid S \subseteq \{1, 2, \ldots, n-2\}\}$ be a fooling set for the language G^c [5, Theorem 5]. Notice that the strings x_S and y_S have the following properties:

(1) by x_S, the initial state goes to the set S;
(2) the string y_S is rejected by N from every state in S and it is accepted by N from every state in $\{1, 2, \ldots, n-2\} \setminus S$.

Then $\mathcal{F}' = \{(cx_S, y_Sc) \mid S \subseteq \{1, 2, \ldots, n-2\}\}$ is a fooling set for L^c. Let $\mathcal{A} = \{(cx_S, y_Sc) \mid S \subseteq \{1, 2, \ldots, n-2\} \text{ and } S \neq \emptyset\}$, $v = y_\emptyset \cdot c$, $u = ca^{n-3}c$. Let us show that L^c, $\mathcal{A}$, u and v satisfy the conditions of Lemma 14.

First, we have $\varepsilon \in L^c$ and $u \notin L^c$. Next, the string $\varepsilon \cdot y_\emptyset \cdot c$ is in L^c since it does not begin with c. The string $uv = ca^{n-3}c \cdot y_\emptyset c$ is in L^c since it contains three c's. The set $\mathcal{A}$ is a fooling set for L^c since $\mathcal{A} \subseteq \mathcal{F}'$. Notice that the string $y_\emptyset c$ is accepted by A from each state in $\{1, 2, \ldots, n-2\}$ since $y_\emptyset$ is accepted by N from each state in $\{1, 2, \ldots, n-2\}$ [5, Theorem 5]. Thus, if S is non-empty, then $cx_S y_\emptyset c \notin L^c$ since by cx_S the NFA A reaches the non-empty set S, from which it accepts $y_\emptyset c$. It follows that the conditions in Lemma 14 are satisfied, and therefore we have $\mathrm{nsc}(L^c) \geq |\mathcal{A}| + 2 = 2^{n-2} + 1$. This completes our proof. $\qquad \square$

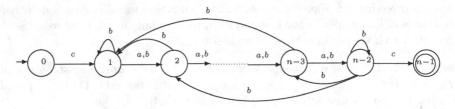

Fig. 3. An NFA of a ternary factor-free language L with $\mathrm{nsc}(L^c) = 2^{n-2} + 1$

It remains to find the bounds for the binary case.

Let us start with the upper bound. Let L be a binary factor-free language with $\mathrm{nsc}(L) = n$, accepted by an n-state NFA N. The NFA N has to have properties as an automaton for a prefix- or suffix-free language, so there is just one final state with no outgoing transition and no transition goes to the initial state. We obtain a similar lemma as in the case of binary prefix-free languages in [7, Lemma 9].

Lemma 16. *There is a positive integer n_0 such that for every $n > n_0$, if L is a binary factor-free language with $nsc(L) = n$ then $\mathrm{nsc}(L^c) \leq 2^{n-2} - 2^{n-4} + 1$.*

For the lower bound, let us consider the language $L = cGc$, where G is accepted by the $(n-2)$-state NFA shown in Fig. 1. Then L is accepted by an n-state NFA A shown in Fig. 3. By a similar strategy as in the binary case of prefix- or suffix-free language, we apply the homomorphism h on the language L. Every string w in $h(L)$ has the form $001u1100$ or $001u1000$ and the string u does not contain the string 00. So in the first case, any proper factor belonging to $h(L)$ does not exist. In the second case, every proper factor belonging to $h(L)$ has to have the form $001u100$ but it has an odd length, and since every string in $h(L)$ has an even length, such a string is not in $h(L)$. So $h(L)$ is factor-free. We get an NFA A for $h(L)$ in a similar way as in the suffix-free or prefix-free cases. The NFA A is minimal and has $2n$ states, so $\mathrm{nsc}(h(L)) = 2n$.

We deal with odd values of n similarly as before. Thus we get the following result.

Lemma 17. *Let $n \geq 8$. There is a binary factor-free language L such that $\mathrm{nsc}(L) = n$ and $\mathrm{nsc}(L^c) \geq \Omega(2^{\frac{n}{2}})$.*

We summarize our results about binary factor-free languages in the following theorem.

Theorem 18. *There is a positive integer n_0 such that for every $n > n_0$, if L is a binary factor-free language with $\mathrm{nsc}(L) = n$ then $\mathrm{nsc}(L^c) \leq 2^{n-2} - 2^{n-4} + 1$. The lower bound is $\Omega(2^{\frac{n}{2}})$.*

Let w be a string such that $w = u_0 v_1 u_1 v_2 u_2 \cdots v_m u_m$, where every u_i and u_j are strings in Σ^*. We say that the string $v = v_1 v_2 \cdots v_m$ is a *subword* of the w. Moreover if $v \neq w$, we say that v is a *proper subword* of w.

For example let $w = abbacb$. Strings $abac, bbb, bc$ are subwords of w, but the string aca is not a subword of w.

Let L be a language. We say L is *subword-free* iff there are no two strings u, v in L such that u is a proper subword of v.

Proposition 19. *Let L be a language. If L is subword-free, then L is finite.*

Theorem 20. *Let $n \geq 4$. Let L be a subword-free language over an alphabet Σ such that $\mathrm{nsc}(L) = n$. Then $\mathrm{nsc}(L) \leq 2^{n-2} + 1$, and the bound is tight if $|\Sigma| \geq 2^{n-2}$.*

Proof. The upper bound is the same as for factor-free languages. To prove tightness, let $\Sigma = \{a_S \mid S \subseteq \{1, 2, \ldots, n-2\}\}$ be an alphabet with 2^{n-2} symbols.

Consider the language L accepted by the NFA $A = (Q, \Sigma, \delta, 0, \{n-1\})$, where $Q = \{0, 1, \ldots, n-1\}$, and the transition function δ is defined as follows: for each symbol a_S in Σ, $\delta(0, a_S) = S$; $\delta(i, a_S) = \emptyset$ if $1 \leq i \leq n-2$ and $i \in S$; $\delta(i, a_S) = \{n-1\}$ if $1 \leq i \leq n-2$ and $i \notin S$; and $\delta(n-1, a_S) = \emptyset$. Notice that each string in L is of length 2, so L is subword-free. Consider the set of pairs $\mathcal{F} = \{(a_S, a_S) \mid S \subseteq \{1, 2, \ldots, n-2\}\}$. Let us show that the set $\mathcal{F}$ is a fooling set for L^c.

(F1) For each S, the string $a_S a_S$ is in L^c, since A goes to S by a_S and a_S is rejected by A from each state in S.

(F2) Let $S \neq T$. Then, without loss of generality, there is a state q in $\{1, 2, \ldots, n-2\}$ such that $q \in S$ and $q \notin T$. Then $a_S a_T$ in not in L^c since A goes to the state q by a_S, and then to the accepting state $n-1$ by a_T.

Hence $\mathcal{F}$ is a fooling set for L^c.

Let $\mathcal{A} = \{(a_S, a_S) \mid \emptyset \neq S \subseteq \{1, 2, \ldots, n-2\}\}$, $u = a_{\{1\}} a_{\{2\}}$, $v = a_{\emptyset}$. Let us show that L^c, $\mathcal{A}$, u, and v satisfy the condition in Lemma 14. First, we have $\varepsilon \in L^c$ and $u \notin L^c$. Next, we have $\varepsilon \cdot v \in L^c$ since it is a one-symbol string, and $uv \in L^c$ since it is of length 3. Finally, notice that $a_{\emptyset}$ is accepted from each state in $\{1, 2, \ldots, n-2\}$. It follows that if $S \neq \emptyset$, then $a_S a_{\emptyset}$ is accepted by A, so it is not in L^c. Hence $\mathcal{A} \cup \{(\varepsilon, v)\}$ and $\mathcal{A} \cup \{(u, v)\}$ are fooling sets for L^c. By Lemma 14, we have $\mathrm{nsc}(L^c) \geq 2^{n-2} + 1$. □

Let us now consider the case for unary alphabets. An arbitrary free language L can contain only one string. We have $L = \{a^n\}$ for some fixed natural number $n \geq 0$. The complement of L consists of every string with length different from n. We can extend the theorem in [7, Theorem 4] by a more general theorem about every free language.

Theorem 21. *Let L be a unary prefix-free or suffix-free or factor-free or subword-free language with $\mathrm{nsc}(L) = n$. Then $\mathrm{nsc}(L^c) = \Theta(\sqrt{n})$.*

Proof. The proof is the same as in [7, Lemma 6]. □

4   Complement on Ideal Languages

Definition 22. *Let L be a language over an alphabet Σ. Then we have four classes of ideals.*

(1) The language L is a right ideal iff $L = L\Sigma^*$.
(2) The language L is a left ideal iff $L = \Sigma^* L$.
(3) The language L is two-sided ideal iff $L = \Sigma^* L\Sigma^*$.
(4) The language L is all-sided ideal iff $L = L \sqcup\!\sqcup \Sigma^*$, where operation $\sqcup\!\sqcup$ is shuffle operation.

The next proposition describes the form of a minimal NFA for some right ideal language.

Proposition 23. *Let L be a language over Σ and let A be a minimal NFA such that $L(A) = L$. The language L is a right ideal if and only if A contains just one final state with a loop on every letter of alphabet Σ.*

Theorem 24. *Let $n \geq 3$. Let L be a right ideal over an alphabet Σ such that $\mathrm{nsc}(L) = n$. Then $\mathrm{nsc}(L) \leq 2^{n-1}$, and the bound is tight if $|\Sigma| \geq 2$.*

Proof. Let $A = (Q, \Sigma, \delta, s, F)$ be a minimal n-state NFA for a right ideal L. Then by Proposition 23 the NFA A has a unique final state f which goes to itself on every input symbol; that is, we have $\delta(f, a) = \{f\}$ for each a in Σ. It follows that in the subset automaton of the NFA A, all final states are equivalent since they accept all the strings in Σ^*. Hence the subset automaton has at most $2^{n-1} + 1$ reachable and pairwise distinguishable states. By interchanging the final and non-final states, we get a DFA B for L^c. The DFA B has a dead state. After removing the dead state, we get an NFA N for L^c of at most 2^{n-1} states.

To prove tightness, let $L = G \cdot b \cdot (a + b)^*$, where G is the language accepted by the binary $(n-1)$-state NFA N shown in Fig. 1. Then L is a right-ideal. The NFA N is minimal because $\mathcal{F} = \{(a^i, a^{n-2-i}b) \mid 0 \le i \le n-2\} \cup \{(a^{n-2}b, \varepsilon)\}$ is a fooling set for L.

Let $\mathcal{F} = \{(u_S, v_S) \mid S \subseteq \{1, 2, \ldots, n-1\}\}$ be a fooling set for G^c as described in [5, Theorem 5]. We prove that the set $\mathcal{F}' = \{(u_S, v_S \cdot b) \mid S \subseteq \{1, 2, \ldots, n-1\}\}$ is a fooling set for L^c.

(F1) For each S, the string $u_S v_S$ is in G^c, so it is not accepted by N. It follows that the string $u_S v_S b$ is not accepted by A. Thus $u_S v_S b$ is in L^c.

(F2) Let $S \ne T$. Then $u_S v_T \notin G^c$ or $u_T v_S \notin G^c$. In the former case, the string $u_S v_T$ is accepted by the NFA N, and therefore the string $u_S v_T b$ is accepted by A. Hence $u_S v_T b \notin L^c$. The latter case is symmetric.

Hence $\mathcal{F}'$ is a fooling set for L^c, which means that $\mathrm{nsc}(L) = 2^{n-1}$.   □

The next proposition describes the form of a minimal NFA for some left ideal languages.

Proposition 25. *Let L be a language over Σ and let A be a minimal NFA such that $L(A) = L$. The language L is a left ideal if and only if there is a minimal NFA A in which the initial state has a loop on every input.*

Theorem 26. *Let $n \ge 3$. Let L be a left ideal over an alphabet Σ such that $\mathrm{nsc}(L) = n$. Then $\mathrm{nsc}(L) \le 2^{n-1}$, and the bound is tight if $|\Sigma| \ge 2$.*

Proof. Let $A = (Q, \Sigma, \delta, s, F)$ be a minimal n-state NFA for a left ideal L. By Proposition 25 we can add a loop on the initial state s on every input symbol, we get an NFA N which is equivalent to A. Since the initial state s of N goes to itself on every input symbol, each reachable subset of the subset automaton of N contains the initial state s, so the number of all reachable subsets is at most 2^{n-1}.

To prove tightness, let the language L be accepted by NFA A in Fig. 4. Then L is a binary left ideal by Proposition 25. The NFA A is minimal because $\mathcal{F} = \{(a^i, a^{n-1-i}) \mid 0 \le i \le n-1\}$ is a fooling set for L.

We are going to consider L^c. Let $\mathcal{F} = \{(u_S, v_S) \mid S \subseteq \{1, 2, \ldots, n-1\}\}$, where string u_S is such that the state 1 goes to the set S after reading u_S in NFA A and the string v_S is such that it is rejected by the NFA from every state $p \in S$ and it is accepted by the NFA from every state $p \notin S$ for any subset S.

Now, we prove that the set $\mathcal{F}' = \{(a \cdot u_S, v_S) \mid S \subseteq \{1, 2, \ldots, n-1\}\}$ is a fooling set for L^c.

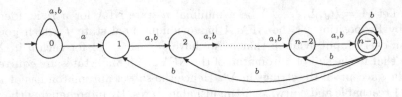

Fig. 4. An NFA of a binary left ideal language L with $\mathrm{nsc}(L^c) = 2^{n-1}$

(F1) For each S, the string $u_S v_S$ is not accepted from state 1, so it follows that the string $a u_S v_S$ is not accepted by A. Thus $a u_S v_S$ is in L^c.

(F2) Let $S \neq T$. Then $u_S v_T \notin L^c$ or $u_T v_S \notin L^c$. Let $u_S v_T$ be accepted by the NFA A, and therefore the string $a u_S v_T$ is accepted by A. Hence $a u_S v_T \notin L^c$. The latter case is symmetric.

Hence $\mathcal{F}'$ is a fooling set for L^c, which means that $\mathrm{nsc}(L) = 2^{n-1}$. $\qquad\square$

Proposition 27. *Let L be a language over Σ and let A be a minimal NFA such that $L(A) = L$. The language L is a two-sided ideal if and only if there is a minimal NFA A with an initial state with a loop on every input and just one final state with a loop on every input.*

Proof. A language L is two-sided ideal if and only if it is left ideal and right ideal; therefore, the proposition follows from Propositions 23, 25. $\qquad\square$

Theorem 28. *Let $n \geq 3$. Let L be a two-sided ideal over an alphabet Σ such that $\mathrm{nsc}(L) = n$. Then $\mathrm{nsc}(L) \leq 2^{n-2}$, and the bound is tight if $|\Sigma| \geq 2$.*

Proposition 29. *Let L be a language over Σ. The language L is an all-sided ideal if and only if there is a minimal NFA A with just one final state and with a loop in every state on every letter of an alphabet Σ, such that $L(A) = L$.*

We can notice that it is not necessary to have a loop for every state on every input symbol. For example a minimal NFA for the binary language L with strings of length at least 3 does not need to have loops on every states except the final one.

Theorem 30. *Let $n \geq 3$. Let L be an all-sided ideal over an alphabet Σ such that $\mathrm{nsc}(L) = n$. Then $\mathrm{nsc}(L^c) \leq 2^{n-2}$, and the bound is tight if $|\Sigma| \geq 2^{n-2}$.*

Proof. The upper bound is the same as for two-sided ideals. To prove tightness, let $\Sigma = \{a_S \mid S \subseteq \{1, 2, \ldots, n-2\}\}$ be an alphabet with 2^{n-2} symbols. Consider the language L accepted by the NFA $A = (\{0, 1, \ldots, n-1\}, \Sigma, \delta, 0, \{n-1\})$ where for each symbol a_S, we have $\delta(0, a_S) = \{0\} \cup S$; $\delta(i, a_S) = \{i\}$ if $i \in S$; $\delta(i, a_S) = \{i, n-1\}$ if $i \in \{1, 2, \ldots, n-2\} \setminus S$; $\delta(n-1, a_S) = \{n-1\}$.

Since in each state of A, we have a loop on every input symbol, the language L is an all-sided ideal by Proposition 29.

Let $\mathcal{F} = \{(a_S, a_S) \mid S \subseteq \{1, 2, \ldots, n-2\}\}$. Let us show that $\mathcal{F}$ is a fooling set for L^c.

(F1) For each S, the NFA A reaches the set $\{0\} \cup S$ by a_S. By the next a_S, the NFA A remains in the set $\{0\} \cup S$, and rejects. Thus $a_S a_S \in L^c$.

(F2) Let S and T be two subsets of $\{1, 2, \ldots, n-2\}$ with $S \neq T$. Without loss of generality, there is a state i with $i \in S$ and $i \notin T$. By a_S, the NFA A goes to $\{0\} \cup S$. Since $i \in S$, the NFA A goes to i by a_S. Then it goes to the state $n-1$ by a_T since $i \notin T$. Hence A accepts $a_S a_T$, and therefore $a_S a_T \notin L^c$.

Thus $\mathcal{F}$ is a fooling set for L^c. It follows that $\mathrm{nsc}(L^c) \geq 2^{n-2}$.   □

Let us consider a unary alphabet. Every type of ideal language has the form $L = \{a^k \mid k \geq n\}$, where n is some fixed natural number. Thus, every minimal NFA A for every type of an ideal language L has a tail of $n-1$ states ending by final state with a loop (see the example in Fig. 5). Such an automaton A is a DFA, so after exchanging of finality and nonfinality of states we get the DFA A' with every state final except one, which is the dead state. After leaving the dead state we get the NFA B with $n-1$ states accepting a complement L^c.

These considerations can be summarized in the following theorem.

Theorem 31. *Let L be ideal over an unary alphabet, such that $\mathrm{nsc}(L) = n$. Then $\mathrm{nsc}(L^c) = n - 1$.*

Fig. 5. Minimal NFA for language $a^k, k \geq n$

5   Conclusions

Let us summarize our results. Let L be a language such that $\mathrm{nsc}(L) = n$.

Firstly, let us review the case for alphabets of size 3 or more. For the suffix-free and prefix-free cases the results come from [7]. The bounds 2^{n-1} are tight in both cases. For the case of factor-free, the bounds $2^{n-2} + 1$ are tight. For the case of subword-free the upper bound is $2^{n-2} + 1$ and it is tight when $|\Sigma| \geq 2^{n-2}$.

Secondly, let us review the case for binary free languages. For suffix-free languages, the lower bound is $2^{\lfloor \frac{n}{2} \rfloor - 1}$ and upper bound is $2^{n-1} - 2^{n-3} + 2$, for prefix-free languages the lower bound is the same as in case of suffix-free and the upper bound [7, Lemma 9] is $2^{n-1} - 2^{n-3} + 1$, for factor-free the lower bound is $\Omega(2^{\frac{n}{2}})$ and the upper bound is $2^{n-2} - 2^{n-4} + 1$.

For right and left ideals the bounds 2^{n-1} are tight. For two-sided ideals the bounds 2^{n-2} are tight. For all-sided languages the upper bound is 2^{n-2} and it is tight when $|\Sigma| \geq 2^{n-2}$.

Finally, we will discuss the case for unary alphabets. In this case, the situation is the same for every class is : the lower bound is $\Theta(\sqrt{n})$ and the upper bound is $\Theta(\sqrt{n})$.

For ideals, the situation is the same: $\text{nsc}(L^c) = n - 1$.

The possibility of improving the bounds for binary cases for prefix-, suffix- and factor-free languages remains open. Also in the case for subword-free languages it remains to solve the binary case. Also the possibility of finding non-exponential alphabets for witness languages for the lower bound in the case of subword-free languages remains open. The possibility of finding non-exponential alphabet for witness language for lower bound in case of all-sided ideal languages remains open.

Acknowledgements. I would like to thank to Galina Jirásková for many helpful discussions which led to writing this article.

References

1. Birget, J.C.: Partial orders on words, minimal elements of regular languages, and state complexity. Theoret. Comput. Sci. **119**, 267–291 (1993). ERRATUM: Partial orders on words, minimal elements of regular languages, and state complexity (2002). http://clam.rutgers.edu/birget/papers.html
2. Cmorik, R., Jirásková, G.: Basic operations on binary suffix-free languages. In: Kotásek, Z., Bouda, J., Černá, I., Sekanina, L., Vojnar, T., Antoš, D. (eds.) MEMICS 2011. LNCS, vol. 7119, pp. 94–102. Springer, Heidelberg (2012)
3. Chrobak, M.: Finite automata and unary languages. Theoret. Comput. Sci. **47**, 149–158 (1986)
4. Holzer, M., Kutrib, M.: Nondeterministic descriptional complexity of regular languages. Int. J. Found. Comput. Sci. **14**, 1087–1102 (2003)
5. Jirásková, G.: State complexity of some operations on binary regular languages. Theoret. Comput. Sci. **330**, 287–298 (2005)
6. Jirásková, G.: Note on minimal automata and uniform communication protocols. In: Grammars and Automata for String Processing: From Mathematics and Computer Science to Biology, and Back: Essays in Honour of Gheorghe Paun, pp. 163–170. Taylor and Francis (2003)
7. Jirásková, G., Mlynárčik, P.: Complement on prefix-free, suffix-free, and non-returning NFA languages. In: Jürgensen, H., Karhumäki, J., Okhotin, A. (eds.) DCFS 2014. LNCS, vol. 8614, pp. 222–233. Springer, Heidelberg (2014)
8. Glaister, I., Shallit, J.: A lower bound technique for the size of nondeterministic finite automata. Inform. Process. Lett. **59**, 75–77 (1996)
9. Sakoda, W.J., Sipser, M.: Nondeterminism and the size of two-way finite automata. In: Proceedings of the 10th Annual ACM Symposium on Theory of Computing, pp. 275–286 (1978)
10. Sipser, M.: Introduction to the Theory of Computation. PWS Publishing Company, Boston (1997)

Universal Disjunctive Concatenation and Star

Nelma Moreira[1], Giovanni Pighizzini[2]([✉]), and Rogério Reis[1]

[1] Centro de Matemática e Faculdade de Ciências da, Universidade do Porto,
Porto, Portugal
{nam,rvr}@dcc.fc.up.pt
[2] Dipartimento di Informatica, Università degli Studi di Milano, Milan, Italy
pighizzini@di.unimi.it

Abstract. Two language operations that can be expressed by suitably combining complement with concatenation and star, respectively, are introduced. The state complexity of those operations on regular languages is investigated. In the deterministic case, optimal exponential state gaps are proved for both operations. In the nondeterministic case, for one operation an optimal exponential gap is also proved, while for the other operation an exponential upper bound is obtained.

1 Introduction

In a recent paper, we investigated automata with partially specified behaviors, shortly called *don't care automata* (dcFA) [5]. These devices are defined as standard nondeterministic finite state automata, with the only difference that they have two sets of final states: the set of *accepting states* and the set of *rejecting states*. In this way, a dcFA A defines two languages: the accepted language $\mathcal{L}^{\oplus}(A)$ and the rejected language $\mathcal{L}^{\ominus}(A)$. It is required that these two languages are disjoint.

Don't care automata can be interesting in situations where it is not necessary to fix the behavior on each possible string, because, for instance, some strings will never be received by the automaton (e.g., when the input of the automaton is generated by a source which produces strings in a certain format), or because for other reasons the answer of the automaton on some strings is not interesting. In the same paper, we studied the optimal reductions, in terms of states, of such devices to compatible deterministic automata; namely to standard deterministic automata which "agree" with the behavior of the given don't care automata.

Triggered by a paper published in 1994 [7], a lot of work has been conducted in the last 20 years to study the state complexity of operations on finite automata. Inspired by this research, we started to investigate how standard operations (boolean, concatenation and Kleene star) could be extended to dcFAs.

N. Moreira and R. Reis—Authors partially funded by the European Regional Development Fund through the programme COMPETE and by the Portuguese Government through the FCT under project UID/MAT/00144/2013.
G. Pighizzini—Author partially supported by MIUR under the project PRIN "Automi e Linguaggi Formali: Aspetti Matematici e Applicativi", code H41J120001 90001.

J. Shallit and A. Okhotin (Eds.): DCFS 2015, LNCS 9118, pp. 197–208, 2015.
DOI: 10.1007/978-3-319-19225-3_17

An obvious requirement is that an operation extended to dcFAs matches the original operation, if the behavior of the automaton is fully specified. For instance, considering union, a string should be accepted when it is accepted by at least one of the two given automata and should be rejected when it is rejected by both automata. Considering don't care values, it is quite obvious how to extend this behavior, as depicted in the table to the right, where *yes* and *no* represent acceptance and rejection, respectively, and *?* represents an unspecified behavior. Similar tables can be filled in for intersection and complement. Notice that this is related to *three-valued logic* [6].

	no	*?*	*yes*
no	no	?	yes
?	?	?	yes
yes	yes	yes	yes

For the other regular operations, concatenation and star, the situation is slightly more complicated and, probably, more interesting. Let us consider two dcFAs A and B on the same input alphabet Σ that have completely specified behaviors, i.e., $\mathcal{L}^{\oplus}(A) \cup \mathcal{L}^{\ominus}(A) = \mathcal{L}^{\oplus}(B) \cup \mathcal{L}^{\ominus}(B) = \Sigma^{\star}$. A dcFA C for concatenation should accept all the strings which can be obtained by concatenating strings accepted by A and B, i.e., $\mathcal{L}^{\oplus}(C) = \mathcal{L}^{\oplus}(A)\mathcal{L}^{\oplus}(B)$, and should reject all the strings which cannot be obtained in this way; namely, all the strings w such that for each factorization $w = uv$ either $u \notin \mathcal{L}^{\oplus}(A)$ or $v \notin \mathcal{L}^{\oplus}(B)$. Since in this case $\mathcal{L}^{\ominus}(A)$ and $\mathcal{L}^{\ominus}(B)$ are the complement of $\mathcal{L}^{\oplus}(A)$ and $\mathcal{L}^{\oplus}(B)$, respectively, this is equivalent to saying that for each factorization $w = uv$ either $u \in \mathcal{L}^{\ominus}(A)$ or $v \in \mathcal{L}^{\ominus}(B)$.

This leads to consider a new language operation, that we call *universal disjunctive concatenation* and, in a similar way, starting from the star, another new operation called *universal disjunctive star*. This paper is devoted to investigating the state complexity of these two operations in both deterministic and nondeterministic cases. Using the fact that these two operations can be expressed by combining complement with concatenation and star, respectively, we prove that in the deterministic case their state complexity is exponential.

We deepen this investigation by considering the nondeterministic case. We prove that given two nondeterministic automata (NFAs) A and B with m and n states, there exists an NFA accepting the universal disjunctive concatenation of the languages accepted by A and B, with at most 2^{m+n} states. Furthermore, the exponential gap cannot be reduced in the worst case.

We also prove that for each NFA A with n states there exists an NFA with no more than 2^n states accepting the universal disjunctive star of $\mathcal{L}(A)$.

In the final part of the paper, we shortly discuss the state complexity of operations on don't care automata.

2   Universal Disjunctive Concatenation

Given a language L over an alphabet Σ, let us denote by $\mathsf{pref}(L)$ the language consisting of *all prefixes* of strings in L. The set of all nonempty prefixes of L, i.e., $\mathsf{pref}(L) \setminus \{\varepsilon\}$, is denoted by $\mathsf{pref}_+(L)$. By $\mathsf{p}(L)$ we denote the *prefix-closed interior* of L; namely the largest subset of L which is prefix closed.

Similar definitions can be given by considering suffixes. Hence, we let $\mathsf{suff}(L)$ denote the set of all suffixes of strings in L, $\mathsf{suff}_+(L)$ denote the set $\mathsf{suff}(L) \setminus \{\varepsilon\}$, and $s(L)$ denote the *suffix-closed interior* of L, namely the largest subset of L which is suffix closed.

The complement of L will be denoted by $\overline{L}$.

Definition 1. *Let L_1 and L_2 be languages over an alphabet Σ. The* universal disjunctive concatenation *of L_1 and L_2 is the language $L_1 \odot L_2$ defined as*

$$L_1 \odot L_2 = \{w \in \Sigma^* \mid \forall x_1, x_2, \; w = x_1 x_2 \; \Rightarrow \; (x_1 \in L_1 \lor x_2 \in L_2)\}.$$

Example 2. Given $\Sigma = \{a, b\}$, consider L_1 the set of all strings containing an even number of occurrences of a and $L_2 = a(a + b)^\star$. Then, $L_1 \odot L_2 = (aa + b)^\star$. Given a string w, let us number the occurrences of the letter a in w, starting from 1. Each prefix of w containing an even number of a's belongs to L_1. If a prefix ends with an odd numbered a then it does not belong to L_1. Thus, in order to have $w \in L_1 \odot L_2$, the remaining suffix should belong to L_2; hence, it should start with an a. Hence each odd numbered a should be immediately followed by another a.

We point out the role of the alphabet Σ we are considering in the definition of the operation $\odot$. For instance, given $L = a^\star$, if $\Sigma = \{a\}$ then $L \odot L = a^\star = L$. However, if the alphabet we are considering is $\Sigma = \{a, b\}$, then $L \odot L = a^\star + a^\star ba^\star$; namely, L is the set of all strings which contain at most one occurrence of the letter b. This is due to the fact that this operation involves, in some sense, complementation, as we will explain below. We can observe that, for $\Sigma = \{a, b\}$, $\overline{L}$ is the set of strings that contain at least one occurrence of the letter b and, hence, $\overline{L}\,\overline{L}$ is the set of all strings that contain at least two occurrences of b. Thus, its complement coincides with $L \odot L$.

Actually, this is a general property. In fact, just by using the definition of $\odot$, we can observe that given two languages L_1 and L_2, $w \notin L_1 \odot L_2$ if and only if there are two strings u, v such that $w = uv$, $u \notin L_1$, and $v \notin L_2$. Thus,

$$L_1 \odot L_2 = \overline{\overline{L_1}\,\overline{L_2}}. \tag{1}$$

As a consequence, the operation $\odot$ preserves regularity; namely, if L_1 and L_2 are regular, then $L_1 \odot L_2$ is regular too. A special case of this operation when $L_1 = \emptyset$ was studied by Birget [2]. We now study some basic properties of the $\odot$ operation.

Proposition 3. *The operation $\odot$ is associative, i.e., $L_1 \odot (L_2 \odot L_3) = (L_1 \odot L_2) \odot L_3$, for all languages L_1, L_2, L_3.*

Because $\odot$ is associative it makes sense to write $L_1 \odot L_2 \odot L_3$ and it is not difficult to realize that

$$L_1 \odot L_2 \odot L_3 = \{w \mid \forall x, y, z, \; w = xyz \; \Rightarrow \; x \in L_1 \lor y \in L_2 \lor z \in L_3\}.$$

Proposition 4. *The $\odot$ operation has an identity element. For any language L, $\Sigma^+ \odot L = L \odot \Sigma^+ = L$.*

Proof. Immediate consequence of Eq. (1), observing that $\overline{\Sigma^+} = \{\varepsilon\}$.   □

We note there are no languages L_1, L_2, apart from Σ^+, such that $L_1 \odot L_2 = \Sigma^+$, i.e. $\odot$ has no nontrivial inverses. In general, it is easy to see that

Proposition 5. *Let L_1, L_2 be languages. If $\varepsilon \notin L_1$, then $L_1 \odot L_2 \subseteq L_2$ and $L_2 \odot L_1 \subseteq L_2$.*

Proposition 6. *Let L be a language. Then $\emptyset \odot L = s(L)$ and $L \odot \emptyset = p(L)$.*

Proof. From the definition of $\odot$, it easily follows that a string w belongs to $\emptyset \odot L$ only if each suffix of w belongs to L. Hence, $\emptyset \odot L \subseteq s(L)$. Conversely, if $w \in s(L)$ then each suffix of w belongs to L, which would imply that $w \in \emptyset \odot L$. In a similar way, it can be proved that $L \odot \emptyset = p(L)$.   □

Proposition 7. *Let $L, X \subseteq \Sigma^\star$ be languages.*

(a) If L is prefix closed then $L \subseteq L \odot X$.
(b) $\mathsf{pref}(L) \subseteq \mathsf{pref}(L) \odot L$.
(c) If $\varepsilon \notin L$ then $\mathsf{pref}(L) \odot L = \mathsf{pref}(L)$.
(d) $\mathsf{pref}_+(L) \odot L = L$.

Proof. (a) is trivial.
(b) Immediately follows from (a).
(c) Since $w = w\varepsilon$ and $\varepsilon \notin L$, from $w \in \mathsf{pref}(L) \odot L$, we obtain $w \in \mathsf{pref}(L)$. Hence $\mathsf{pref}(L) \odot L \subseteq \mathsf{pref}(L)$. The converse inclusion is given in (b).
(d) Let $w \in L$, then for every x, y such that $xy = w$, $x \in \mathsf{pref}_+(L)$ unless $x = \varepsilon$, but in that case $y = w \in L$. On the other hand, if $w \in \mathsf{pref}_+(L) \odot L$ then, as $\varepsilon w = w$, necessarily $w \in L$.   □

Notice that if $\varepsilon \in L$ then $\mathsf{pref}(L) \odot L$ could differ from both L and $\mathsf{pref}(L)$. For instance, given $\Sigma = \{a\}$, for $L = \{\varepsilon, aa\}$, we have $\mathsf{pref}(L) \odot L = \{\varepsilon, a, aa, aaa\}$.

We can prove properties similar to those in Proposition 7, considering suffixes:

Proposition 8. *Let $L, X \subseteq \Sigma^*$ be languages.*

(a) If L is suffix closed then $L \subseteq X \odot L$.
(b) $\mathsf{suff}(L) \subseteq L \odot \mathsf{suff}(L)$.
(c) If $\varepsilon \notin L$ then $L \odot \mathsf{suff}(L) = \mathsf{suff}(L)$.
(d) $L \odot \mathsf{suff}_+(L) = L$.

Proposition 9. *If L_1 is prefix closed and L_2 is suffix closed, then $L_1 L_2 \subseteq L_1 \odot L_2$.*

Proof. Suppose $w \in L_1 L_2$. Let $x \in L_1$ and $y \in L_2$ be such that $w = xy$. Then for all strings u, v verifying $w = uv$ either u is a prefix of x, thus implying $u \in L_1$, or v is a suffix of y, thus implying $v \in L_2$. This allows to conclude that $w \in L_1 \odot L_2$.   □

Our previous example with $\Sigma = \{a, b\}$ and $L_1 = L_2 = a^\star$ shows that the inclusion proved in Proposition 9 can be proper.

Now we are going to study the state complexity of the operation ⊚. First of all, by using results on the state complexity of concatenation [7], we can obtain the following bound:

Theorem 10. *For all integers* $m, n \geq 2$, *let* A' *be an* m-*state DFA and* A'' *an* n-*state DFA. Then any DFA that accepts* $\mathcal{L}(A') \oplus \mathcal{L}(A'')$ *needs at most* $m2^n - (m-f)2^{n-1}$ *states, where* f *is the number of* final *states of* A'. *Futhermore, this bound is tight.*

Proof. It follows immediately from Eq. (1) that the state complexity of ⊚ coincides with the state complexity of concatenation because the state complexity of the complement of a language L coincides with the state complexity of L. Hence, the upper bound follows from Theorem 2.3 in [7], after switching the role of final and nonfinal states in the automaton A', due to the complementation. The lower bound also derives from a result in the same paper (Theorem 2.1).   □

The investigation of the state complexity of ⊚ is now deepened by proving that the bound in Theorem 10 cannot be reduced if we allow the resulting automaton to be nondeterministic. To this aim, for each integer $n \geq 1$, let us consider the following language over $\Sigma = \{a, b\}$:

$$L_n = (a(a+b)^{n-1})^*(\varepsilon + a(a+b)^{<n}) + (b(a+b)^{n-1})^*(\varepsilon + b(a+b)^{<n}), \quad (2)$$

where $(a+b)^{<n}$ denotes *less than* n repetitions of $a+b$. In other words, a string w belongs to L_n if and only if the same symbol occurs in all positions $in+1$ of w, with $i \geq 0$ and $in+1 \leq |w|$.

Theorem 11. *Let* L_n *be the language defined in (2). Then:*

(a) *The minimum DFA accepting* L_n *has* $2n+2$ *states.*
(b) $s(L_n) = \{x^k y \mid \text{ for some } x \in \{a, b\}^n, k \geq 0, y \in \text{pref}(x)\}$.
(c) *Each NFA accepting* $s(L_n)$ *requires at least* 2^n *states.*

Proof. (a) A DFA accepting L_n is depicted in Fig. 1. By a standard distinguishability argument, it can be proved that it is minimal.
(b) From the definition of L_n, we can observe that a string $w \in s(L_n)$ if and only if each two symbols of w at distance n, i.e., with $n-1$ symbols in between, are equal. This implies that w consists of a prefix x of length n which is repeated a certain number of times and a suffix y of length $< n$, which is a prefix of x.
(c) Consider the set $S = \{(x, x) \mid x \in \{a, b\}^n\}$. From (b), it follows that for $x, y \in \{a, b\}^n$, $xx \in s(L_n)$ and $xy \notin s(L_n)$. Hence, S is a *fooling set* for $s(L_n)$ and each NFA accepting it requires at least $\#S$ states [1].   □

As a consequence of Theorem 11 we can now conclude that the exponential upper bound given in Theorem 10 cannot be reduced, even if the resulting automaton is nondeterministic.

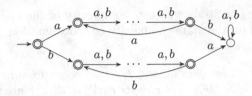

Fig. 1. The minimum DFA accepting L_n.

Theorem 12. *Let A be the 1-state automaton accepting the empty language. For each integer n there exists an n-state DFA B accepting a language defined over a binary alphabet such that each NFA accepting $\mathcal{L}(A) \odot \mathcal{L}(B)$ requires at least $2^{\lfloor (n-2)/2 \rfloor}$ states.*

Proof. Given $k = \lfloor (n-2)/2 \rfloor$, consider the language L_k, defined according to (2). If n is even, we choose as DFA B the minimum automaton accepting L_k. If n is odd, we obtain B by splitting the trap state of the minimum automaton accepting L_k in two states (this can be done with a simple change in the transition graph in Fig. 1). This adds one extra state, without changing the accepted language.

In both cases, the DFA B has n states and recognizes $L_{\lfloor (n-2)/2 \rfloor} = L_k$. From Proposition 6, it turns out that $\emptyset \odot L_k = \mathsf{s}(L_k)$. Hence, by Theorem 11, each NFA accepting $\emptyset \odot L_k$ requires 2^k states. $\qquad \square$

We point out that Theorem 12 improves a similar gap proved in [2], by considering a three-letter alphabet.

Now, we further investigate the nondeterministic case, by studying the state complexity of $\odot$ in the case of nondeterministic automata. The upper bound in Theorem 10 is derived from (1) which uses a double complementation. This leads to a double exponential upper bound on the state complexity, when the given automata are nondeterministic. However, it produces a deterministic automaton. It could be interesting to see if the state complexity can be reduced, if we want to derive a *nondeterministic* automaton. This could be done using *alternating automata* [4]. However, in the next result we present a direct construction.

Theorem 13. *For each integers $m, n \geq 1$, let A' be an m-state NFA and A'' an n-state NFA, then there is an NFA that accepts $\mathcal{L}(A') \odot \mathcal{L}(A'')$ with at most 2^{m+n} states.*

Proof. Let $A' = \langle Q', \Sigma, \delta', I', F' \rangle$, $A'' = \langle Q'', \Sigma, \delta'', I'', F'' \rangle$, $L' = \mathcal{L}(A')$, and $L'' = \mathcal{L}(A'')$. We define an NFA $A = \langle Q, \Sigma, \delta, i, F \rangle$ which accepts the language $L' \odot L''$. First we informally explain how A works, in the case the two given automata A' and A'' are deterministic. Let i' and i'' be their initial states. Given a string $w \in \Sigma^\star$, in order to test if $w \in L' \odot L''$, the automaton A has to check that for each prefix of w which is rejected by A' the corresponding suffix is accepted by A''. To this aim, while reading w, A simulates the deterministic control of A'. Each time that in the simulation A' reaches a nonfinal state — namely the prefix read so far does not belong to L' — the automaton A starts the simulation of a computation of A'' to check if the remaining suffix belongs to L''. In this way, A works by simulating in parallel a computation of A' on the given input and, for each suffix corresponding to a prefix not in L', one computation of A''. At the end, A must verify that all the computations are accepting.

The automaton A is implemented using states (q, α) where $q \in Q'$ and $\alpha \subseteq Q''$. The first component is used for the simulation of A', the second one keeps track of the states reached by A'' on the suffixes under examination. So, if the initial state i' of A' is final, then A starts its computation in $(i', \emptyset)$. Otherwise, since $\varepsilon \notin L'$, A needs to verify that all the input belongs to L'' and hence it starts the computation in $(i', \{i''\})$. When in the state (q, α) the automaton A reads a symbol σ, it moves to the state (p, β) where $p = \delta'(q, \sigma)$ and β contains all the states that are reached by states in α reading σ. In this way, the second component continues the inspection of input suffixes. However, if $p \notin F'$ then A needs to simulate A'' on the incoming input suffix. To this aim, in this case, β also contains the state i''. At the end of the computation, A has to verify that all the suffixes under examination are accepted by A''. Hence, if (q, α) is the state reached at the end of the computation, all states in α should belong to F''. However, since we also have to verify that either the input w belongs to L' or $\varepsilon \in L''$, we should additionally ask if either $q \in F'$ or $i'' \in F''$. Notice that the resulting automaton A is deterministic. The resulting DFA A is formally defined with $Q = Q' \times 2^{Q''}$; $i = (i', \emptyset)$ if $i' \in F'$ and $i = (i', \{i''\})$ otherwise; $F = \{(q, \alpha) \mid \alpha \subseteq F'' \wedge (q \in F' \vee i'' \in F'')\}$ and

$$\delta((q, \alpha), \sigma) = \begin{cases} (\delta'(q, \sigma), \delta''(\alpha, \sigma)), & \text{if } \delta'(q, \sigma) \in F'; \\ (\delta'(q, \sigma), \delta''(\alpha, \sigma) \cup \{i''\}), & \text{otherwise.} \end{cases}$$

Furthermore, in the construction above, we can observe that all the states (q, α) with $q \notin F'$ and $i'' \notin \alpha$ are not reachable in A. Then the total number of states of A is at most $m2^n - (m - f)2^{n-1}$, where f is the number of final states of A', which is exactly the same number derived in Theorem 10.

When A'' is nondeterministic, the construction is slightly more complicated. In fact, on each suffix we could have different computations. We need to verify that at least one of them is accepting. To do that, we simply use nondeterministic choices. Hence, when in the state (q, α) the automaton reads a symbol σ, each possible next state (p, β) is obtained by taking $p = \delta'(q, \sigma)$ and by nondeterministically choosing a state $s \in \delta''(r, \sigma)$ to be in β for each state $r \in \alpha$. When $p \notin F'$, the automaton A needs to start a computation of A'' on the incoming suffix. Hence, a nondeterministically chosen state $i'' \in I''$ is added to β. The formal definition of A, in the case of A' deterministic and A'' nondeterministic is the following:

- $Q = Q' \times 2^{Q''}$,
- $I = \begin{cases} \{(i', \emptyset)\}, & \text{if } i' \in F'; \\ \{(i', \{i''\}) \mid i'' \in I''\}, & \text{otherwise;} \end{cases}$
- for $(q, \alpha) \in Q' \times 2^{Q''}$, $\sigma \in \Sigma$, let us consider the following set

$$\text{next}(\alpha, \sigma) = \{\gamma \in 2^{Q''} \mid \exists f : \alpha \to \gamma \text{ s.t. } f \text{ is surjective and}$$
$$f(r) = s \Rightarrow s \in \delta''(r, \sigma)\},$$

the set $\delta((q, \alpha), \sigma)$ contains all the states $(p, \beta) \in Q' \times 2^{Q''}$ such that $p = \delta'(q, \sigma)$ and $\beta = \begin{cases} \gamma, & \text{if } p \in F'; \\ \gamma \cup \{i''\}, & \text{otherwise;} \end{cases}$ for some $\gamma \in \text{next}(\alpha, \sigma), i'' \in I''$,

$$- F = \begin{cases} \{(q, \alpha) \mid q \in F', \alpha \subseteq F''\}, & \text{if } I'' \cap F'' = \emptyset; \\ \{(q, \alpha) \mid q \in Q', \alpha \subseteq F''\}, & \text{otherwise.} \end{cases}$$

Finally, when even A' is nondeterministic, we can preliminary convert it into an equivalent DFA applying the subset construction, and then proceed as above described. In this case, the set of states of the resulting automaton is a subset of $2^{Q'} \times 2^{Q''}$. Hence, its cardinality is bounded by 2^{m+n}. □

From Theorem 12, it follows that the exponential upper bound in Theorem 13 cannot be reduced.

3   Universal Disjunctive Star

In this section we study the other operation we are interested in, the universal disjunctive star, defined in the following way:

Definition 14. *Let $L \subseteq \Sigma^\star$ be a language. Let $L^{\odot 0} = \Sigma^+$ and $L^{\odot k} = L^{\odot k-1} \odot L$, for each integer $k > 0$. Then we define the* universal disjunctive star *as*

$$L^\circledast = \bigcap_{k \geq 0} L^{\odot k}.$$

Notice that by this definition, it turns out that a string $w \in L^\circledast$ if and only if for each factorization of w as $w = x_1 x_2 \cdots x_k$, with $k \geq 1$, at least one factor x_i belongs to the language L. We now show that we can restrict our attention to the nonempty factors. Due to space limitations, we omit the proof of the following propositions.

Proposition 15. *For each integer $i \geq 0$:*

(a) *If $\varepsilon \in L$, then $\Sigma^{<i} \subseteq L^{\odot i}$.*
(b) *If $w \in \Sigma^\star$, $|w| = i$, and $w \in L^{\odot i}$, then for each $j > i$, $w \in L^{\odot j}$.*
(c) *If $\varepsilon \notin L$ and $|w| = i$, $w \in L^{\odot i}$ if and only if for each $j > i$, $w \in L^{\odot j}$.*

As a consequence we obtain:

Proposition 16. *Given $L \subseteq \Sigma^\star$ and $w \in \Sigma^\star$, $w \in L^\circledast$ if and only if, for each $0 \leq i \leq |w|$, $w \in L^{\odot i}$.*

As a consequence of the previous proposition, we get that a string $w \in L^\circledast$ if and only if for each decomposition of w in at most $|w|$ factors, at least one of them belongs to L. Hence, we can express $L^\circledast$ as:

$$L^\circledast = \{w \in \Sigma^\star \mid \forall k \leq |w| \, \forall x_1, \ldots, x_k \in \Sigma^+, w = x_1 \cdots x_k, \exists i \leq k \; x_i \in L\}. \quad (3)$$

We can also observe that

$$L^{\circledast} = \overline{(\overline{L})^{\star}} \tag{4}$$

that implies that the class of regular languages is closed under of this operation. Considering Eq. (4), $\circledast$ is exactly the Kleene interior studied by Brzozowski et al. [3] when characterising the number of different languages that can occur by successive application of star and complement to a given regular language. Furthermore, using the results about the state complexity of the star [7, Corollary 3.2, Theorem 3.3], we immediately obtain the following result:

Theorem 17. *For any n-state DFA A, $n \geq 1$, there exists a DFA A' of at most $2^{n-1} + 2^{n-2}$ states such that $\mathcal{L}(A') = (\mathcal{L}(A))^{\circledast}$. Furthermore, this bound cannot be reduced in the worst case.*

We now consider the state complexity of $\circledast$ in the nondeterministic case. We prove that the upper bound remains exponential.

Theorem 18. *For any n-state NFA A, $n \geq 1$, there exists an NFA A' with at most 2^n states such that $\mathcal{L}(A') = (\mathcal{L}(A))^{\circledast}$.*

Proof. To make clearer the main argument used to define A' from A, first we discuss the construction for the deterministic case. Subsequently, we will describe the generalization to the nondeterministic case.

Let us start by supposing $A = \langle Q, \Sigma, \delta, I, F \rangle$, with $I = \{i\}$, is deterministic. We also suppose that $\varepsilon \notin L$, i.e., $i \notin F$. We describe a DFA $A' = \langle Q', \Sigma, \delta', I', F' \rangle$ which accepts the language $L^{\circledast}$.

First of all, we remind the reader that, by definition, $\varepsilon \notin L^{\circledast}$. So let us consider an input $w \in \Sigma^+$. The automaton A' has to verify that for each factorization of w in $k \geq 1$ nonempty factors, at least one of them belongs to the language L. In the following, a factorization satisfying such a property will be said to be *accepted*.

A' works by exploring all input factorizations in parallel computation branches that are generated while reading the input in the following way. Suppose that a string u has been read and consider a computation branch corresponding to a factorization $u = u_1 u_2 \cdots u_h$ of the input in $h \geq 1$ nonempty strings. Before reading the next input symbol γ, the computation branch is split into two branches according to the following possible factorizations of $w\gamma$:

(a) $u_1, \ldots, u_h\gamma$; namely, γ will be considered as a further symbol of the hth factor,

(b) $u_1, \ldots, u_h, \gamma$; namely, γ will be considered as a new factor.

Now suppose that the string u is the prefix of the input w that has been read so far, namely, $w = uv$, for some $v \in \Sigma^{\star}$. Let $w = w_1 \cdots w_k$ be a factorization of w in a computation branch which is obtained, after inspecting the suffix v, from the computation branch on the factorization $u = u_1 \cdots u_h$. Then, $w_j = u_j$ for $j = 1, \ldots, h-1$ (however u_h could be a *proper* prefix of w_h).

Suppose $u_j \in L$, for some $j < h$. In this case, the factorization of w is accepted regardless the suffix v and all factors $u_{j'}$, with $j' > j$, namely, *each* input factorization which begins by $u_1, \ldots, u_j$ is accepted and, thus, the input symbols after the factor u_j do not need to be inspected.

On the other hand, if $u_j \notin L$ for each $j < h$, then the computation branch has to test the membership to L of the factor u_h. To do this, it remembers the state $q = \delta(i, u_h)$. We observed that before reading the next input symbol, the computation branch is split in two. In the computation branch corresponding to (a), the simulation of A is continued from the state q. For (b) there are two possibilities. If $q \in F$, i.e., $u_h \in L$, then all factorizations beginning by $u_1, \ldots, u_h$ are accepted and A' does not need to consider the remaining part of the input. Thus, the second computation branch is not needed. Otherwise a computation branch corresponding to (b) is generated in order to inspect, from the initial state, a factor which begins with the next input symbol.

The automaton has to accept when each computation branch discovers that its corresponding factorization is accepted. This can happen either by testing during the computation that a factor is in L or at the end of the input by reaching a final state, so proving that the last factor is in L.

For each computation branch, the automaton A' needs only to remember the current state. This allows to implement A' by keeping in its finite state control the set of states which are reached by computation branches. More precisely, the formal definition of $A' = \langle Q', \Sigma, \delta', i', F' \rangle$ is as follows

- $Q' = 2^Q$,
- $i' = \{i\}$,
- for $\alpha \in Q'$, $\sigma \in \Sigma$, $\delta'(\alpha, \sigma) = \begin{cases} \delta(\alpha, \sigma), & \text{if } \delta(\alpha, \sigma) \subseteq F; \\ \delta(\alpha, \sigma) \cup \{i\}, & \text{otherwise}; \end{cases}$
- $F' = \{\alpha \in Q' \mid \alpha \subseteq F\}$.

In particular, in the definition of $\delta'(\alpha, \sigma)$, the part $\delta(\alpha, \sigma)$ corresponds to the computation branch (a), while the part $\{i\}$ corresponds to (b) and it is not added when after reading a symbol σ all the states are final, namely, all factors that end in σ (after the already inspected input prefix) are in L. See Fig. 2.

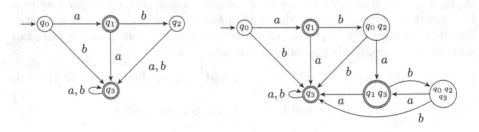

Fig. 2. Let $\Sigma = \{a, b\}$ and $L = \Sigma^* \backslash \{ab\}$. The DFA depicted on the left recognizes $L \backslash \{\varepsilon\}$. Applying to it the construction presented in the proof of Theorem 18, the DFA depicted on the right is obtained, which accepts $L^\circledast = \Sigma^* \backslash \{(ab)^*\}$.

Now, we switch to the nondeterministic case supposing that $A = \langle Q, \Sigma, \delta, I, F \rangle$ is nondeterministic, where $I = \{i\}$ with $i \notin F$. We build an NFA A' which accepts the language $L^\circledast$. The general working strategy of A' is similar to that described for the deterministic case: in parallel computation branches, A' inspects all different factorizations of the input. However, in this case A' needs also to simulate the nondeterministic choices of A.

In the construction for the operation $\circledcirc$, the purpose of the nondeterministic simulation was to check the membership of an input suffix to the language L''. In this construction the situation is more delicate, because we have to check input factors instead of suffixes. For the initial state there is only one choice. However, from a state r we can have a nondeterministic choice which leads to the acceptance of a certain factor x and to the rejection of another factor y, and a different choice which leads to reject x and to accept y. Since A' has to inspect all different factorizations, both choices have to be considered. Thus, each time A' needs to simulate a transition from a state r on a symbol σ, a nonempty set of transitions from r on σ is nondeterministically selected, guessing that this set will lead to test that each factorization inspected in computation branches that visit the state r at that point is accepted.

The formal definition of $A' = \langle Q', \Sigma, \delta', I', F' \rangle$, obtained along these lines, is the following:

- $Q' = 2^Q$,
- $I' = I = \{i\}$,
- for $\alpha \in Q$, $\sigma \in \Sigma$, let us consider the following set:

$$\mathsf{next}(\alpha, \sigma) = \{\gamma \in 2^Q \mid \exists f : \alpha \to 2^\gamma \text{ s.t. } \gamma = \bigcup_{r \in \alpha} f(r) \wedge$$
$$\wedge \forall r \in \alpha (\emptyset \neq f(r) \subseteq \delta(r, \sigma))\}; \quad (5)$$

the set $\delta'(\alpha, \sigma)$ contains all $\beta \in 2^Q$ such that $\beta = \begin{cases} \gamma, & \text{if } \gamma \subseteq F; \\ \gamma \cup \{i\}, & \text{otherwise;} \end{cases}$

for some $\gamma \in \mathsf{next}(\alpha, \sigma)$,
- $F' = \{\alpha \in Q' \mid \alpha \subseteq F\}$.

Finally, we study an upper bound on the state complexity. Given an NFA with n states accepting a language L, by adding one more state i, we can obtain an NFA A in the form required by our last construction, accepting the language $L \setminus \{\varepsilon\}$. Notice that $L^\circledast = (L \setminus \{\varepsilon\})^\circledast$. Thus, the number of states of the resulting automaton A' is at most 2^{n+1}. However, inspecting the construction, we can see that only at most one-half of them can be reached. In fact, from the definition of δ' we observe that each state β in the range of δ' is defined by a subset $\gamma \subseteq Q \setminus \{i\}$, taking $\beta = \gamma$ when all the states in γ are final, and $\beta = \gamma \cup \{i\}$ otherwise. This gives 2^n possibilities. However, $\gamma = \emptyset$ should not be considered because it is not reachable (see the definition of $\mathsf{next}(\alpha, \sigma)$) and, on the other hand, the state $\{i\}$ does not belong the range of δ' but it is used at the beginning of the computation. Hence, we conclude that the total number of reachable states is at most 2^n. $\square$

4   Conclusion

As mentioned in the Introduction, the investigation of universal disjunctive concatenation and star was initially motivated by the study of state costs of boolean and regular operations on dcFAs. To this respect, we now sketch some results that can be easily derived. All automata we consider in this discussion are nondeterministic.

First of all, the *complementation* of a dcFA can be trivially done, without increasing the number of its states, just switching the set of accepting states with the set of rejecting states. For the other operations, we can proceed as follows. From each dcFA A, we can get two NFAs $A^{\oplus}$ and $A^{\ominus}$ with at most the same number of states as A, recognizing the languages $\mathcal{L}^{\oplus}(A)$ and $\mathcal{L}^{\ominus}(A)$. Hence, given two dcFAs A and B we can easily build a dcFA C, corresponding to the *union*, by combining $A^{\oplus}$ and $B^{\oplus}$, according to the standard construction for the union, and $A^{\ominus}$ and $B^{\ominus}$, according to the standard construction for the intersection. The number of states of C is polynomial with respect to those of A and B. We can proceed in a similar way for the *intersection*.

In the case of the *concatenation*, the accepting part of C is obtained by combining $A^{\oplus}$ and $B^{\oplus}$, as in the standard construction for the concatenation. Since we are interested in a nondeterministic automaton, this uses a number of states which is bounded by the sum of the states in A and B. The rejecting part of C should recognize the language $A^{\ominus} \odot B^{\ominus}$, so according to Theorems 12 and 13, it uses an exponential number of states. In a similar way, for the *star* we have a polynomial number of states for the accepting part, but an exponential upper bound for the rejecting part, according to Theorem 18.

References

1. Birget, J.C.: Intersection and union of regular languages and state complexity. Inf. Process. Lett. **43**(4), 185–190 (1992)
2. Birget, J.C.: The state complexity of $\overline{\Sigma^{*}\overline{L}}$ and its connection with temporal logic. Inf. Process. Lett. **58**, 185–188 (1996)
3. Brzozowski, J.A., Grant, E., Shallit, J.: Closures in formal languages and Kuratowski's theorem. Int. J. Found. Comput. Sci. **22**(2), 301–321 (2011)
4. Chandra, A.K., Kozen, D., Stockmeyer, L.J.: Alternation. J. ACM **28**(1), 114–133 (1981). http://doi.acm.org/10.1145/322234.322243
5. Moreira, N., Pighizzini, G., Reis, R.: Optimal state reductions of automata with partially specified behaviors. In: Italiano, G.F., Margaria-Steffen, T., Pokorný, J., Quisquater, J.-J., Wattenhofer, R. (eds.) SOFSEM 2015-Testing. LNCS, vol. 8939, pp. 339–351. Springer, Heidelberg (2015). http://dx.doi.org/10.1007/978-3-662-46078-8_28
6. Putnam, H.: Three-valued logic. Philos. Stud. **8**(5), 73–80 (1957)
7. Yu, S., Zhuang, Q., Salomaa, K.: The state complexities of some basic operations on regular languages. Theor. Comput. Sci. **125**(2), 315–328 (1994)

Quasi-Distances and Weighted Finite Automata

Timothy Ng, David Rappaport, and Kai Salomaa[✉]

School of Computing, Queen's University, Kingston, ON K7L 3N6, Canada
{ng,daver,ksalomaa}@cs.queensu.ca

Abstract. We show that the neighbourhood of a regular language L with respect to an additive quasi-distance can be recognized by an additive weighted finite automaton (WFA). The size of the WFA is the same as the size of an NFA (nondeterministic finite automaton) for L and the construction gives an upper bound for the state complexity of a neighbourhood of a regular language with respect to a quasi-distance. We give a tight lower bound construction for the determinization of an additive WFA using an alphabet of size five. The previously known lower bound construction needed an alphabet that is linear in the number of states of the WFA.

Keywords: Regular languages · Weighted finite automata · State complexity · Distance measures

1 Introduction

In many applications it is crucial to measure the similarity between data. How we define the distance between objects depends on what the objects we want to compare are and why we want to compare them [5]. One of the most commonly used similarity measures for words is the Levenshtein distance [13], also called the edit distance [4,11,12,15]. By the edit distance between languages L_1 and L_2 we mean the smallest distance between a word of L_1 and of L_2, respectively. This definition is natural for error correction applications; however, other definitions such as the relative distance or Hausdorff distance have also been considered [3,5].

The edit distance is additive with respect to concatenation of words in the sense defined by Calude et al. [2]. Pighizzini [15] has shown that the edit distance between a word and a language recognized by a one-way nondeterministic auxiliary pushdown automaton is computable in polynomial time. Konstantinidis [12] showed that the edit distance of a regular language, that is, the smallest edit distance between two distinct words in the language can be computed in polynomial time. Han et al. [8] gave a polynomial time algorithm to compute the edit distance between a regular language and a context-free language. Error/edit systems for error correction have been studied by Kari and Konstantinidis [10], and the error correction capabilities of regular languages with respect to edit operations were recently investigated by Benedikt et al. [1].

© Springer International Publishing Switzerland 2015
J. Shallit and A. Okhotin (Eds.): DCFS 2015, LNCS 9118, pp. 209–219, 2015.
DOI: 10.1007/978-3-319-19225-3_18

A quasi-distance is a generalization of the notion of distance in that it allows the possibility of distinct elements having distance zero. Calude et al. [2] showed that the neighbourhood of a regular language with respect to an additive distance or quasi-distance is regular. The neighbourhood of radius r of a language L consists of all words that have distance at most r from some word of L.

In an additive weighted finite automaton (WFA) [17] the weight of a path is the sum of the weights of the individual transitions that make up the path and the weight of an accepted word w is the minimum weight of a path from the start state to a final state that spells out w. Note that this differs significantly from weighted automata used, for example, in image processing applications [6,7].

For a given nondeterministic finite automaton (NFA) A, an additive distance d and radius r, Salomaa and Schofield [17] gave a construction for an additive weighted finite automaton (WFA) which recognizes the neighbourhood of radius r of the language recognized by A. The construction relies on the fact that additive distances are finite, that is, the neighbourhood of any word is always finite. This makes the construction not suitable for quasi-distances, since neighbourhoods of additive quasi-distances are not guaranteed to be finite [2].

Here we show that neighbourhoods of a regular language with respect to an additive quasi-distance can be recognized by a WFA. Given an NFA A, the WFA recognizing a constant radius neighbourhood of $L(A)$ can be constructed in polynomial time. The construction relies on the property that the neighbourhoods with respect to a quasi-distance are regular and a finite automaton for the neighbourhood can be constructed effectively. The construction yields also an upper bound for the size of a deterministic finite automaton (DFA) needed to recognize the neighbourhood of radius r of a regular language (given by an NFA) with respect to a quasi-distance. The upper bound is significantly better than the bound obtained by constructing an NFA for the neighbourhood [2] and then determinizing the NFA.

We study also the state complexity of additive WFAs. A WFA A within a given weight bound R recognizes a regular language, and Salomaa and Schofield [17] gave an upper bound for the size of a DFA for this language. They also gave a matching lower bound construction; however, the WFAs used for the lower bound construction needed an alphabet of size linear in the number of states of the WFA. Here we give a tight lower bound construction for the "determinization of WFAs" using a five-letter alphabet.

The paper concludes with a discussion of open problems on the state complexity of neighbourhoods of a regular language with respect to an additive distance or quasi-distance.

2   Preliminaries

We assume that the reader is familiar with the basics of finite automata and regular languages [9,19,20]. A general reference for weighted finite automata is [6].

In the following Σ is always a finite alphabet, Σ^* is the set of words over Σ and ε is the empty word. The length of a word w is $|w|$. When there is no danger of confusion, a singleton set $\{w\}$ is denoted simply as w. The set of non-negative integers (respectively, rationals) is $\mathbb{N}_0$ (respectively, $\mathbb{Q}_0$).

A *nondeterministic finite automaton* (NFA) is a tuple $A = (Q, \Sigma, \delta, q_0, F)$ where Q is a finite set of states, Σ is an alphabet, δ is a multi-valued transition function $\delta : Q \times \Sigma \to 2^Q$, $q_0 \in Q$ is the initial state, and $F \subseteq Q$ is a set of final states. We extend the transition function δ to $Q \times \Sigma^* \to 2^Q$ in the usual way. A word $w \in \Sigma^*$ is *accepted* by A if $\delta(q_0, w) \cap F \neq \emptyset$ and the language recognized by A consists of all strings accepted by A.

The automaton A is a *deterministic finite automaton* (DFA) if, for all $q \in Q$ and $a \in \Sigma^*$, $\delta(q, a)$ either consists of one state or is undefined. A DFA A is *complete* if δ is defined for all $q \in Q$ and $a \in \Sigma$. Two states p and q of a DFA A are equivalent if $\delta(p, w) \in F$ if and only if $\delta(q, w) \in F$ for every string $w \in \Sigma^*$. A DFA A is *minimal* if each state of Q is reachable from the initial state and no two states are equivalent.

The (right) Kleene congruence of a language $L \subseteq \Sigma^*$ is the relation $\equiv_L \subseteq \Sigma^* \times \Sigma^*$ defined by setting, for $x, y \in \Sigma^*$,

$$x \equiv_L y \text{ iff } [(\forall z \in \Sigma^*) \, xz \in L \Leftrightarrow yz \in L].$$

A language L is regular if and only if the index of $\equiv_L$ is finite and, in this case, the index of $\equiv_L$ is equal to the size of the minimal complete DFA for L [19,20]. The minimal DFA for a regular language L is unique. The *state complexity* of L, $\mathrm{sc}(L)$, is the size of the minimal complete DFA recognizing L.

Definition 1 [17]. *An additive weighted finite automaton (WFA) is a 6-tuple $A = (Q, \Sigma, \gamma, \omega, q_0, F)$ where Q is a finite set of states, Σ is an alphabet, $\gamma : Q \times \Sigma \to 2^Q$ is the transition function, $\omega : Q \times \Sigma \times Q \to \mathbb{Q}_0$ is a partial weight function where $\omega(q_1, a, q_2)$ is defined if and only if $q_2 \in \gamma(q_1, a)$, $q_0 \in Q$ is the initial state, and $F \subseteq Q$ is the set of accepting states.*

Strictly speaking, the transitions of γ are also determined by the domain of the partial function β. In the following by a WFA we always mean an additive weighted finite automaton as in Definition 1. By a transition of A on symbol $a \in \Sigma$ we mean a triple (q_1, a, q_2) such that $q_2 \in \gamma(q_1, a)$, $q_1, q_2 \in Q$. A computation path α of a WFA A along a word $w = a_1 a_2 \cdots a_m$, $a_i \in \Sigma$, $i = 1, \ldots, m$, from state p_1 to p_2 is a sequence of transitions that spell out the word w,

$$\alpha = (q_0, a_1, q_1)(q_1, a_2, q_2) \cdots (q_{m-1}, a_m, q_m),$$

where $p_1 = q_0$, $p_2 = q_m$, and $q_i \in \gamma(q_{i-1}, a_i)$, $1 \le i \le m$. The weight of a computation path is

$$\omega(\alpha) = \sum_{i=1}^{m} \omega(q_{i-1}, a_i, q_i).$$

We let $\Theta(p_1, w, p_2)$ denote the set of all computation paths along a word w from p_1 to p_2. The *language recognized by A within the weight bound* $r \ge 0$ is the set of words for which there exists a computation path that is accepted by A and has weight at most r, defined as

$$L(A, r) = \{w \in \Sigma^* : (\exists f \in F)(\exists \alpha \in \Theta(q_0, w, f)) \ \omega(\alpha) \leq r\}.$$

Proposition 1 [17]. *If A is a WFA with n states where all transition weights are integers and $r \in \mathbb{N}_0$, then $L(A, r)$ can be recognized by a DFA with at most $(r + 2)^n$ states.*

3   WFA Construction for a Quasi-Distance Neighbourhood

We construct a WFA to recognize the neighbourhood of a regular language with respect to a quasi-distance. First we recall some definitions concerning additive distances and quasi-distances between words [2].

A function $d : \Sigma^* \times \Sigma^* \to \mathbb{Q}_0$ is a *distance* if it satisfies, for all $x, y, z \in \Sigma^*$,

1. $d(x, y) = 0$ if and only if $x = y$,
2. $d(x, y) = d(y, x)$,
3. $d(x, z) \leq d(x, y) + d(y, z)$.

The function d is a *quasi-distance* if it satisfies conditions 2 and 3 and $d(x, y) = 0$ always when $x = y$, that is, a quasi-distance allows the possibility that distinct word may have distance zero. The *neighbourhood* of radius r of a language L is the set

$$E(L, d, r) = \{x \in \Sigma^* : (\exists y \in L) \ d(x, y) \leq r\}.$$

A distance d is said to be *finite* if the neighbourhood of any given radius of an individual word with respect to d is finite. A (quasi-)distance d is *additive* if for every factorization $w = w_1 w_2$ and radius $r \geq 0$,

$$E(w, d, r) = \bigcup_{r_1 + r_2 = r} E(w_1, d, r_1) \cdot E(w_2, d, r_2).$$

It is known that the neighbourhood of a regular language with respect to a quasi-distance is regular [2]. The next lemma constructs a WFA for this language. The construction is inspired by related constructions in [2, 18].

An additive (quasi-)distance d is determined by the finite number of values $d(a, b)$, $d(a, \varepsilon)$, where $a, b \in \Sigma$. For the complexity estimate of the lemma we assume that d is a fixed additive quasi-distance that is given by listing the values $d(a, b)$, $d(a, \varepsilon)$, $a, b \in \Sigma$.

Lemma 1. *Let $N = (Q, \Sigma, \delta, q_0, F)$ be an NFA with n states, d an additive quasi-distance, and $R \geq 0$ is a constant. There exists an additive WFA A with n states such that for any $0 \leq r \leq R$,*

$$L(A, r) = E(L(N), d, r)$$

Furthermore, the WFA A can be constructed in time $O(n^3)$.

Proof. We define an additive WFA $A = (Q, \Sigma, \gamma, \omega, q_0, F)$ as follows. The transition function γ is defined by setting, for $p \in Q$, $a \in \Sigma$,

$$\gamma(p, a) = \{q : (\exists x \in \Sigma^*)\ q \in \delta(p, x) \text{ and } d(a, x) \leq R\}.$$

That is, for each pair of states p, q, we add a transition from p to q on a in the WFA A if there is a word $x \in \Sigma^*$ with $d(a, x) \leq R$ that takes p to q in the NFA N. The transition (p, a, q) in A has weight

$$\omega((p, a, q)) = \min_{x \in \Sigma^*} \{d(a, x) : q \in \delta(p, x)\}. \tag{1}$$

We claim that a word w spells out a path in A with weight r $(\leq R)$ from the start state q_0 to a state q_1 if and only if some word u with $d(w, u) \leq r$ takes the state q_0 to q_1 in the NFA B.

We prove the "only if" direction of the claim using induction on the length of w. If $w = \varepsilon$, then $q_1 = q_0$ and there is nothing to prove. For the inductive step consider $w = ub$, $u \in \Sigma^*$, $b \in \Sigma$, where the claim holds for u. Since w takes state q_0 to q_1 by a path with weight r in the WFA A, the word u takes q_0 to a state p by a path of weight r_1 where $r_1 + \omega(p, b, q_1) = r$.

By the inductive assumption, there exists $u_p \in \Sigma^*$, $d(u, u_p) \leq r_1$ such that u_p in the NFA N takes q_0 to the state p. By the definition of the transition weights of A in (1), there exists a word $v_{p,b}$ with $d(b, v_{p,b}) = \omega(p, b, q_1)$ such that in the NFA N the word $v_{p,b}$ takes state p to state q_1.

Since d is additive and $r_1 + \omega(p, b, q_1) = r$, we have

$$E(u, d, r_1) \cdot E(b, d, \omega(p, b, q_1)) \subseteq E(w, d, r).$$

Thus, $d(w, u_p v_{p,b}) \leq r$ and in the NFA N the word $u_p v_{p,b}$ takes the start state q_0 to q_1. This concludes the proof of the "only if" direction of the claim.

An analogous argument establishes the "if" direction of the claim. Since the start states of A and N coincide and A and N have the same set of final states, the claim implies that, for any $r \leq R$, $L(A, r) = E(L(N), d, r)$.

It remains to give an upper bound for the time complexity of finding the weights (1) in order to verify the claim concerning the time bound for constructing A. Since d is additive, for given $p, q \in Q$ and $a \in \Sigma$, the set of words x such that $d(a, x) \leq R$ and x takes p to q in the NFA N is regular. This means that, for $p \in Q$ and $a \in \Sigma$, the set $\gamma(p, a)$ can be efficiently constructed and the weights of the transitions of N are computed as follows.

A word $x = b_1 b_2 \cdots b_m$, $b_i \in \Sigma$ is in the neighbourhood of a of radius R if and only if there exists an index $i \in \{1, \ldots, m\}$ such that

$$d(a, b_i) + \sum_{j \in \{1, \ldots, m\}, j \neq i} d(\epsilon, b_j) \leq R.$$

For the radius R neighbourhood of a, $a \in \Sigma$, we define the two-state WFA $B_a = (\{I_0, I_1\}, \Sigma, \eta, \rho, I_0, \{I_1\})$, shown in Fig. 1. The states of B_a are $\{I_0, I_1\}$. For each symbol $\sigma \in \Sigma$, we define self-loop transitions $\eta(q, \sigma) = q$ with weight

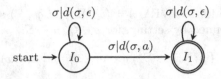

Fig. 1. The WFA B_a recognizing the language $\{x \in \Sigma : d(a, x) \le R\}$

$d(\sigma, \epsilon)$ for both states and the transition $\eta(I_0, \sigma) = I_1$ with weight $d(\sigma, a)$ for the transition which consumes the symbol a.

Let $M_a = (\{I_0, I_1\} \times Q, \Sigma, \delta_a, \omega_a, (I_0, q_0), I_1 \times F)$ be the WFA obtained as a cross product of the WFA B_a and the NFA N. The states of M_a are of the form (P, q), where $P \in \{I_0, I_1\}$ and $q \in Q$. The transitions of M_a are defined by setting, for $q \in Q$, $\sigma \in \Sigma$,

$$\delta_a((I_0, q), \sigma) = \{(I_0, \delta(q, \sigma)), (I_1, \delta(q, \sigma))\},$$
$$\delta_a((I_1, q), \sigma) = \{(I_1, \delta(q, \sigma))\}.$$

The weights of transitions $((P_1, q_1), \sigma, (P_2, q_2))$ defined in δ_{M_a} are defined

$$\omega_a((P_1, q_1), \sigma, (P_2, q_2)) = \begin{cases} d(\sigma, \epsilon), & \text{if } P_1 = P_2; \\ d(\sigma, a), & \text{if } P_1 \neq P_2. \end{cases}$$

For states $p, q \in Q$, paths from states (I_0, p) to (I_1, q) are labelled by words x with weight $d(a, x)$.

We compute the paths with the least weight for every pair of states of M_a. There are $2n$ states in the product machine and minimal weight paths for every pair of states can be computed in time $O(n^3)$ via the Floyd-Warshall algorithm [4]. A transition from p to q on a is added if there is a path from (I_0, p) to (I_1, q) with weight at most R. $\square$

Lemma 1 gives the following result.

Theorem 1. *Suppose that L has an NFA with n states and d is a quasi-distance. The neighbourhood of L of radius R can be recognized by an additive WFA having n states within weight bound R.*

As a consequence of Theorem 1 and Proposition 1 we get in Corollary 1 an upper bound for the state complexity of the neighbourhood of a regular language with respect to an additive quasi-distance d where all values $d(u, v)$, $u, v \in \Sigma^*$ are integers.

We note that if a quasi-distance d associates a non-negative integer value withany pair of words, then the weights of the WFA A constructed in the proof of Lemma 1 are integral. Furthermore, a neighbourhood with respect to a quasi-distance d with rational values can be converted to a neighbourhood with respect to a quasi-distance with integral values by multiplying the radius and the values of d by a suitably chosen constant. This can be done since the distance between any two words is determined by distances between two alphabet symbols and alphabet symbols and the empty word.

Corollary 1. *Let N be an NFA with n states, $R \in \mathbb{N}_0$, and d a quasi-distance $\Sigma^* \times \Sigma^* \to \mathbb{N}_0$. Then the neighbourhood $E(L(N), d, R)$ can be recognized by a DFA with $(R + 2)^n$ states.*

The upper bound $(R + 2)^n$ is significantly better than what is obtained by first constructing an NFA for $E(L(N), d, R)$ as in [2] and then determinizing the NFA. If the set of states of N is Q, Theorem 8 of [2][1] constructs an NFA for $E(L(N), d, R)$ with set of states $Q \times D$ where $D \subseteq \mathbb{N}$, roughly speaking, consists of all integers at most R that can be represented as a sum of distances between an element of Σ and an element of Σ^*.

We do not have a lower bound corresponding to the upper bound of Corollary 1, and the state complexity of neighbourhoods of regular languages with respect to an additive distance or quasi-distance remains an open question. Povarov [16] has given a lower bound for the radius-one Hamming neighbourhood of a regular language that is tight within an order of magnitude.

In the next section we will give a lower bound construction for the size of a DFA needed to simulate an additive WFA that matches the upper bound of Proposition 1. However, this does not necessarily shed light on the state complexity of neighbourhoods of regular languages because an arbitrary additive WFA need not recognize a neighbourhood of a (regular) language.

4   State Complexity of Weighted Finite Automata

Salomaa and Schofield [17] have given a matching lower bound construction for Proposition 1 using a family of WFAs over an alphabet of size $2n - 1$ where n is the number of states of the WFA. Here, we define a family of WFAs over a five-letter alphabet which reaches the upper bound $(r + 2)^n$.

Let $A_n = (Q_n, \Sigma, \gamma, \omega, 1, n)$ be an additive WFA with $Q_n = \{1, 2, \ldots, n\}$ and $\Sigma = \{a, b, c, d, e\}$. The transition function γ with $q \in Q$ and $\sigma \in \Sigma$ is defined

$$\gamma(q, \sigma) = \begin{cases} \{1, 2\}, & \text{if } q = 1, \sigma = a \text{ or } q = 2, \sigma = b; \\ \{3\}, & \text{if } q = 1, \sigma = b \text{ or } q = 2, \sigma = a; \\ \{q + 1\}, & \text{if } q = 3, \ldots, n - 1 \text{ and } \sigma = a, b; \\ \{q\}, & \text{if } q = 1, \ldots, n \text{ and } \sigma = c, d, e. \end{cases}$$

The weight function ω for a transition $\alpha \in Q_n \times \Sigma \times Q_n$ is defined

$$\omega(\alpha) = \begin{cases} 1, & \text{if } \alpha = (1, c, 1); \\ 1, & \text{if } \alpha = (2, d, 2); \\ 1, & \text{if } \alpha = (q, e, q) \text{ for all } q \in Q; \\ 0, & \text{for all other transitions defined by } \gamma. \end{cases}$$

---

[1] Theorem 8 of [2] assumes that N is deterministic. However, the construction used in the proof works also for an NFA.

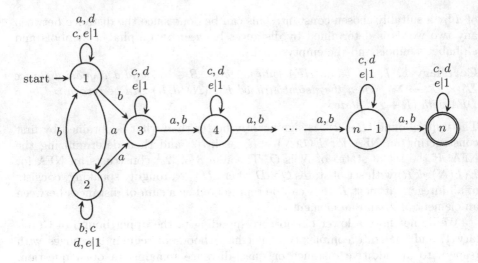

Fig. 2. The weighted finite automaton A_n used in the proof of Lemma 2.

The transition diagram for A_n is shown in Fig. 2 with the non-zero weights of each transition marked after the alphabet symbols labeling the transition. For example, state 1 has self-loops on a and d with weight zero and self-loops on c and e with weight one.

We will use the WFAs A_n to give a lower bound for the size of DFAs for a language recognized by a WFA within a given weight bound. First in Lemma 2 we establish a technical property of the weights of computations of A_n reaching a particular state and for this purpose we introduce the following notation.

For $0 \le k_i \le r+1$ and $1 \le i \le n$, we define the words

$$w(k_1, \ldots, k_n) = \begin{cases} ac^{k_n}bd^{k_{n-1}}ac^{k_{n-2}} \cdots ac^{k_3}bd^{k_2}c^{k_1}, & \text{if n is odd;} \\ abd^{k_n}ac^{k_{n-1}}bd^{k_{n-2}} \cdots ac^{k_3}bd^{k_2}c^{k_1}, & \text{if n is even.} \end{cases}$$

Lemma 2. *Let $n \in \mathbb{N}$. The WFA A_n after processing the input $w(k_1, \ldots, k_n)$ can reach the state s, $1 \le s \le n$, on a path with weight k_s. Furthermore, any computation of A_n on input $w(k_1, \ldots, k_n)$ that reaches state s, $1 \le s \le n$, has weight k_s.*

Proof. In the string $w(k_1, \ldots, k_n)$ occurrences of symbols a and b alternate. Thus the computation of A can exit states 1 and 2 after making a self-loop on a in state 1 or a self-loop on b in state 2 and, furthermore, this is the only way for the computation to get out of the "binary cycle" of states 1 and 2.

Below using a case analysis we verify that, for $1 \le s \le n$, A_n has a computation with weight k_s that ends in state s and, furthermore, any computation ending in s has weight k_s.

(i) First consider the case where n is even. Consider a computation of A_n that reaches a state s where $s \geq 2$ is even. Note that after exiting the cycle of states 1 and 2, only the symbols a or b move the computation to the next state. Thus, the only way to reach s is that the computation must make a self-loop on b in state 2 directly before reading the substring d^{k_s}. After that the following k_s symbols d are read via the weight one transitions. This also applies for the case $s = 2$.

If $s \geq 3$ is odd, in order to reach state s, directly before reading the substring c^{k_s} the computation must on input a make a self-loop in state 1 and then the following k_s symbols c are read with transitions of weight one in state 1.

Finally consider the case $s = 1$. In order to end in state 1, the computation must not have made any self-loops on a in state 1 or b in state 2. If this is done the computation ends in a state z with $z \geq 2$. Thus, reading the final b takes the computation from state 2 to state 1, where the transition on d is taken k_2 times. The computation remains in state 1 and reads the rest of the word c^{k_1} on the transition of weight 1 exactly k_1 times.

(ii) Next consider the case where n is odd. The above argument remains the same, almost word for word. The only minor difference is in the case $s = n$. In order to reach state n, the computation must read the first symbol a using a self-loop and then the following k_n symbols c using transitions of weight 1. (Note that when n is odd, in $w(k_1, \ldots, k_n)$ the first symbol a is followed by k_n symbols c.)

□

Lemma 3. *Let A_n be the WFA defined above and $r \in \mathbb{N}$. Then the minimal DFA for $L(A_n, r)$ needs $(r + 2)^n$ states.*

Proof. It is sufficient to show that all words $w(k_1, \ldots, k_n)$, $0 \leq k_i \leq r + 1$, $i = 1, \ldots, n$, belong to distinct classes of $\equiv_{L(A_n, r)}$.

Consider two distinct words $w(k_1, \ldots, k_n)$ and $w(k'_1, \ldots, k'_n)$ with $0 \leq k_i, k'_i \leq r + 1, i = 1, \ldots, n$. There exists an index j such that $k_j \neq k'_j$. Without loss of generality, we assume that $k_j < k'_j$. Choose

$$z = e^{r - k_j} a^{n - j}.$$

Since $k_j < k'_j \leq r + 1$, it follows that $r - k_j \geq 0$ and z is a well-defined word. We claim that

$$w(k_1, \ldots, k_n) \cdot z \in L(A, r), \quad w(k'_1, \ldots, k'_n) \cdot z \notin L(A, r).$$

By Lemma 2, A has a computation on input $w(k_1, \ldots, k_n)$ that ends in state j with weight k_j. In state j, A reads the first $r - k_j$ symbols e of z, after which the total weight is $k_j + (r - k_j) = r$. The zero weight transitions on the suffix a^{n-j} take the automaton from state j to the final state n.

Now consider from which states q the WFA A can reach the accepting state n on input z. On any state of A, the symbols c, d, e define self-loops. On states

$3 \leq q \leq n-1$, transitions to state $q+1$ only occur on a, b. For states $q = 1, 2$, a transition to state $q+1$ occurs only on a. Thus, A can reach the accepting state n from a state q on input z only if $q = j$.

Thus, the only possibility for A to accept $w(k'_1, \ldots, k'_n) \cdot z$ would be that the computation has to reach state j on the prefix $w(k'_1, \ldots, k'_n)$. By Lemma 2, the weight of this computation can only be k'_j. But when continuing the computation on z from state j, A has to read the first $r - k_j$ symbols e, each with a self-loop transition having weight one. After this, the weight of the computation will be $k'_j + r - k_j > r$. Thus, $w(k'_1, \ldots, k'_n) \cdot z \notin L(A, r)$.

Thus, the equivalence relation $\equiv_{L(A,r)}$ has index at least $(r+2)^n$.   □

As a consequence of Lemma 3 and Proposition 1 we have:

Theorem 2. *If A is an n state WFA with integer weights for transitions and $r \in \mathbb{N}$, then*

$$\mathrm{sc}(L(A,r)) \leq (r+2)^n.$$

For $n, r \in \mathbb{N}$, there exists an n state WFA A with integral weights defined over a five-letter alphabet such that $\mathrm{sc}(L(A,r)) = (r+2)^n$.

5   Conclusion

For the state complexity of a language recognized by an additive WFA with a given weight we have established a tight lower bound using a constant size alphabet. The earlier known lower bound construction [17] used a variable alphabet that has size linear in the number of states of the WFA.

We have also constructed a WFA recognizing the neighbourhood of a regular language with respect to an additive quasi-distance. This yields an upper bound $(r+2)^n$ for the state complexity of a neighbourhood of radius r of an n state NFA language with respect to an additive quasi-distance. The upper bound is significantly better than a bound obtained by directly constructing an NFA for the neighbourhood [2] and then determinizing the NFA. The same upper bound $(r+2)^n$ has been known previously for neighbourhoods with respect to an additive distance.

The precise state complexity of neighbourhoods with respect to a distance or a quasi-distance remains open. Povarov [16] gives an upper bound $n \cdot 2^{n-1} + 1$ for the Hamming neighbourhood of radius one of an n-state regular language and an almost matching lower bound. For neighbourhoods of radius $r \geq 2$ no good lower bounds are known. Finding such lower bounds will be a topic of a forthcoming paper [14].

References

1. Benedikt, M., Puppis, G., Riveros, C.: Bounded repairability of word languages. J. Comput. Syst. Sci. **79**, 1302–1321 (2013)
2. Calude, C.S., Salomaa, K., Yu, S.: Distances and quasi-distances between words. J. Univ. Comput. Sci. **8**(2), 141–152 (2002)

3. Choffrut, C., Pighizzini, G.: Distances between languages and reflexivity of relations. Theoret. Comput. Sci. **286**, 117–138 (2002)
4. Cormen, T.H., Leiserson, C.E., Rivest, R.L., Stein, C.: Introduction to Algorithms, 2nd edn. MIT Press, Cambridge (2001)
5. Deza, M.M., Deza, E.: Encyclopedia of Distances. Springer, Heidelberg (2009)
6. Droste, M., Kuich, W., Vogler, H. (eds.): Handbook of Weighted Automata. EATCS Monographs in Theoretical Computer Science. Springer, Heidelberg (2009)
7. Eramian, M.: Efficient simulation of nondeterministic weighted finite automata. J. Automata Lang. Comb. **9**, 257–267 (2004)
8. Han, Y.-S., Ko, S.-K., Salomaa, K.: The edit distance between a regular language and a context-free language. Int. J. Found. Comput. Sci. **24**, 1067–1082 (2013)
9. Holzer, M., Kutrib, M.: Descriptional and computational complexity of finite automata — a survey. Inf. Comput. **209**, 456–470 (2011)
10. Kari, L., Konstantinidis, S.: Descriptional complexity of error/edit systems. J. Automata Lang. Comb. **9**, 293–309 (2004)
11. Konstantinidis, S.: Transducers and the properties of error detection, error-correction, and finite-delay decodability. J. Univ. Comput. Sci. **8**, 278–291 (2002)
12. Konstantinidis, S.: Computing the edit distance of a regular language. Inf. Comput. **205**, 1307–1316 (2007)
13. Levenshtein, V.I.: Binary codes capable of correcting deletions, insertions, and reversals. Sov. Phys. Dokl. **10**(8), 707–710 (1966)
14. Ng, T., Rappaport, D., Salomaa, K.: State complexity of neighbourhoods and approximate pattern matching (March 2015, Submitted for publication)
15. Pighizzini, G.: How hard is computing the edit distance? Inf. Comput. **165**, 1–13 (2001)
16. Povarov, G.: Descriptive complexity of the Hamming neighborhood of a regular language. In: Proceedings of the 1st International Conference Language and Automata Theory and Applications, LATA 2007, pp. 509–520 (2007)
17. Salomaa, K., Schofield, P.: State complexity of additive weighted finite automata. Int. J. Found. Comput. Sci. **18**(6), 1407–1416 (2007)
18. Schofield, P.: Error Quantification and Recognition Using Weighted Finite Automata. MSc thesis, Queen's University, Kingston, Canada (2006)
19. Shallit, J.: A Second Course in Formal Languages and Automata Theory. Cambridge University Press, Cambridge (2009)
20. Yu, S.: Regular languages. In: Rozenberg, G., Salomaa, A. (eds.) Handbook of Formal Languages, vol. 1, pp. 41–110. Springer, Heidelberg (1997)

The State Complexity of Permutations on Finite Languages over Binary Alphabets

Alexandros Palioudakis[1], Da-Jung Cho[1], Daniel Goč[2], Yo-Sub Han[1], Sang-Ki Ko[1], and Kai Salomaa[2(✉)]

[1] Department of Computer Science, Yonsei University, 50 Yonsei-Ro, Seodaemum-Gu, Seoul 120–749, Republic of Korea
{alex,dajung,emmous,narame7}@cs.yonsei.ac.kr
[2] School of Computing, Queen's University, Kingston, Ontario K7L 3N6, Canada
{goc,ksalomaa}@cs.queensu.ca

Abstract. We investigate the state complexity of the permutation operation over finite binary languages. We first give an upper bound of the state complexity of the permutation operation for a restricted case of these languages. We later present a general upper bound of the state complexity of permutation over finite binary languages, which is asymptotically the same as the previous case. Moreover, we show that there is a family of languages that the minimal DFA recognizing each of these languages needs at least as many states as the given upper bound for the restricted case. Furthermore, we investigate the state complexity of permutation by focusing on the structure of the minimal DFA.

Keywords: Finite automata · State complexity · Finite languages · Permutation · Parikh equivalence

1 Introduction

Finite automata are well studied in the theory of computation. McCulloch and Pitts [14] first introduced this model of computation. Following on these ideas Kleene [10] wrote the first paper on finite automata and regular expressions. Later, Rabin and Scott [18] first studied the nondeterministic version of finite automata, for which they received the Turing Award, the highest award in computer science.

Since then, much work has been done in the descriptional complexity of finite automata [12,13,15,16]. The descriptional complexity of finite automata is usually measured in the number of transitions or the number of states that a finite automaton requires in order to accept a given language. Most researchers have focused on the state complexity of finite automata [6,9].

A widely studied topic in the state complexity of finite automata is the state complexity of language operations. Yu et al. [20] studied the state complexity of some basic operations. Han and Salomaa [7] studied the state complexity of union and intersection of finite languages. Holzer and Kutrib [8]

© Springer International Publishing Switzerland 2015
J. Shallit and A. Okhotin (Eds.): DCFS 2015, LNCS 9118, pp. 220–230, 2015.
DOI: 10.1007/978-3-319-19225-3_19

studied the nondeterministic state complexity of some basic language operations. Câmpeanu et al. [1] studied the state complexity of basic language operation for finite languages. Domaratzki [2] studied the state complexity of proportional removals and recently, Goč et al. [5] studied the nondeterministic state complexity of proportional removals. For more information on the state complexity of language operations, the reader can consult the recent review by Gao et al. [4].

Here we investigate the operational state complexity of the permutation operation. The family of regular languages is not closed under permutation and, hence, in this paper we focus on finite languages. We first compute an upper bound of the state complexity of the permutation on a restricted case of regular languages over binary alphabets. We show an upper bound for the state complexity of permutation for general binary finite languages. We mention that the permutation operation is related to the Parikh mapping, which maps each string over n letters to an n-dimensional vector whose components give the number of occurrences of the letters in the string [11,17]. Ellul et al. [3] have given strong lower bounds for the size of NFAs or regular expressions recognizing permutations of symbols of a growing alphabet.

In Sect. 2, we briefly present definitions and notation used throughout the paper. In Sect. 3, we give the state complexity of permutation of binary languages recognized by DFAs that form a chain, and present a general upper bound of the state complexity of permutation of binary finite languages. In Sect. 4, we give an upper bound on the state complexity of permutation for languages that recognize strings with equal length. Moreover, we give lower bounds that are tight in the restricted cases and asymptotically tight in the general case.

2   Preliminaries

We assume that the reader is familiar with the basic definitions concerning finite automata [19,21] and descriptional complexity [6,9]. Here we just fix some notation needed in the following.

The set of strings over a finite alphabet Σ is Σ^*, the length of $w \in \Sigma^*$ is $|w|$ and ε is the empty string. Moreover, for a letter $a \in \Sigma$ and a string $w \in \Sigma^*$, we denote the numbers of occurrences of the letter a in the string w by $|w|_a$. The set of positive integers is denoted by $\mathbb{N}$. The cardinality of a finite set S is $\#S$.

A deterministic finite automaton (DFA) is a 5-tuple $A = (Q, \Sigma, \delta, q_0, F)$, where Q is a finite set of states, Σ is a finite alphabet, $\delta : Q \times \Sigma \to Q$ is the transition function (partial function), $q_0 \in Q$ is the start state and $F \subseteq Q$ is the set of accepting states. The function δ is extended in the usual way as a function $Q \times \Sigma^* \to Q$ and the language recognized by A consists of strings $w \in \Sigma^*$ such that $\delta(q_0, w) \in F$. By the *size of A*, we mean the number of states of A, $\text{size}(A) = \#Q$.

The minimal size of a DFA recognizing a regular language L is called the state complexity of L and denoted by $\text{sc}(L)$. Note that we allow DFAs to be incomplete and, consequently, the deterministic state complexity of L may differ by one from the definition using complete DFAs.

An important relation of languages is the Myhill-Nerode relation R_L of a language L. The relation R_L contains pairs of strings x and y if and only if for every $z \in \Sigma^*$ both strings $x \cdot z$ and $y \cdot z$ belong in L or both strings $x \cdot z$ and $y \cdot z$ do not belong in L. It is well known that the MyhillNerode relation of the language L has finite number of equivalence classes if and only if the language L is regular. Moreover, the unique minimal DFA for L has the same number of states as the number of equivalence classes of R_L. Hence, when we want to find a lower bound on the state complexity of a regular language L, it is sufficient to find a set of strings S such that, for every strings $w, w' \in S$, there is a string $u \in \Sigma^*$ with $w \cdot u \in L$ and $w' \cdot u \notin L$, or $w \cdot u \notin L$ and $w' \cdot u \in L$. Then, we have that $sc(L) \geq \#S$.

We consider the state complexity of the operation of permutation on finite languages. We now define the permutation $per(L)$ of a regular language L over the alphabet Σ as follows: A string w belongs in $per(L)$ if and only if there is a string $u \in L$ such that the strings w and u have the same number of occurrences of every letter of Σ. Formally, we define

$$per(L) = \{w \in \Sigma^* \mid (\exists u \in L)(\forall a \in \Sigma)(|u|_a = |w|_a)\}.$$

Remark that the family of regular languages is not closed under permutation. For example, for the language $(a \cdot b)^*$, the permutation of this language contains all the strings w such that $|w|_a = |w|_b$, which is not a regular language.

Given an alphabet $\Sigma = \{a_1, a_2, \ldots, a_k\}$, let $\Psi : \Sigma^* \to [\mathbb{N}_0]^k$ be a mapping defined by $\Psi(w) = (|w|_{a_1}, |w|_{a_2}, \ldots, |w|_{a_k})$. This function is called a *Parikh mapping* and $\Psi(w)$ is called the *Parikh vector* of w. The Parikh mapping is extended for a set of strings, with $\Psi : 2^{\Sigma^*} \to 2^{[\mathbb{N}_0]^k}$ be a mapping defined by $\Psi(L) = \{\Psi(w) \mid w \in L\}$. Two languages L_1, L_2 are Parikh equivalent, denoted by $L_1 \equiv_{\text{Parikh}} L_2$, if $\Psi(L_1) = \Psi(L_2)$. Similarly, we say that two DFAs A, B are Parikh equivalent if $\Psi(L(A)) = \Psi(L(B))$ and denote it by $A \equiv_{\text{Parikh}} B$.

3   Permutation Operation for Chain DFAs

We consider the problem of finding the state complexity of the permutation of a binary language L. We start with giving an upper bound for the restricted case of the language L where each string of L has length $sc(L) - 1$.

Lemma 1. *Let n be a positive integer and $L \subseteq \{a, b\}^{n-1}$ be a finite language such that $sc(L) = n$. Then, we have the following inequality for the state complexity of the permutation of L:*

$$sc(per(L)) \leq \frac{n^2 + n + 1}{3}.$$

Proof. Let A be the minimal DFA recognizing L over the alphabet $\{a, b\}$. Since we have that $L \subseteq \{a, b\}^{n-1}$, we know that each string w of L is of length $n - 1$. Hence, the DFA A forms a chain, that is, we can enumerate the states of A such that for all $1 \leq h \leq n - 1$ at least one of the following is true; $\delta(h, a) = h + 1$

or $\delta(h, b) = h + 1$. Additionally, each state h has only one target state $h + 1$ for $1 \leq h \leq n - 1$ and the transition function for h is one the the following three cases:

1. $\delta(h, a) = h + 1$ (a-transition)
2. $\delta(h, b) = h + 1$ (b-transition)
3. $\delta(h, a) = h + 1, \delta(h, b) = h + 1$ ($a\&b$-transition)

By the definition of the language $\text{per}(L)$, we have all the possible permutations of the strings of L. The order of the different types of transitions (a, b, or $a\&b$) of A does not affect $\text{per}(L)$. Hence, we can assume that, without loss of generality, we start with a-transitions, followed by b-transitions, followed by $a\&b$-transitions. From this assumption, we have that L is of the form $a^i b^j (a+b)^k$ for some non-negative integers i, j, k such that $i + j + k = n - 1$. It is not difficult to construct a DFA with $(i+1)\cdot(j+1)+k\cdot j+k\cdot i+k$ states recognizing $\text{per}(L)$. Since we search for an upper bound of the state complexity of $\text{per}(L)$, we can find for which values of i, j, k the function $f(i, j, k) = (i+1)\cdot(j+1)+k\cdot j+k\cdot i+k$, and $i + j + k = n - 1$, has a maximal value. It is easy to verify that f is maximized when i, j, k are $i = j = k = \frac{n-1}{3}$. Thus, for integer values of i, j, k

$$\max f(i, j, k) = \begin{cases} \frac{n^2+n+1}{3}, & \text{if } n \equiv 1 \pmod{3}; \\ \frac{n^2+n}{3}, & \text{otherwise.} \end{cases}$$

$\square$

In Lemma 1, we assume that every string w in L has a specific length—$|w| = \text{sc}(L) - 1$. This restriction ensures that the states of the minimal DFA recognizing L form a chain. Now, we move one step further and show an upper bound of the state complexity of the permutation of L without such restrictions.

Lemma 2. *Let L be a binary finite language and $m = \max\{|w| \mid w \in L\}$ for some positive integer m. Then, we have*

$$\text{sc}(\text{per}(L)) \leq \frac{m^2 + m + 2}{2}.$$

Proof. We construct a DFA A that recognizes $\text{per}(L)$ over the binary alphabet $\{a, b\}$. We keep track at each state of A what is the number of occurrences of a's and what is the number of occurrences of b's that we have already read. Then A has states of the form (i, j), for $0 \leq i, j \leq m$, where i (and j) keeps track of the occurrences of a's (and b's, respectively). Since m is the length of the longest string in L, we know that there is no computation path in A with more than $m + 1$ states. Thus, for all states (i, j) of A we have $i + j \leq m$. Moreover, it is immediate that all states (i, j) with $i + j = m$ are final and equivalent—we can merge them to one final state.

Now we counter the number of states. The total number of states of the resulting DFA is $1 + 2 + \cdots + m + 1$(the merged final state) $= \frac{m\cdot(m+1)}{2} + 1$. The final states of A are all states (i, j) such that there is a string $w \in L$ with $|w|_a = i$ and $|w|_b = j$.

$\square$

We notice that the maximum length of all the strings of the language L can be at most the state complexity of L minus one; in other words, we have $1 + \max\{|w| \mid w \in L\} \leq \mathrm{sc}(L)$. By this observation and Lemma 2, we have the following corollary.

Corollary 1. *Let L be a binary finite language and $\mathrm{sc}(L) = n$ for some positive integer n. Then we have*

$$\mathrm{sc}(\mathrm{per}(L)) \leq \frac{n^2 - n + 2}{2}.$$

In the following theorem, we give a lower bound on the state complexity of the permutation of a language. This bound is asymptotically tight with the general upper bound in Corollary 1.

Theorem 1. *For any $n_0 \in \mathbb{N}$, there exists a regular language L with $\mathrm{sc}(L) = n$, for $n \geq n_0$, such that*

$$\mathrm{sc}(\mathrm{per}(L)) \geq \frac{n^2 + n + 1}{3}.$$

Proof. Let $n = 3k + 1 \geq n_0$, $k \in \mathbb{N}$ and $L_n = L(a^k b^k (a + b)^k)$. In Fig. 1, for $k = 3$ and $n = 10$, we see that L_n can be accepted by an incomplete DFA with n states.

Fig. 1. The state minimal DFA recognizing the language L_{10}.

We prove a lower bound for the state complexity of $\mathrm{per}(L_n)$.

$$\mathrm{per}(L_n) = \{w \in \Sigma^{3 \cdot k} \mid |w|_a, |w|_b \geq k, n = 3 \cdot k + 1\}.$$

Let X and Y be the sets of strings as follows:

$$X = \{a^i b^j : 0 \leq i \leq 2k, 0 \leq j \leq k\} \text{ and } Y = \{a^i b^j : 0 \leq i < k, k < j \leq 2k\}.$$

We show that all strings of $X \cup Y$ are pairwise inequivalent with respect to the Myhill-Nerode congruence of $\mathrm{per}(L_n)$. Let $u = a^i b^j$ and $u' = a^{i'} b^{j'}$ be two arbitrary distinct strings from $X \cup Y$. We consider first the case where $|u| \neq |u'|$ and later we consider three separate cases $u, u' \in X$, $u, u' \in Y$, and, $u \in X$ and $u' \in Y$ (same case as $u' \in X$ and $u \in Y$):

1. We have that $|u| \neq |u'|$. It is straightforward to verify that u and u' are inequivalent since one can easily find a string z such that $uz \in \mathrm{per}(L_n)$ and $|u'z| \neq 3 \cdot k$—$u'z \notin \mathrm{per}(L_n)$.
2. We have $|u| = |u'|$ and $u, u' \in X$. Since $u \neq u'$, either $|u|_a < |u'|_a$ or $|u'|_a < |u|_a$. Without loss of generality, we assume that $|u|_a < |u'|_a$. For $z = a^{2 \cdot k - i} b^{k - j}$, we have $uz \in \mathrm{per}(L_n)$. However, for the string $u'z$, we have $|u'z|_a > 2 \cdot k$, which means, since $|uz| = |u'z| = 3 \cdot k$, that $|u'z|_b < k$ and $u'z \notin \mathrm{per}(L_n)$.

3. We have $|u| = |u'|$ and $u, u' \in Y$. Similar with the second case above, we assume that, without loss of generality, $|u|_b < |u'|_b$. For $z = a^{k-i}b^{2 \cdot k-j}$, we have $uz \in \mathrm{per}(L_n)$. However, we have $|u'z|_b > 2 \cdot k$, which implies that $u'z \notin \mathrm{per}(L_n)$.

4. We have that $u \in X$ and $u' \in Y$ and $|u| = |u'|$. Since $u' \in Y$ and $u \in X$, we know that $|u'|_b > k$ and $|u|_b \leq k$. This implies that $|u|_a > |u'|_a$ because $|u| = |u'|$. Now for the string $z = a^{k-i}b^{2 \cdot k-j}$, we have $uz \in \mathrm{per}(L_n)$. However, for the string $u'z$, we have that $|u'z|_a < k$ and, thus, $u'z \notin \mathrm{per}(L_n)$.

An example of the minimal DFA recognizing the language $\mathrm{per}(L_{10})$ is presented in Fig. 2.

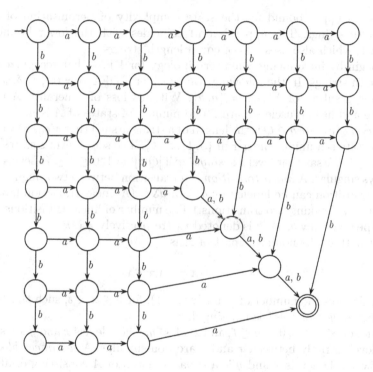

Fig. 2. The minimal DFA recognizing the language $\mathrm{per}(L_{10})$).

Hence, the number of states of the minimal DFA recognizing the language $\mathrm{per}(L_n)$ has at least $(2 \cdot k + 1) \cdot (k + 1) + k^2 = 3 \cdot k^2 + 3 \cdot k + 1$ states. We know that $n = 3 \cdot k + 1$ (namely, $k = \frac{n-1}{3}$) and, thus, the minimal DFA for $\mathrm{per}(L_n)$ has at least

$$3 \cdot \left(\frac{n-1}{3} \right)^2 + 3 \cdot \left(\frac{n-1}{3} \right) + 1 = \frac{n^2 - 2 \cdot n + 1}{3} + (n-1) + 1 = \frac{n^2 + n + 1}{3}$$

states.   □

From the simple case studied in Lemma 1 and Theorem 1, we have the following corollary.

Corollary 2. *Let n be a positive integer and $L \subseteq \{a, b\}^{n-1}$ be a finite language such that* $\mathrm{sc}(L) = n$. *Then the state complexity of the permutation of L is bounded by the inequality,*

$$\mathrm{sc}(\mathrm{per}(L)) \leq \frac{n^2 + n + 1}{3}.$$

Moreover, sometimes $\frac{n^2+n+1}{3}$ states are necessary for the minimal DFA recognizing $\mathrm{per}(L)$.

4   Upper Bound for Sets of Equal Length Strings

We prove an upper bound for the state complexity of permutation of sets of equal length strings. The upper bound coincides with the lower bound from Theorem 1, which also uses sets of equal length strings.

We begin by introducing some terminology for DFAs that recognize sets of equal length strings. In the following, we consider a DFA $A = (Q, \Sigma, \delta, q_0, \{q_f\})$ recognizing a subset of Σ^ℓ, $\Sigma = \{a, b\}$. Without loss of generality A has one final state and has no useless states. The number of states of A is n.

The *level of a state* $q \in Q$ is the length of a string w such that $\delta(q_0, w) = q$. The level of a state is a unique integer in $\{0, 1, \ldots, \ell\}$. The set of level z states is $Q[z]$ for $0 \leq z \leq \ell$. We say that level z is *singular* if $|Q[z]| = 1$, $0 \leq z \leq \ell$. Levels 0 and ℓ are always singular. A *linear transition* is a transition between two singular levels. A linear transition can be labeled by a, b or $a\&b$. (A linear transition labeled by $a\&b$ is strictly speaking two transitions.) The number of linear transitions labeled by a (respectively, by b, $a\&b$) is denoted i_A (respectively, j_A, k_A).

The length of the nonlinear part of A is

$$h_A = \ell - (i_A + j_A + k_A). \tag{1}$$

Thus h_A denotes the number of pairs $(z, z+1)$, for $0 \leq z < \ell$, such that at least one of the levels z or $z+1$ is not singular.

Consider $0 \leq x \leq \ell$, $0 \leq y \leq \ell$, and $x+1 < y$, where levels x and y are singular and all levels strictly between x and y are non-singular. A *nonlinear block* $B_{x,y}$ of A between the levels x and y is a subautomaton of A consisting of all states of $\cup_{x \leq z \leq y} Q[z]$ and the transitions between them. The initial (respectively, final) state of the subautomaton is the state having level x (respectively, y). The length of the nonlinear block $B_{x,y}$ is $y - x$. The length of a block is always at least two.

Note that a nonlinear block begins and ends in a singular level and all levels between these are non-singular. In the following, nonlinear blocks are called simply *blocks*. Examples of blocks are illustrated in Fig. 3.

The sum of the lengths of the blocks of A equals to h_A. The estimation of the length of accepted strings ℓ in terms of the number of states n depends on the types of blocks that A has.

Assume that the total length h_A of the blocks of A is fixed. Then the maximal value of ℓ can be reached if all blocks have length two (and h_A is even). Note that a block of length two has always exactly 4 states. Thus, we have

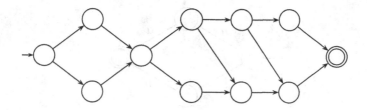

Fig. 3. A DFA with a block of length 2 and a block of length 4.

$$\ell \leq n - 1 - \frac{1}{2}h_A.\tag{2}$$

Example of the worst-case situation where $\ell = n - 1 - \frac{1}{2}h_A$ is illustrated in Fig. 4.

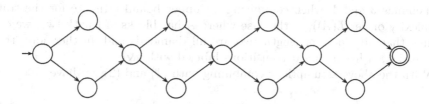

Fig. 4. $n = 13$ and $h_A = 8$, $\ell = 8$.

4.1 Estimate for DFAs Having Blocks of Length Two

We begin by providing an upper bound in the case where a DFA A includes blocks of length two and none of bigger length. As observed in the following subsection the same upper bound holds for arbitrary DFAs recognizing sets of equal length strings. The proof of the general case is based on similar ideas but is considerably more complicated. In this extended abstract we include the proof only for the case where the DFA has blocks of length at most two.

A block of length two that recognizes the language $\{aa, bb\}$ is called a *diamond* (see Fig. 5). There are a total of 9 different blocks of length two and it is easy to see that any block of length two that is not a diamond is "redundant" in the sense that it can be replaced by linear transitions and the modified DFA is Parikh-equivalent to A. This is stated in the following lemma.

Lemma 3. *Assume that A has a block of length two that is not a diamond. Then there exists a DFA A_1 having $n - 1$ states such that $L(A_1) \equiv_{\text{Parikh}} L(A)$.*

Next we observe that if A has one or more diamonds, then without loss of generality A can be assumed to have no linear transitions with label $a\&b$.

Lemma 4. *Assume that A has $r \geq 1$ diamonds and $k_A \geq 1$. Then there exists a DFA A_2 with $n - r$ states such that $L(A_2) \equiv_{\text{Parikh}} L(A)$.*

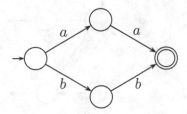

Fig. 5. A diamond.

Proof. This follows from the observation that when $k_A \geq 1$,

$$(aa + bb)^r (a + b)^{k_A} \equiv_{\text{Parikh}} (a + b)^{2r + k_A}.$$

□

By Lemmas 3 and 4, when computing an upper bound estimate for the state complexity of $\text{per}(L(A))$, in the case where A has blocks of length two, we can assume that all blocks of length two are all diamonds and, furthermore, that $k_A = 0$ (i.e., A has no linear transitions labeled with $a\&b$).

With the above assumptions combining with (1) and (2), we have

$$\frac{3}{2} \cdot h_A + i_A + j_A \leq n - 1.$$

We construct a DFA B recognizing $\text{per}(L(A))$. Note that it is sufficient for B to count a's up to $i_A + h_A$ and count b's up to $j_A + h_A$ with the further restriction that the sum of the counts is at most $i_A + j_A + h_A$. The states of B consist of pairs (x, y), where x is the a-count and y is the b-count. The states can be listed as follows:

- $(i_A + 1) \cdot (j_A + 1)$ pairs, where a-count is at most i_A and b-count is at most j_A.
- When a-count is $i_A + z$, for $1 \leq z \leq h_A$, b-count can be between 0 and $j_A + h_A - z$. This results in $\frac{1}{2} h_A (2j_A + h_A + 1)$ states. (The number of states comes from calculating, for some positive integers m and n, the cardinality of the following set $\{(i_0, j_0) \mid 1 \leq i_0 \leq m, 0 \leq j_0 \leq n + m - i\}$. After some calculations we conclude that the set has $\frac{1}{2} m(2n + m + 1)$ elements.)
- Additionally, for each b-count greater than j_A, we need to count up to i_A a's, which results in $h_A \cdot (i_A + 1)$ added states. (The situation where also the a-count is above i_A was included already in states listed above.)

In total, B has

$$i_A j_A + h_A i_A + h_A j_A + \frac{1}{2} h_A^2 + i_A + j_A + \frac{3}{2} h_A + 1$$

states.

This number is maximized whenever $i_A = \frac{2}{7}(n - 1)$, $j_A = \frac{2}{7}(n - 1)$, and $h_A = \frac{2}{7}(n-1)$ leading to a value of $\frac{2}{7}(n-1)^2 + n$ (in these cases $\frac{3}{2} \cdot h_A + i_A + j_A = n - 1$). (This maximization can be easily checked by mathematics software such as Maple.) This polynomial is bounded by $\frac{n^2 + n + 1}{3}$ and only reaches that bound in the trivial case where $i_A = 0$, $j_A = 0$, $h_A = 0$, and $n = 1$.

Above we have verified the following:

Proposition 1. *If A is a DFA with n states that recognizes a set of equal length strings over $\{a, b\}$ and the nonlinear part of A has only blocks of length two, then*

$$\text{sc}(\text{per}(L(A)) \leq \frac{n^2 + n + 1}{3}.$$

4.2   Estimate for General DFAs for Equal Length Languages

The result of the following theorem extends the result of Proposition 1 to all DFAs recognizing sets of equal length strings. Due to the limit on the number of pages the proof of Theorem 2 is omitted in this extended abstract.

Theorem 2. *Let A be an n-state DFA accepting a language $L \subseteq \Sigma^\ell$. Then there exists a DFA C accepting $\text{per}(L)$ with no more than $\frac{n^2+n+1}{3}$ states.*

From Theorem 1, we already know that the upper bound of Theorem 2 can be reached.

Corollary 3. *For every $n_0 \in \mathbb{N}$, there is a positive integer ℓ and a regular language $L \subseteq \Sigma^\ell$, with $\text{sc}(L) = n$, for $n \geq n_0$, such that every DFA accepting $\text{per}(L)$ needs at least $\frac{n^2+n+1}{3}$ states.*

5   Conclusions

We have studied the deterministic state complexity of permutation of finite languages over binary alphabets. More specifically, we have presented asymptotically tight upper and lower bounds for the general case. We have also established the matching upper and lower bound on the restricted cases when the given language recognizes strings with equal length. Matching bounds of the general case remain open. Moreover, the state complexity of permutation over non-binary languages remains open, as well as, the nondeterministic state complexity of finite languages.

Acknowledgment. This research was supported by the Basic Science Research Program through NRF funded by MEST (2012R1A1A2044562), the International Cooperation Program managed by NRF of Korea (2014K2A1A2048512) and the Natural Sciences and Engineering Research Council of Canada Grant OGP0147224.

References

1. Câmpeanu, C., Culik, K., Salomaa, K., Yu, S.: State complexity of basic operations on finite languages. In: Boldt, O., Jürgensen, H. (eds.) WIA 1999. LNCS, vol. 2214, pp. 60–70. Springer, Heidelberg (2001)
2. Domaratzki, M.: State complexity of proportional removals. J. Autom. Lang. Comb. **7**(4), 455–468 (2002)
3. Ellul, K., Krawetz, B., Shallit, J., Wang, M.: Regular expressions: new results and open problems. J. Autom. Lang. Comb. **10**(4), 407–437 (2005)

4. Gao, Y., Moreira, N., Reis, R., Yu, S.: A review of state complexity of individual operations. Technical report, Universidade do Porto, Technical Report Series DCC-2011-08, Version 1.1, September (2012). www.dcc.fc.up.pt/Pubs (To appear in Computer Science Review, 2015)
5. Goč, D., Palioudakis, A., Salomaa, K.: Nondeterministic state complexity of proportional removals. In: Jurgensen, H., Reis, R. (eds.) DCFS 2013. LNCS, vol. 8031, pp. 102–111. Springer, Heidelberg (2013)
6. Goldstine, J., Kappes, M., Kintala, C.M.R., Leung, H., Malcher, A., Wotschke, D.: Descriptional complexity of machines with limited resources. J. UCS **8**(2), 193–234 (2002)
7. Han, Y.S., Salomaa, K.: State complexity of union and intersection of finite languages. Int. J. Found. Comput. Sci. **19**(03), 581–595 (2008)
8. Holzer, M., Kutrib, M.: State complexity of basic operations on nondeterministic finite automata. In: Champarnaud, J.-M., Maurel, D. (eds.) CIAA 2002. LNCS, vol. 2608, pp. 148–157. Springer, Heidelberg (2003)
9. Holzer, M., Kutrib, M.: Descriptional and computational complexity of finite automata – a survey. Inf. Comput. **209**(3), 456–470 (2011)
10. Kleene, S.C.: Representation of events in nerve nets and finite automata. Technical report, DTIC Document (1951)
11. Lavado, G.J., Pighizzini, G., Seki, S.: Operational state complexity under Parikh equivalence. In: Jürgensen, H., Karhumäki, J., Okhotin, A. (eds.) DCFS 2014. LNCS, vol. 8614, pp. 294–305. Springer, Heidelberg (2014)
12. Lupanov, O.: A comparison of two types of finite sources. Problemy Kibernetiki **9**, 328–335 (1963)
13. Maslov, A.: Estimates of the number of states of finite automata. In: Soviet Mathematics Doklady, Translation from Doklady Akademii Nauk SSSR 194, vol. 11, pp. 1266–1268, 1373–1375 (1970)
14. McCulloch, W.S., Pitts, W.: A logical calculus of the ideas immanent in nervous activity. Bull. Math. Biophys. **5**(4), 115–133 (1943)
15. Meyer, A.R., Fischer, M.J.: Economy of description by automata, grammars, and formal systems. In: SWAT (FOCS), pp. 188–191. IEEE Computer Society (1971)
16. Moore, F.: On the bounds for state-set size in the proofs of equivalence between deterministic, nondeterministic, and two-way finite automata. IEEE Trans. Comput. **C–20**(10), 1211–1214 (1971)
17. Parikh, R.J.: On context-free languages. J. ACM (JACM) **13**(4), 570–581 (1966)
18. Rabin, M.O., Scott, D.S.: Finite automata and their decision problems. IBM J. Res. Dev. **3**(2), 114–125 (1959)
19. Shallit, J.: A Second Course in Formal Languages and Automata Theory. Cambridge University Press, Cambridge (2008)
20. Yu, S., Zhuang, Q., Salomaa, K.: The state complexities of some basic operations on regular languages. Theor. Comput. Sci. **125**(2), 315–328 (1994)
21. Yu, S.: Handbook of Formal Languages, Volume 1, Chap. Regular Languages, pp. 41–110. Springer, Heidelberg (1998)

Star-Complement-Star on Prefix-Free Languages

Matúš Palmovský[1] and Juraj Šebej[2(✉)]

[1] Mathematical Institute, Slovak Academy of Sciences,
Grešákova 6, 040 01 Košice, Slovakia
palmovsky@saske.sk
[2] Institute of Computer Science, Faculty of Science, P.J. Šafárik University,
Jesenná 5, 040 01 Košice, Slovakia
juraj.sebej@gmail.com

Abstract. We study the star-complement-star operation on prefix-free languages. We get a tight upper bound $2^{n-3}+2$ for the state complexity of this combined operation on prefix free languages. To prove tightness, we use a binary alphabet. Then we present the results of our computations concerning star-complement-star on binary prefix-free languages. We also show that state complexity of star-complement-star of every unary prefix-free language is one, except for the language $\{a\}$, where it is two.

1 Introduction

If we apply the operations of star and complement in any order and any number of times on a formal language, then we can get only a finite number of distinct languages [1]. To get the state complexities of the resulting languages, it is enough to know the complexity of languages L^* and L^{*c*}; notice that a language and its complement have the same complexity [10,14].

The star-complement-star operation on regular languages was studied by Jirásková and Shallit in [8]. They obtained an upper bound of $2^{3n \log n}$ and a lower bound of $2^{\frac{1}{8}n \log n}$ on the state complexity of this combined operation. Hence the state complexity of star-complement-star on regular languages is $2^{\Theta(n \log n)}$. However since we have Θ in an exponent, the gap between the upper bound and the lower bound is large.

On the class of prefix-free languages, investigated, for example, in [2–5,9], the star is an easy operation [5]. If a prefix-free language is accepted by a deterministic finite automaton (DFA) of n states, then its star is accepted by a DFA of at most n states. The question that arises is whether we can use this fact to get a tight upper bound on the state complexity of star-complement-star on prefix-free languages; cf. the result on cyclic shift of prefix-free languages in [6].

In this paper we give a positive answer. We prove that if a prefix-free language L is recognized by an n-state DFA, then the language $((L^*)^c)^*$, which we denote by L^{*c*}, is recognized by a DFA of at most $2^{n-3} + 2$ states. We prove that this upper bound is tight already in the binary case.

M. Palmovský—Research supported by VEGA grant 2/0084/15.
J. Šebej—Research supported by VEGA grant 1/0142/15.

J. Shallit and A. Okhotin (Eds.): DCFS 2015, LNCS 9118, pp. 231–242, 2015.
DOI: 10.1007/978-3-319-19225-3_20

Then we present the results of our computations. Using the lists of minimal n-state binary prefix-free non-isomorphic DFAs, $n = 3, 4, 5, 6, 7, 8$, we computed the state complexity of star-complement-star for each automaton in the list. We also computed the frequencies of the resulting complexities and average complexities. We also observed that for $n = 4, 5, 6, 7, 8$, there is exactly one n-state DFA A (up to renaming the input symbols) such that state complexity of the star-complement-star of the language $L(A)$ is two. In the second part of our paper we prove that this is true for every $n \geq 4$.

We conclude the paper by discussing the unary case. Here the state complexity of star-complement-star of every unary prefix-free language is one, except for the case $L = \{a\}$, where it is two.

2   Preliminaries

Let Σ be a finite alphabet and Σ^* the set of all strings over Σ. The empty string is denoted by ε. The length of a string w is $|w|$. A *language* is any subset of Σ^*. We denote the size of a set A by $|A|$, and its power-set by 2^A.

A *deterministic finite state automaton* is a quintuple $A = (Q, \Sigma, \delta, s, F)$, where Q is a finite set of states; Σ is a finite alphabet; $\delta \colon Q \times \Sigma \to Q$ is the transition function, $s \in Q$ is the initial state; $F \subseteq Q$ is the set of final states (or accepting states). A non-final state q is a *dead state* if $\delta(q, a) = q$ for each a in Σ. The language accepted or recognized by the DFA A is defined to be the set $L(A) = \{w \in \Sigma^* \mid \delta(s, w) \in F\}$.

A *nondeterministic finite automaton* is a quintuple $A = (Q, \Sigma, \delta, s, F)$, where Q, Σ, s, and F are the same as for a DFA, and $\delta \colon Q \times \Sigma \to 2^Q$ is the transition function. Through the paper we use the notation (p, a, q) to mean that there is a transition from p to q on input a, that is, $q \in \delta(p, a)$. The language accepted by the NFA A is defined to be the set $L(A) = \{w \in \Sigma^* \mid \delta(s, w) \cap F \neq \emptyset\}$.

Two automata are *equivalent* if they recognize the same language.

A DFA A is *minimal* if every equivalent DFA has at least as many states as A. It is known that every regular language has a unique minimal DFA (up to isomorphism), and that a DFA $A = (Q, \Sigma, \delta, s, F)$ is minimal if and only if all its states are reachable and distinguishable.

The *state complexity* of a regular language L, denoted by $sc(L)$, is the number of states in the minimal DFA accepting the language L.

Every NFA can be converted to an equivalent DFA by the subset construction [12,13] as follows. Let $A = (Q, \Sigma, \delta, s, F)$ be an NFA. Construct the DFA $A' = (2^Q, \Sigma, \delta', \{s\}, F')$, where $F' = \{R \subseteq Q \mid R \cap F \neq \emptyset\}$, and $\delta'(R, a) = \bigcup_{r \in R} \delta(r, a)$ for each R in 2^Q and each a in Σ. The DFA A' is called the *subset automaton* of the NFA A. The subset automaton may not be minimal since some of its states may be unreachable or equivalent.

For languages K and L the *concatenation* $K \cdot L$ is defined to be $K \cdot L = \{uv \mid u \in K, v \in L\}$. The language L^k with $k \geq 0$ is defined inductively by $L^0 = \{\varepsilon\}$, $L^1 = L$, $L^{i+1} = L^i \cdot L$. The *Kleene closure (star)* of a language L is the language $L^* = \bigcup_{i \geq 0} L^i$. We denote by L^{*c*} the star-complement-star of L.

For a language L accepted by a DFA A, the following construction gives an NFA A^* for the language L^* [7].

Construction 1. *Let $A = (Q, \Sigma, \delta, s, F)$ be a DFA accepting a language L. Construct an NFA A^* for the language L^* from the DFA A as follows:*

- *For each state q in Q and each symbol a in Σ such that $\delta(q, a) \in F$, add the transition on a from q to s.*
- *If $s \notin F$, then add a new initial state q_0 to Q and make this state accepting. For each symbol a in Σ, add the transition on a*
  from q_0 to $\delta(s, a)$ if $\delta(s, a) \notin F$, and
  from q_0 to $\delta(s, a)$ and from q_0 to s if $\delta(s, a) \in F$.

If $w = uv$ for some strings u and v, then u is a *prefix* of w. If, moreover, the string v is non-empty, then u is a *proper prefix* of w. A language is *prefix-free* if it does not contain two distinct strings one of which is a prefix of the other. It is known that L is prefix-free if and only if the minimal DFA for L contains exactly one final state that goes to the dead state on every input symbol [5,9].

For a prefix-free language L accepted by a minimal DFA A, the following construction gives a DFA B for the language L^*.

Construction 2. *Let $A = (Q, \Sigma, \delta, s, \{f\})$ be a DFA accepting a prefix-free language L. Construct a DFA $B = (Q, \Sigma, \delta_B, f, \{f\})$ for L^*, where*

$$\delta_B(q, a) = \begin{cases} \delta(q, a), & \text{if } q \neq f; \\ \delta(s, a), & \text{if } q = f. \end{cases}$$

3 State Complexity of Star-Complement-Star on Prefix-Free Languages

The aim of this section is to prove that the tight bound on the state complexity of the star-complement-star operation on prefix-free languages is $2^{n-3} + 2$. We start with an upper bound. Recall that if a prefix-free language L is accepted by an n-state DFA, then the language L^* is accepted by a DFA of at most n states. The DFA for L^* is described in Construction 2.

Lemma 3. *Let $n \geq 3$. Let L be a binary prefix-free language with $sc(L) = n$. Then $sc(L^{*c*}) \leqslant 2^{n-3} + 2$.*

Proof. Let $A = (Q, \Sigma, \delta, s, \{f\})$ be a minimal DFA for a prefix-free language L, where $Q = \{s, 1, 2, \ldots, n-3, f, d\}$ and $\delta(f, a) = \delta(d, a) = d$ for each a in Σ. Now we describe a DFA B for L^*, a DFA C for L^{*c}, and an NFA C^* for L^{*c*}. Then we illustrate these constructions on a simple example in Fig. 1. First, construct a DFA $B = (Q, \Sigma, \delta_B, f, \{f\})$ for L^* as described in Construction 2. Next, construct a DFA C for L^{*c} from the DFA B by interchanging the final and non-final states. Finally, construct an NFA C^* for L^{*c*} as described in Construction 1.

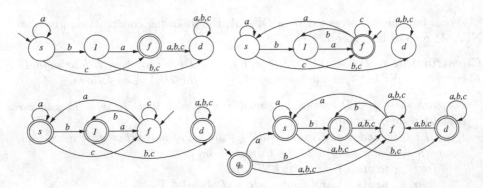

Fig. 1. A DFA A for L (top-left); a DFA B for L^* (top-right); a DFA C for L^{*c} (bottom-left); an NFA C^* for L^{*c*} (bottom-right).

Our aim is to show that the subset automaton of the NFA C^* has at most $2^{n-3} + 2$ reachable and distinguishable states. Notice that if a reachable subset contains a final state of C^*, then it also contains the state s. Next, the empty set is unreachable since A is deterministic and complete. Let S be a reachable subset. Consider four cases:

(1) $d \notin S, f \notin S$. Then S is a non-empty subset of $\{s, 1, 2, \ldots, n-3\}$. Since S contains a final state of NFA C^*, it must also contain the state S. Hence we have at most 2^{n-3} subsets in this case.

(2) $d \notin S, f \in S$. Then either $S = \{f\}$ or $S = \{f\} \cup S'$ where $\emptyset \neq S' \subseteq \{s, 1, 2, \ldots, n-3\}$. In the second case, S' contains a final state of NFA C^*. It follows that S' also contains the state s. Then $\{f\} \cup S'$ is equivalent to S' which was considered in case (1). Thus we get only one new subset, that is, the subset $\{f\}$.

(3) $d \in S$. Then S is equivalent to $\{d\}$ since every string is accepted from $\{d\}$ in the subset automaton.

(4) The initial set $\{q_0\}$ is equivalent to $\{s\}$.

It follows that the subset automaton has at most $2^{n-3} + 2$ reachable and distinguishable states. $\qquad\square$

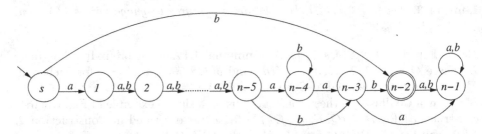

Fig. 2. The DFA of a binary witness prefix-free language for star-complement-star meeting the upper bound $2^{n-3} + 2$.

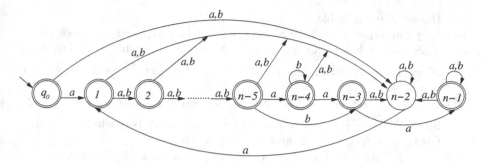

Fig. 3. The NFA C^* for L^{*c*}, where L is accepted by the DFA in Fig. 2.

Now we prove that the upper bound $2^{n-3} + 2$ can be met by a binary prefix-free language.

Lemma 4. *Let $n \geq 3$. There exists a binary regular prefix-free language L with $sc(L) = n$ such that $sc(L^{*c*}) = 2^{n-3} + 2$.*

Proof. If $n = 3$, then $2^{n-3} + 2 = 3$. Consider the language $L = \{a\}$ over the binary alphabet $\{a, b\}$. Then $L^{*c*} = \{a, b\}^* \setminus a^+$, so $sc(L^{*c*}) = 3$.

Now let $n \geq 4$. Let L be the binary prefix-free language accepted by the DFA A shown in Fig. 2. Construct a DFA B for L^* from the DFA A by removing the transitions $(n - 2, a, n - 1)$ and $(n - 2, b, n - 1)$, and by adding the transitions $(n - 2, a, 1)$ and $(n - 2, b, n - 2)$.

Then interchange the final and non-final states to get a DFA C for L^{*c}. Finally, construct an NFA C^* for L^{*c*} from C by adding a new initial and final state q_0 going to $\{1, n - 2\}$ by a and to $\{n - 2\}$ by b; next, for each state q with $q \neq n - 3$ add the transitions $(q, a, n - 2)$ and $(q, b, n - 2)$, and finally, add the transition $(n - 3, a, n - 2)$. The NFA C^* is shown in Fig. 3. Our aim is to show that the subset automaton of the NFA C^* has $2^{n-3} + 2$ reachable and pairwise distinguishable states.

First, by induction on $|S|$, we prove that every subset S of $\{1, 2, \ldots, n - 2\}$ containing $n - 2$ is reachable.

The basis, $|S| \leq 2$, holds since we have

$$\{q_0\} \xrightarrow{b} \{n - 2\} \xrightarrow{a} \{1, n - 2\} \xrightarrow{b^{i-1}} \{i, n - 2\}, \text{ if } 1 \leq i \leq n - 5,$$

$$\{n - 5, n - 2\} \xrightarrow{a} \{1, n - 4, n - 2\} \xrightarrow{b^n} \{n - 4, n - 2\},$$

$$\{n - 5, f\} \xrightarrow{b} \{n - 3, n - 2\}.$$

Assume that $2 \leq k \leq n - 3$, and that each set S, where $|S| = k$ and $n - 2 \in S$, is reachable. Let $S = \{i_1, i_2, \ldots, i_k, n - 2\}$, where $1 \leq i_1 < i_2 < \cdots < i_k \leq n - 3$, be a set of size $k + 1$ and containing the state $n - 2$. Consider several cases:

(i) $i_1 = 1$. Take $S' = \{i_2 - 1, \ldots, i_k - 1, n - 2\}$. Then $|S'| = k$ and $n - 2 \in S'$. Hence S' is reachable by the induction hypothesis. Since

$$S' \xrightarrow{a} \{1, i_2, \ldots, i_k, n - 2\} = S,$$

the set S is reachable.

(ii) $i_1 \geq 2$ and $i_k = n-4$. Take $S' = \{1, i_2-i_1+1, \ldots, i_{k-1}-i_1+1, n-4, n-2\}$. Then $|S'| = k+1$ and $n-2 \in S'$. Hence S' is reachable by (i). Since $S' = \{1, i_2 - i_1 + 1, \ldots, i_{k-1} - i_1 + 1, n-4, n-2\} \xrightarrow{b^{i_1-1}} S$, the set S is reachable.

(iii) $i_1 \geq 2$ and $i_k \leq n-5$. Take $S' = \{1, i_2 - i_1 + 1, \ldots, i_k - i_1 + 1, n-2\}$. Since $S' \xrightarrow{b^{i_1-1}} \{i_1, i_2, \ldots, i_k, n-2\} = S$, the set S is reachable.

(iiiA) $i_1 \geq 2$, $i_k = n-3$, and $i_k - 1 \leq n - 5$.

Take $S' = \{i_1-1, i_2-1, \ldots, i_{k-1}-1, n-5, n-2\}$. Then $i_{k-1}-1 \leq n-6$, $|S'| = k+1$, $n-5 \in S'$ and $n-2 \in S'$. Hence S' is reached by (iii). Since $S' \xrightarrow{b} \{i_1, i_2, \ldots, i_{k-1}, n-3, n-2\} = S$, the set S is reachable.

(iiiB) $i_1 \geq 2$, $i_k = n-3$, and $i_k - 1 = n-4$.

Take $S' = \{i_1 - 1, i_2 - 1, \ldots, i_{k-2} - 1, n-5, n-4, n-2\}$. Since $S' \xrightarrow{b} \{i_1, i_2, \ldots, i_{k-2}, n-3, n-4, n-2\} = S$, the set S is reachable.

We proved that each subset of $\{1, 2, \ldots n-2\}$ containing $n-2$ is reachable. The set $\{q_0\}$ is the initial state and $\{n-2, n-1\}$ is reached from $\{n-3, n-2\}$ by a. This gives $2^{n-3} + 2$ reachable subsets.

Let us show that all the reachable subsets are pairwise distinguishable. Let S and T, two distinct subsets of $\{1, 2, \ldots, n-2\}$ containing $n-2$. Let i be the greatest state in $\{1, 2, \ldots, n-2\}$ which is in exactly one of S and T. Without loss of generality, we can assume that $i \in S$ and $i \notin T$. Consider three cases:

(1) $i = n-3$. Then the string $ab^{n-4}ab^{n-4}$ is accepted from S. This string is rejected from T since T goes to a subset of $\{1, n-3\}$ by $ab^{n-4}a$, and then to $\{n-2\}$ by b^{n-4}.

(2) $i = n-4$. Then we have $S \xrightarrow{b^{n-4}} \{n-4, n-2\}$ and $T \xrightarrow{b^{n-4}} \{n-2\}$, so b^{n-4} is accepted from S and rejected from T.

(3) $i \leq n-5$. Then we have $S \xrightarrow{b^{n-4-i}} S'$, where $n-3 \in S'$, $T \xrightarrow{b^{n-4-i}} T'$ where $n-3 \notin T'$. Since S' and T' are distinguishable by (1), the sets S and T are distinguishable as well.

Next we distinguish state $\{q_0\}$ from all other subsets $S = \{1, 2, \ldots, n-2\}$, such that $n-2 \in S$. If $n-3 \in S$, then we can distinguish S from $\{q_0\}$ by ab^n. If $n-3 \notin S$, then we can distinguish S from $\{q_0\}$ by b.

We also need to distinguished $\{n-2, n-1\}$ from all subsets S and $\{q_0\}$. The state $\{q_0\}$ can be distinguished from $\{n-2, n-1\}$ by b. The subsets of $\{1, 2, \ldots, n-2\}$ can be distinguished from $\{n-2, n-1\}$ by $b^n ab^n$.   □

We summarize the results of the two lemmas above in the following theorem.

Theorem 5. *Let $n \geq 3$. Let L be a prefix-free language over an alphabet Σ with $sc(L) = n$. Then $sc(L^{*c*}) \leq 2^{n-3} + 2$, and the bound is tight if $|\Sigma| \geq 2$.*   □

Table 1. The frequencies of the complexities and the average complexity of star-complement-star on prefix-free languages in the binary case; $n = 3, 4, 5, 6, 7$.

n\sc(L^{*c*})	1	2	3	4	5	6	7	8	9	10	Average
3	-	2	1	-	-	-	-	-	-	-	2.333
4	18	1	7	2	-	-	-	-	-	-	1.75
5	374	1	83	37	24	2	-	-	-	-	1.737
6	10374	1	1638	353	482	359	172	42	26	6	1.71
7	356623	1	47123	5259	7501	8194	8044	4450	2663	1867	1.738

Table 2. The frequencies of the complexities 11–18 of star-complement-star on prefix-free languages in the binary case; $n = 7$.

n\sc(L^{*c*})	11	12	13	14	15	16	17	18	19	20
7	896	447	608	174	-	-	164	26	-	-

4 Computations

We did some computations concerning the star-complement-star operation on binary prefix-free languages. Having the lists of all the minimal prefix-free and pairwise non-isomorphic n-state deterministic automata, $n = 3, 4, 5, 6, 7, 8$, we computed the state complexity of the star-complement-star of corresponding languages. We computed the frequency of the resulting complexities and an average complexities. Our results are summarized in Tables 1 and 2. The results for $n = 8$ can be found at http://im.saske.sk/~palmovsky/StarComplementStar.

Notice that for $n = 4, 5, 6, 7$, there is exactly one n-state DFA A such that the state complexity of the star-complement-star of the language $L(A)$ is two. In the remaining part of this section, we prove that this is true for every $n \geq 4$. Our first lemma shows that for every $n \geq 4$, there exist at least one such language.

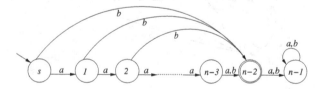

Fig. 4. The DFA of a binary prefix-free language L such that sc(L^{*c*}) = 2.

Lemma 6. *Let* $n \geq 4$. *Let* L *be the binary prefix-free language accepted by the DFA shown in Fig. 4. Then* sc(L^{*c*}) = 2.

Proof. Construct the DFA for L^{*c} as shown in Fig. 5. Next, construct the NFA N for L^{*c*} as shown in Fig. 6. Let us determinize NFA N. The reachable states of the subset automaton of N is shown in Fig. 7. Notice that all the final subsets are equivalent. It follows that after minimization we get the two-state DFA shown in Fig. 8(D). Hence $\mathrm{sc}(L^{*c*}) = 2$.   □

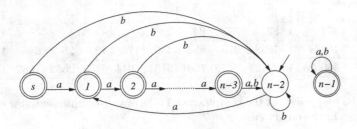

Fig. 5. A DFA for L^{*c}, where L is accepted by the DFA in Fig. 4.

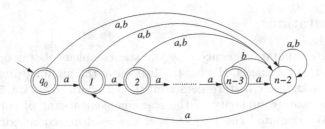

Fig. 6. An NFA N for L^{*c*}, where L is accepted by the DFA in Fig. 4.

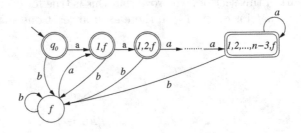

Fig. 7. The reachable states of the subset automaton of the NFA N from Fig. 6.

The next lemma shows how a minimal DFA for L^{*c*} look if we know that $\mathrm{sc}(L^{*c*}) = 2$ and L is a binary prefix-free language.

Lemma 7. *Let $n \geq 4$. Let L be a binary prefix-free language with $\mathrm{sc}(L) = n$ such that $\mathrm{sc}(L^{*c*}) = 2$. Then the minimal DFA for L^{*c*}, up to renaming the input symbols, must look like the one shown in Fig. 8(D).*

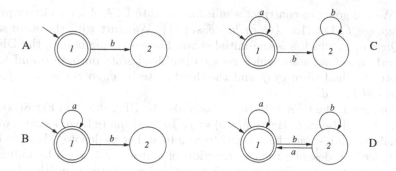

Fig. 8. The transitions in a two-state DFA for L^{*c*}.

Proof. Let B be a minimal 2-state DFA for the language L^{*c*} with the state set $\{1, 2\}$, in which the state 1 is the initial state. Let δ be the transition function of the DFA B.

(a) Since B accepts the star of some language, the initial state 1 must be accepting. As B is minimal, the state 2 must be non-final, and it must be reachable. Without loss of generality, we can assume that it is reached from the initial state 1 by letter b, so $\delta(1, b) = 2$ as shown in Fig. 8(A). As B is deterministic, we have $b \notin L^{*c*}$, and therefore $b \notin L^{*c}$. It follows that $b \in L^*$, which means that $b \in L$.

(b) Now we show that $\delta(1, a) = 1$. Assume for a contradiction, that $\delta(1, a) = 2$; see Fig. 8(B). Then $a \notin L^{*c*}$, and therefore $a \notin L^{*c}$. But then $a \in L^*$, which means that $a \in L$. As $a, b \in L$, and L is a binary prefix-free language, there are no other words in L, so $sc(L) = 3$. This is a contradiction with $n \geq 4$. Therefore $\delta(1, a) = 1$.

(c) Next we show that $\delta(2, b) = 2$; see Fig. 8(C). Since $b \in L$ as shown in (a), we have $b \in L^*$ and $bb \in L^*$. It follows that $b \notin L^{*c}$ and $bb \notin L^{*c}$. This means that $bb \notin L^{*c*}$, and therefore $\delta(2, b) = 2$.

(d) Finally, we show that $\delta(2, a) = 1$; see Fig. 8(D). Assume, in order to get a contradiction, that $\delta(2, a) = 2$. This means that $ba \notin L^{*c*}$. Therefore $ba \notin L^{*c}$, so $ba \in L^*$. It follows that either $ba \in L$ or $a, b \in L$. As shown in (b) the latter case cannot occur. The former case contradicts (a) and the prefix-free property because $b \in L$. Hence $\delta(2, a) = 1$, and our proof is complete. □

Now we prove that, up to renaming input symbols, the language accepted by the DFA in Fig. 4 is a unique binary prefix-free language whose star-complement-star is of complexity two.

Lemma 8. *Let $n \geq 4$. Let L be a binary prefix-free language such that $sc(L) = n$ and $sc(L^{*c*}) = 2$. Then, up to renaming the input symbols, a minimal n-state DFA for L must look like the one shown in Fig. 4.*

Proof. We are going to construct a minimal n-state DFA A for a binary prefix-free language L. Let $A = \{Q, \{a, b, \}, \delta, s, \{f\}\}$. We start with the state set Q with $|Q| = n$, in which s is the initial state. Since L is prefix-free, the DFA A has exactly one final state which goes to the dead state on both a and b. Let us denote the final state by f and the dead state by d, so $\delta(f, a) = \delta(f, b) = \delta(d, a) = \delta(d, b) = d$.

By Lemma 7, the DFA B for L^{*c*} looks like the DFA shown in Fig. 8(D), and therefore b must be in L. Hence $\delta(s, b) = f$. To reach more then 3 states, one of the remaining states must be reached from s by a. Let us denote this state by 1.

Next, we will describe the construction of states $1, \ldots, n - 3$ by induction. Let us assume that for every i such that $1 \leq i \leq n - 4$, the transition function δ is defined as follows:

$\delta(i, b) = f$, and
$\delta(i, a) = i + 1$.

Since $a^{i+1}b \notin L^{*c*}$, we have $a^i b \notin L^{*c}$ and $a^i b \in L^*$. It follows that either $a^i b \in L$, or we can split $a^i b$ into more strings which are in L. But in A, we cannot accept any a^j with $j \leq i$. This means that $a^i b \in L$, and therefore $\delta(i + 1, b) = f$.

Now, we are going to define $\delta(i + 1, a)$. First, we will discuss the case when $i + 1 \neq n - 3$. The same arguments as in definition of $\delta\{s, a\}$ can be used, so we need to reach more states, and therefore $\delta(i + 1, a) = i + 2$.

Next, let $i + 1 = n - 3$, that is, we have already reached all n states. If $\delta(n - 3, a) \in \{s, 1, 2, \ldots, n - 3\}$, then the states in $\{s, 1, 2, \ldots, n - 3\}$ are equivalent. Therefore $n - 3$ goes by a either to d, or to f. Let $\delta(i + 1, a) = d$, so a^{n-2} is not in L.

We know that $a^{n-2}b \in L^{*c*}$, and therefore $a^{n-2}b \notin L^{*c}$ and $a^{n-2}b \in L^*$. This means that $a^{n-2}b \in L$ or we can split $a^{n-2}b$ into more strings from L. Since $\delta(n - 3, a) = d$, the string $a^{n-2}b$ is not in L. Moreover, no a^i with $i \leq n - 2$ is in L. It follows that $a^{n-2}b$ is not L^*, which is a contradiction. Hence $\delta(n-3, a) = f$, and our proof is complete. $\square$

As a corollary of the three lemmas above, we get the following theorem.

Theorem 9. *Let $n \geq 4$. There exists exactly one binary prefix-free language (up to renaming the input symbols) such that $sc(L) = n$ and $sc(L^{*c*}) = 2$. Such a language L is accepted by the DFA in Fig. 4.* $\square$

5   Unary Alphabet

Recall that a unary language is prefix free if it is empty or if it contains exactly one string. In the following theorem we discuss the state complexity of the star-complement-star operation on unary prefix-free languages.

Theorem 10. *Let L be a unary prefix free language. Then $sc(L^{*c*}) = 1$, except for the case when $L = \{a\}$ in which $sc(L^{*c*}) = 2$.*

Proof. We consider four cases:

(a) $L = \emptyset$. Then $L^* = \varepsilon$ and $L^{*c} = a^* \setminus \varepsilon$. Thus $L^{*c*} = a^*$ and $sc(L^{*c*}) = 1$.
(b) $L = \varepsilon$. Then $L^* = \varepsilon$, and we have already obtained this in case (a). So in this case, we again get $sc(L^{*c*}) = 1$.
(c) $L = a$. Then $L^* = a^*$ and $L^{*c} = \emptyset$. Therefore $L^{*c*} = \varepsilon$ and $sc(L^{*c*}) = 2$.
(d) $L = a^n$, where $n \geq 2$. Then $a \notin L$. Therefore $a \notin L^*$, so $a \in L^{*c}$. It follows that $L^{*c*} = a^*$, so $sc(L^{*c*}) = 1$. Our proof is complete. $\qquad\square$

6   Conclusions

We investigated the star-complement-star operation on prefix-free languages. We proved that if a language L is accepted by an n-state deterministic finite automaton, then the language L^{*c*} is accepted by a deterministic finite automaton of at most $2^{n-3} + 2$ states. We also proved that this upper bound is tight for every alphabet containing at least two symbols.

Our computations showed that if $n \in \{4, 5, 6, 7, 8\}$, then there exist exactly one, up to renaming input symbols, binary prefix-free language with $sc(L) = n$ and $sc(L^{*c*}) = 2$. We proved that this true for all $n \geq 4$.

We showed that the state complexity of the star-complement-star of a unary language is 1, except for the language $\{a\}$, where it is 2. Taking into account that the only prefix-free language with state complexity 1 or 2 is the empty language or the language $\{\varepsilon\}$, respectively, and that $\emptyset^{*c*} = \varepsilon^{*c*} = \Sigma^*$, we can summarize the results of our paper in the following theorem.

Theorem 11. *Let $f_k(n)$ be the state complexity of the star-complement-star operation on prefix-free regular languages over a k-letter alphabet defined to be*

$$f_k(n) = \max\{sc(L^{*c*}) \mid L \subseteq \Sigma^*, |\Sigma| = k, \, sc(L) = n, \, \text{and } L \text{ is prefix-free}\}.$$

Then

1. *if $k \geq 2$, then* $f_k(n) = \begin{cases} 1, & \text{if } n = 1 \text{ or } n = 2; \\ 2^{n-3} + 2, & \text{if } n \geq 3; \end{cases}$
2. *if $n \geq 4$, then there exists exactly one, up to renaming input symbols, binary prefix-free language such that $sc(L) = n$ and $sc(L^{*c*}) = 2$;*
3. $f_1(n) = \begin{cases} 2, & \text{if } n = 3; \\ 1, & \text{otherwise.} \end{cases}$ $\qquad\square$

References

1. Brzozowski, J.A., Grant, E., Shallit, J.: Closures in formal languages and Kuratowski's theorem. Int. J. Found. Comput. Sci. **22**(2), 301–321 (2011). doi:10.1142/S0129054111008052
2. Eom, H.-S., Han, Y.-S., Salomaa, K.: State complexity of k-union and k-intersection for prefix-free regular languages. In: Jurgensen, H., Reis, R. (eds.) DCFS 2013. LNCS, vol. 8031, pp. 78–89. Springer, Heidelberg (2013)

3. Eom, H., Han, Y., Salomaa, K., Yu, S.: State complexity of combined operations for prefix-free regular languages. In: Paun, G., Rozenberg, G., Salomaa, A. (eds.) Discrete Mathematics and Computer Science. In: Memoriam Alexandru Mateescu (1952–2005), pp. 137–151. The Publishing House of the Romanian Academy (2014)
4. Han, Y., Salomaa, K., Wood, D.: Nondeterministic state complexity of basic operations for prefix-free regular languages. Fundamenta Informaticae **90**(1–2), 93–106 (2009). doi:10.3233/FI-2009-0008
5. Han, Y., Salomaa, K., Wood, D.: Operational state complexity of prefix-free regular languages. In: Ésik, Z., Fülöp, Z. (eds.) Automata, Formal Languages, and Related Topics – Dedicated to Ferenc Gécseg on the occasion of his 70th birthday, pp. 99–115. Institute of Informatics, University of Szeged, Hungary (2009)
6. Jirásek, J., Jirásková, G.: Cyclic shift on prefix-free languages. In: Bulatov, A.A., Shur, A.M. (eds.) CSR 2013. LNCS, vol. 7913, pp. 246–257. Springer, Heidelberg (2013)
7. Jirásková, G., Palmovský, M.: Kleene closure and state complexity. In: Vinar, T. (ed.) Proceedings of the Conference on Information Technologies – Applications and Theory, Slovakia, 11–15 September 2013. CEUR Workshop Proceedings, vol. 1003, pp. 94–100 (2013). CEUR-WS.org, http://ceur-ws.org/Vol-1003/94.pdf
8. Jirásková, G., Shallit, J.: The state complexity of star-complement-star. In: Yen, H.-C., Ibarra, O.H. (eds.) DLT 2012. LNCS, vol. 7410, pp. 380–391. Springer, Heidelberg (2012)
9. Krausová, M.: Prefix-free regular languages: closure properties, difference, and left quotient. In: Kotásek, Z., Bouda, J., Černá, I., Sekanina, L., Vojnar, T., Antoš, D. (eds.) MEMICS 2011. LNCS, vol. 7119, pp. 114–122. Springer, Heidelberg (2012)
10. Maslov, A.: Estimates of the number of states of finite automata. Sov. Math. Dokl. **11**, 1373–1375 (1970)
11. Salomaa, A., Salomaa, K., Yu, S.: State complexity of combined operations. Theor. Comput. Sci. **383**(2–3), 140–152 (2007). doi:10.1016/j.tcs.2007.04.015
12. Sipser, M.: Introduction to the Theory of Computation. PWS Publishing Company, Boston (1997)
13. Yu, S.: Regular languages. In: Rozenberg, G., Salomaa, A. (eds.) Word, Language, Grammar, Handbook of Formal Languages, vol. 1, pp. 41–110. Springer, New York (1997)
14. Yu, S., Zhuang, Q., Salomaa, K.: The state complexities of some basic operations on regular languages. Theor. Comput. Sci. **125**(2), 315–328 (1994). doi:10.1016/0304-3975(92)00011-F

Groups Whose Word Problem
is a Petri Net Language

Gabriela Aslı Rino Nesin and Richard M. Thomas[✉]

Department of Computer Science, University of Leicester, Leicester LE1 7RH, UK
garn1@le.ac.uk, rmt@mcs.le.ac.uk

Abstract. There has been considerable interest in exploring the connections between the word problem of a finitely generated group as a formal language and the algebraic structure of the group. However, there are few complete characterizations that tell us precisely which groups have their word problem in a specified class of languages. We investigate which finitely generated groups have their word problem equal to a language accepted by a Petri net and give a complete classification, showing that a group has such a word problem if and only if it is virtually abelian.

Keywords: Finitely generated group · Word problem · Petri net language

1 Introduction

There has been considerable interest in exploring the connections between the word problem of a finitely generated group as a formal language and the algebraic structure of the group. Whilst the seminal work of Boone and Novikov in the 1950's showed that a finitely presented group could have a word problem that is not recursive, it was not really until the 1970's that languages lower down in the Chomsky hierarchy were investigated. Anisimov showed in 1971 that a group has a regular word problem if and only if it is finite [1]. Whilst this result is not difficult to prove, asking such a question was an innovative idea, and naturally led to an investigation as to what happens with other classes of languages.

Muller and Schupp showed in [17] (modulo a subsequent result of Dunwoody [3]) that a group has a context-free word problem if and only if it is virtually free. Indeed, the word problem of such a group must be deterministic context-free, and even an NTS language [2]. Apart from Dunwoody's result, this characterization uses other deep group-theoretical results (such as Stallings' classification of groups with more than one end).

Within the context-free languages there are essentially not many other possibilities if we assume certain natural conditions on the class of languages. Herbst [8] showed that, if $\mathcal{F}$ is a cone that is a subset of the context-free languages, then the class of groups whose word problem lies in $\mathcal{F}$ is either the class of groups with a regular word problem, the class of groups with a one-counter

© Springer International Publishing Switzerland 2015
J. Shallit and A. Okhotin (Eds.): DCFS 2015, LNCS 9118, pp. 243–255, 2015.
DOI: 10.1007/978-3-319-19225-3_21

word problem[1] or the class of groups with a context-free word problem. He also classified the groups with a one-counter word problem as being the virtually cyclic groups. This was extended in [10], where it was shown that a group has a word problem that is a finite intersection of one-counter languages if and only if it is virtually abelian. There is also an interesting result in [4], where it is shown that a group has a word problem that is accepted by a blind counter machine if and only it is virtually abelian; see Sect. 7 for the definition of such an automaton and for some further discussion.

Whilst other classes of languages have been investigated, there are very few complete characterizations. We investigate groups whose word problem is a terminal Petri net language and establish the following:

Theorem 1. *A finitely generated group G has word problem that is a terminal Petri net language if and only if G is virtually abelian.*

Whilst this gives a correspondence between an important family of languages and a natural class of groups, there are many variations on Petri net languages which could potentially give rise to different classes of groups. Many of these modifications are so powerful that the class of languages is found to be equal to the class of recursively enumerable languages, but there are other interesting possibilities, such as the class obtained by allowing λ-transitions in the Petri net.

The structure of this paper is as follows. We recall some basic facts about Petri nets and group theory in Sects. 2 and 4. In Sect. 3 we comment on the equivalence of various definitions for Petri net languages. Given this background material, showing that a finitely generated virtually abelian group has a word problem that is a Petri net language is fairly straightforward, and we do this in Sect. 5. The proof of the converse is rather more involved and we provide that in Sect. 6. We finish in Sect. 7 by commenting how this class of groups relates to certain other classes which have arisen in considering word problems.

2   Petri Nets

In this section we set out our conventions and notation for Petri nets and recall some properties of the class of languages they accept. A *labelled Petri net* is a tuple $P = (S, T, W, m_0, \Sigma, l)$ where:

(i) S is a finite set, called the set of *places*; we will assume that an order is imposed on S and so it will be displayed as a tuple.

(ii) T is a finite set disjoint from S, called the set of *transitions*.

(iii) $W : (S \times T) \cup (T \times S) \to \mathbb{N}$ is the *weight function*, assigning a multiplicity to pairs of places and transitions. If $W(x, y) = n$ then we will write $x \xrightarrow{n} y$. If $W(x, y) = 0$ then we say there is no arrow from x to y.

(iv) The *initial marking* $m_0 \in \mathbb{N}^S$ assigns a natural number to each place.

---

[1] The one-counter languages are those languages accepted by a pushdown automaton where we have a single stack symbol apart from a bottom marker.

(v) Σ is a finite set called the *alphabet* and the *labelling function* $l : T \to \Sigma$ assigns a label to each transition.

The function l can be extended to a function $T^* \to \Sigma^*$ in the natural way (where we define λl to be λ). Note that l does not have to be bijective (if it were we would have a "free Petri net"), but we do assume that l is a (total) function.

As usual, we represent a labelled Petri net by a labelled directed graph, where the places are represented by circles, transitions by rectangles (we will denote transitions by their labels for simplicity), the weight function by arrows and arrow multiplicities by numbers on the arrows (with no arrow drawn if the multiplicity is zero and no number if the multiplicity is one). Markings (i.e., elements of $\mathbb{N}^S$) are represented by tokens or natural numbers in each place.

Now we describe the execution semantics of Petri nets. Let $m \in \mathbb{N}^S$ be a marking and $t \in T$ be a transition. We say that t is *enabled* at m if, for all $s \in S$, we have $W(s,t) \leqslant m(s)$; we denote this by $m[t\rangle$. If t is enabled at m, we can fire t to get a new marking $m' \in \mathbb{N}^S$, defined by $m'(s) = m(s) + W(t,s) - W(s,t)$ for all $s \in S$, and we write $m[t\rangle m'$ noting that asserting this automatically implies that $m[t\rangle$ must hold. We generalize this to sequences w of transitions (i.e., to elements w of T^*) and define $m[w\rangle m'$ in the obvious way.

We will need the notion of a labelled Petri net accepting a language; there are various possibilities and we consider the "terminal language" of a Petri net. We extend the definition of a labelled Petri net $P = (S, T, W, m_0, \Sigma, l)$ to include a finite set of *terminal markings* $M \subset \mathbb{N}^S$ and write $P = (S, T, W, m_0, M, \Sigma, l)$. The *(terminal) language* $L(P)$ recognized by P is the set

$$\{l(w) : m_0[w\rangle m \text{ some } m \in M, w \in T^*\}$$

We say that a language $L \subseteq \Sigma^*$ is a *Petri net language* (PNL for short) if there is a labelled Petri net whose terminal language is L and let $\mathcal{PNL}$ denote the class of all Petri net languages. The class $\mathcal{PNL}$ has several nice closure properties (see references such as [13]), some of which we note here for future reference:

Proposition 2. *(i) $\mathcal{PNL}$ contains all regular languages.*
(ii) $\mathcal{PNL}$ is closed under union.
(iii) $\mathcal{PNL}$ is closed under intersection.
(iv) $\mathcal{PNL}$ is closed under inverse GSM mappings (and, in particular, under inverse homomorphisms).

Some authors define Petri net languages in a slightly different way; several clever constructions (see pages 8–21 in [7]) show that these definitions are equivalent, up to the inclusion of the empty word λ in the language. We will survey some of these approaches in the next section.

3   Equivalence of the Various Definitions

We will now give various different definitions of Petri net languages and note their equivalence. We are not sure that all of this has been explicitly proved in previous papers, but it does appear that these equivalences are already known.

Our definition of a PNL is the same as that given by Jantzen in [13]. We will keep our terminology of labelled Petri net and PNL as above. The following definition is used by Petersen to define CSS (computation sequence sets) in [19]:

Definition 3. *A P-Petri net N is a 5-tuple (P, T, Σ, S, F) where P is a finite set of places, T is a finite set of transitions disjoint from P, Σ is the input alphabet (or the set of labels), $S \in P$ is a designated start place, $F \subseteq P$ is a designated set of final places, and each transition $t \in T$ is a triple consisting of a label in Σ, a multiset (bag) I of input places, and a bag O of output places.*

This is almost the same as our definition, except for the designated start and final places. The labelling implies that there can be more than one transition with the same label, and the multiplicity of a place in a bag is just the multiplicity of its arrow to or from the transition in our original definition. Enabled transitions and so on are defined in the same way. We then have the following:

Definition 4. *The* Computation Sequence Set *of a P-Petri net is the set of all sequences of labels of transitions leading from the start marking (one token in the start place, none anywhere else) to one of the final markings (one token in one of the final places, none in any other place).*

Let $\mathcal{CSS}$ denote the class of languages which are the CSS of a P-Petri net.

Hack's definition of a labelled Petri net in [7] is the same as the one given here (he actually splits the weight function into two separate forwards and backwards incidence functions, but this is not an essential difference). He then has the following:

Definition 5. *The set of H-terminal label sequences of a labelled Petri net N for a final marking $m_f \neq m_0$ is the set labels of sequences of transitions leading from m_0 to m_f.*

Essentially, the difference between our definition and Hack's is twofold: he only allows one final marking, and this final marking cannot be equal to the start marking. His motivation is that one then avoids having any H-terminal languages containing λ: if one keeps the unique final marking condition but allows these languages to contain λ, then the class of H-terminal languages of labelled Petri nets would no longer be closed under union (see page 8 in [7]). Hack calls this class $\mathfrak{L}_0$, and we shall adopt this terminology. It is known (see pages 19–20 of [7]) that $\mathfrak{L}_0$ and $\mathcal{CSS}$ are the same up to inclusion of the empty word:

Theorem 6. *For any language L, we have that $L \in \mathcal{CSS} \iff L - \{\lambda\} \in \mathfrak{L}_0$.*

It is clear that $\mathcal{CSS} \subseteq \mathcal{PNL}$ and the reverse inclusion also holds: one can use a "standardisation" of the Petri nets described in [7] to transform a Petri net into a P-Petri net without changing the terminal language. This gives the following:

Theorem 7. $\mathcal{PNL} = \mathcal{CSS}$.

4   Group Theory

In this section we review the background material we need from group theory and establish some general facts about groups whose word problem is a Petri net language. For general information about group theory, see [14, 20] for example.

Let A be a finite set and let A^{-1} be another set disjoint from, but in a one-to-one correspondence with, A; we write a^{-1} for the element in A^{-1} corresponding to the element a in A. Let $\Sigma = A \cup A^{-1}$. We say that A is a *generating set* for a group G if we have a monoid homomorphism φ from Σ^* onto G such that $(a\varphi)^{-1} = a^{-1}\varphi$ for all $a \in A$; we normally then identify an element $x \in \Sigma$ with the image $x\varphi \in G$, so that A becomes a subset of G. A group with such a finite generating set is said to be *finitely generated*. The groups considered in this paper will all be finitely generated.

With this convention we define the *word problem* $W_A(G)$ of G with respect to the generating set A to be $\{\alpha \in \Sigma^* : \alpha =_G 1_G\}$ where the notation $\alpha =_G \beta$ (where $\alpha, \beta \in \Sigma^*$) denotes the fact that α and β represent the same element of G (i.e., that $\alpha\varphi = \beta\varphi$) and $\alpha =_G g$ (where $\alpha \in \Sigma^*$ and $g \in G$) denotes the fact that α represents the element g of G (i.e., that $\alpha\varphi = g$).

With this definition the word problem $W_A(G)$ is a subset of Σ^* and hence is a language; so we can consider which groups have their word problem in a given class of languages. This would seem to depend on the choice of A but, under certain mild assumptions of $\mathcal{F}$, this does not matter (see [9] for example):

Proposition 8. *If a class of languages $\mathcal{F}$ is closed under inverse homomorphism and the word problem of a group G with respect to some finite generating set lies in $\mathcal{F}$ then the word problem of G with respect to any finite generating set lies in $\mathcal{F}$.*

Given Proposition 2 (iv), we may talk about the word problem of a finitely generated group G being a PNL without reference to the choice of generating set. If $\mathcal{F}$ is any class of languages closed under inverse homomorphism then we will (mildly) abuse notation and write $G \in \mathcal{F}$ if the word problem of G lies in $\mathcal{F}$.

As the class $\mathcal{PNL}$ is closed under inverse homomorphisms and intersection with regular languages (the latter fact following from parts (i) and (iii) of Proposition 2), we have the following immediate consequence of Lemma 2 of [11]:

Proposition 9. *The class of finitely generated groups with word problem a PNL is closed under taking finitely generated subgroups.*

We also have the following:

Proposition 10. *If G and H are finitely generated groups with word problems in $\mathcal{PNL}$ then the word problem of the direct product $G \times H$ is also in $\mathcal{PNL}$.*

Proof. If P_1 and P_2 are Petri nets recognising the word problems of G and H with respect to finite generating sets A and B respectively (where $A \cap B = \emptyset$) then the disjoint union of P_1 and P_2 recognizes the word problem of $G \times H$. $\square$

Of fundamental importance in what follows will be the so-called *Heisenberg group*, which is the group of matrices

$$\left\{ \begin{pmatrix} 1 & a & c \\ 0 & 1 & b \\ 0 & 0 & 1 \end{pmatrix} : a, b, c \in \mathbb{Z} \right\}$$

under multiplication. This is an example of a "nilpotent group". One way of defining this concept is to let $Z(G)$ denote the *centre* of a group G (i.e., the set of elements in G that commute with all the elements of G) and then define a series of normal subgroups $Z_1(G) \leqslant Z_2(G) \leqslant \cdots$ of G as follows:

$$Z_1(G) := Z(G), \quad Z_{i+1}(G)/Z_i(G) := Z(G/Z_i(G)) \text{ for } i \geqslant 1.$$

We say that G is *nilpotent* if $Z_i(G) = G$ for some $i \in \mathbb{N}$.

A generalization of this is to say that a group G is *virtually nilpotent* if G has a nilpotent subgroup H of finite index in G (where the *index* of a subgroup H is the number of distinct right cosets of the form Hg for $g \in G$). In general, if $\wp$ is any property of groups, then we say that G is *virtually $\wp$* if G has a subgroup of finite index with the property $\wp$. It is a standard result that, if H has finite index in G, then H is finitely generated if and only if G is finitely generated. The following fact (see [11] for example) will be important here:

Proposition 11. *A finitely generated torsion-free virtually nilpotent group that does not contain the Heisenberg group is virtually abelian.*

The term "torsion-free" means that the group does not contain any non-trivial elements of finite order.

The notion of finite index will be particularly relevant in this paper. Given that $\mathcal{PNL}$ is closed under union with regular sets and inverse GSM mappings by Proposition 2, we have the following immediate consequence of Lemma 5 in [11]:

Proposition 12. *If H is a finitely generated group with word problem in $\mathcal{PNL}$ and G is a group containing H as a finite index subgroup, then the word problem of G is also in $\mathcal{PNL}$.*

Returning to generating sets, we say that a group G with finite generating set A has *polynomial growth* if there is a polynomial $p(x)$ such that the number of distinct elements of G represented by words in $(A \cup A^{-1})^*$ of length at most n is bounded above by $p(n)$.

5   Virtually Abelian Implies PNL Word Problem

In this section we prove one direction of Theorem 1, showing that a finitely generated virtually abelian group G has its word problem in $\mathcal{PNL}$. We start with the case where G is abelian:

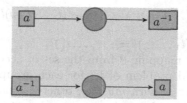

Fig. 1. A labelled Petri net recognizing the word problem of $\mathbb{Z}$. The empty marking is both initial and terminal, and there are no other terminal markings.

Proposition 13. *The word problem of a finitely generated abelian group is always a PNL.*

Proof. Let G be a finitely generated abelian group. According to the structure theorem for finitely generated abelian groups, G is expressible as a direct product

$$\mathbb{Z}^r \times \mathbb{Z}/a_1\mathbb{Z} \times \cdots \times \mathbb{Z}/a_m\mathbb{Z}$$

where $r \geqslant 0$, $m \geqslant 0$ and $a_i = p_i^{n_i}$ for some prime p_i and some natural number $n_i \geqslant 1$. As noted in the introduction, the word problem of a finite group such as $\mathbb{Z}/a\mathbb{Z}$ is regular, and hence a PNL. The word problem of $\mathbb{Z}$ with respect to some generating set $\{a\}$ is a PNL as shown in Fig. 1.

The result now follows from Proposition 10. $\qquad\qquad\qquad\qquad\square$

Propositions 12 and 13 immediately give the following:

Corollary 14. *Any finitely generated virtually abelian group has word problem in $\mathcal{PNL}$.*

6 PNL Word Problem Implies Virtually Abelian

Now we consider the converse to Corollary 14 which (together with Corollary 14) will establish Theorem 1. First we prove the following:

Proposition 15. *A finitely generated group with PNL word problem has polynomial growth.*

Proof. Let G be a group generated by a finite set A, let $\Sigma = A \cup A^{-1}$, and assume that the word problem $W_A(G)$ of G is recognized by a Petri net $P = (S, T, W, m_0, M, \Sigma, l)$ with initial marking m_0 and set of terminal markings M.

We call markings that are reachable from m_0 in P and which allow the possibility of reaching a terminal marking *acceptable* markings. Note that, given an acceptable marking m, any two sequences of transitions reaching m from m_0 must represent the same element of G. This is because, if $m_0[t_1 \ldots t_n\rangle m$ and $m_0[t'_1 \ldots t'_k\rangle m$ and if m is acceptable, then there is a sequence of transitions w from m to some terminal marking m'. But then

$$m_0[t_1 \ldots t_n\rangle m[w\rangle m' \quad \text{and} \quad m_0[t'_1 \ldots t'_k\rangle m[w\rangle m',$$

and hence both sequences of transitions label elements of $W_A(G)$, i.e.,

$$(t_1 \dots t_n w)l =_G 1_G =_G (t_1' \dots t_k' w)l,$$

from which we get that $(t_1 \dots t_n)l =_G (t_1' \dots t_k')l$.

So we have a natural mapping θ from the set of acceptable markings to G. As P recognizes the word problem of G, for each group element g there must be an acceptable marking m with $m\theta = g$; otherwise no word ww^{-1}, where w represents g, can be accepted by P. So the mapping θ is surjective.

In order to show polynomial growth, we want to show that there is a polynomial $p(n)$ such that the number of elements of G represented by a sequence of generators of length n is at most $p(n)$. Since the mapping θ is surjective, it is therefore sufficient to bound the number of acceptable markings reachable by a sequence of transitions of length n by such a polynomial $p(n)$.

If a sequence $t_1 \dots t_n$ reaches an acceptable marking and if $t_{\sigma(1)} \cdots t_{\sigma(n)}$ does as well for some permutation σ of $\{1, 2, \dots, n\}$, then the two sequences reach the same marking[2]; this follows directly from the effect on a marking of firing a transition. In counting the number of acceptable markings, one can therefore ignore the order in which the transitions fire: the only important thing is their multiplicities. If $T = \{u_1, u_2, \dots, u_k\}$ then there are at most as many acceptable markings induced by sequences of n transitions as there are possible values for $\mu(u_1), \dots, \mu(u_k) \in \mathbb{N}$ such that $\mu(u_1) + \cdots + \mu(u_k) = n$, where $\mu(u_i)$ denotes the multiplicity of u_i in the transition sequence. It is now clear that the number of acceptable markings is bounded above by the polynomial $(n + 1)^k$, as there are at most $n + 1$ choices for each of the $\mu(u_i)$. $\square$

Using Gromov's wonderful theorem [6] about groups with polynomial growth we immediately deduce the following:

Corollary 16. *A finitely generated group whose word problem is a PNL is virtually nilpotent.*

We now want to show that a finitely generated group whose word problem is a PNL is virtually abelian. As we will show later, it is enough to show that the Heisenberg group's word problem is not a PNL. To show this, we use Lambert's pumping lemma; we state here a corollary to it (see Theorem 5.1 in [15]):

Theorem 17. *Let $P = (S, T, W, m_0, \Sigma, l)$ be a labelled Petri net and m_f a final marking. Let $a \in \Sigma$. Defining*

$$\mathcal{L}(a) := \{|l(u)|_a : m_0[u\rangle m_f\}$$

we have that $\mathcal{L}(a)$ is infinite if and only if it contains an arithmetic sequence with a non-zero common difference.

---

[2] Recall that the t_i are actual transitions, not labels of transitions (i.e., generators); therefore this argument does not imply that G is abelian as, for example, being able to swap labels a and b in one such sequence does not mean that we would necessarily be able to do so in all such sequences.

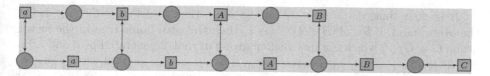

Fig. 2. A labelled Petri net recognizing K. The initial marking is a token in the bottom left place, and the terminal marking is a token in the bottom right place.

Note that the use of only one final marking does not pose a problem, because of Theorems 6 and 7. Our result will follow from Theorem 18 below.

Theorem 18. *Let $\Sigma = \{a, b, A, B, C\}$. Then $L = \{a^i b^j A^i B^j C^{ij} : i, j \in \mathbb{N}\}$ is not a Petri net language.*

Proof. Assume that L is a Petri net language. If so, then we can intersect it with the language $K = \{a^n b^n A^n B^n C^k : n, k \in \mathbb{N}\}$ to get $\{a^n b^n A^n B^n C^{n^2} : n \in \mathbb{N}\}$. K is a Petri net language, recognisable by the Petri net (Fig. 2).

Since Petri nets are closed under intersection, we have that $L \cap K \in \mathcal{PNL}$ as well. By Theorem 17, $\mathcal{L}(C) = \{n^2 : n \in N\}$ would then contain an arithmetical sequence, a contradiction. □

We can now deduce the required result about the Heisenberg group:

Corollary 19. *The word problem of the Heisenberg group H is not a PNL.*

Proof. If a, b and c, respectively, denote the matrices

$$\begin{pmatrix} 1 & 1 & 0 \\ 0 & 1 & 0 \\ 0 & 0 & 1 \end{pmatrix}, \begin{pmatrix} 1 & 0 & 0 \\ 0 & 1 & 1 \\ 0 & 0 & 1 \end{pmatrix} \text{ and } \begin{pmatrix} 1 & 0 & 1 \\ 0 & 1 & 0 \\ 0 & 0 & 1 \end{pmatrix}$$

then a, b and c generate H and every relation in H can be deduced from the relations

$$ac = ca, \ bc = cb \text{ and } a^{-1}b^{-1}ab = c;$$

see [14] for example. Let W denote the word problem of H with respect to $\{a, b, c\}$. To ease clutter, we let A represent a^{-1}, B represent b^{-1} and C represent c^{-1}. We claim that the language L from Theorem 18 is just $W \cap a^* b^* A^* B^* C^*$. To see this, we note that $ab =_G bac$. So we get

$$
\begin{aligned}
a^i b^j &=_G a^{i-1} abb^{j-1} \ =_G a^{i-1} bacb^{j-1} =_G a^{i-1} bab^{j-1}c =_G a^{i-1} babb^{j-2}c \\
&=_G a^{i-1} b^2 ab^{j-2} c^2 =_G \ \cdots \ =_G a^{i-1} b^j a c^j =_G \ \cdots \\
&=_G \ b^j a^i c^{ij}.
\end{aligned}
$$

Now:

$$a^i b^j A^k B^l C^m =_G 1 \iff b^j a^i c^{ij} A^k B^l C^m =_G 1 \iff b^j a^i A^k B^l c^{ij} C^m =_G 1.$$

It is clear that, if $i = k$, $j = l$ and $ij = m$, then $b^j a^i A^k B^l c^{ij} C^m =_G 1$. On the other hand, if $b^j a^i A^k B^l c^{ij} C^m =_G 1$, then this still holds true in the factor group $\overline{G} = G/\langle c \rangle$ which is a free abelian group of rank 2 generated by $\overline{a}$ and $\overline{b}$. So $\overline{b}^j \overline{a}^i \overline{A}^k \overline{B}^l =_{\overline{G}} 1$, which gives that $i = k$ and $j = l$. So $b^j a^i A^k B^l c^{ij} C^m =_G c^{ij} C^m$, and so we must have that $m = ij$ as well. So $a^i b^j A^k B^l C^m =_G 1$ if and only if $i = k$, $j = l$ and $ij = m$, which is what we wanted to establish.

Since the class $\mathcal{PNL}$ is closed under intersection with regular languages, we have that $W \notin \mathcal{PNL}$.   □

Our result now follows:

Proposition 20. *If a finitely generated group G has a PNL word problem, then G is virtually abelian.*

Proof. We know already that G is virtually nilpotent by Corollary 16. Assume that G is not virtually abelian; then G has a nilpotent but not virtually abelian subgroup K of finite index in G. In turn, it is known (see 5.4.15 (i) of [20], for example) that K must have a torsion-free subgroup L of finite index, and L must then be nilpotent but not virtually abelian. By Proposition 11 we have that $H \leqslant L \leqslant K \leqslant G$ where H is the Heisenberg group. So H is a finitely generated subgroup of G. Since G has a PNL word problem, so does H by Proposition 9, contradicting Corollary 19.   □

Taken together with Corollary 14, this completes the proof of Theorem 1.

7   Relation to Other Classes of Languages

In this section, we put our results into some context, comparing the class of groups with word problem in $\mathcal{PNL}$ with those in some other classes of languages. We let $\mathcal{OC}$ denote the class of one-counter languages, $\mathcal{CF}$ the class of context-free languages and $co\mathcal{CF}$ the class of co-context-free languages (i.e., languages that are complements of context-free languages).

The one-counter languages are those languages accepted by a pushdown automaton where we have a single stack symbol apart from a bottom marker. We could think of the stack as a counter where we store a natural number and can test for zero. If we have two such counters, it is well known that we can simulate a Turing machine, and so such machines accept all the recursively enumerable languages; however there are natural variations which restrict the class of languages that can be accepted. If we allow the counters to contain integers and accept a word if we can reach a designated accept state with all the counters equal to zero, then we have a BMM (*blind multicounter machine*). We could strengthen the model to allow the counters to contain natural numbers and, whilst we still cannot test if the counters are empty, transitions that attempt to decrease a counter which currently has value zero are not enabled; we again accept if we can reach a designated accept state with all the counters equal to zero. Such a machine is called a PBMM (*partially blind multicounter machine*).

It was shown in [5] that every Petri net language is accepted by such a machine but that, if L is the language $\{a^n b^n : n \geqslant 0\}$, which is in both $\mathcal{OC}$ and $\mathcal{PNL}$, then L^* is not accepted by a PBMM, and hence is not in $\mathcal{PNL}$. Since $L^* \in \mathcal{OC}$, we have that $\mathcal{OC} \not\subseteq \mathcal{PNL}$. On the other hand, it is well known that $\mathcal{PNL} \not\subseteq \mathcal{CF}$ (let alone $\mathcal{OC}$); for example, the language $\{a^n b^n c^n : n \geqslant 0\}$ is in $\mathcal{PNL}$.

We mentioned in the introduction that the groups whose word problem is a one-counter or context-free language have been classified. As we have just seen, these families of languages are both incomparable with $\mathcal{PNL}$. However, when we turn to word problems of groups, the situation changes:

Proposition 21. *If $G \in \mathcal{OC}$ then $G \in \mathcal{PNL}$.*

Proof. A group with one-counter word problem is virtually cyclic by [8] and hence virtually abelian; the result follows from Corollary 14. □

Of course, since $\mathcal{OC} \not\subseteq \mathcal{PNL}$, finite intersections of one-counter languages are not necessarily in $\mathcal{PNL}$; however this situation also changes when we restrict ourselves to word problems. As we mentioned in the introduction, it was shown [10] that a group has a word problem that is the intersection of finitely many one-counter languages if and only if it is virtually abelian; so we have the following:

Corollary 22. $G \in \mathcal{PNL}$ *if and only if* $G \in \bigcap_{\text{fin}} \mathcal{OC}$.

We mention in passing that, not only is $\bigcap_{\text{fin}} \mathcal{OC}$ not a subset of $\mathcal{PNL}$, but $\mathcal{PNL}$ is not a subset of $\bigcap_{\text{fin}} \mathcal{OC}$ either:

Proposition 23. $L = \{a^n b^m : 1 \leqslant m \leqslant 2^n, 1 \leqslant n\}$ *is in* $\mathcal{PNL}$ *but not* $\bigcap_{\text{fin}} \mathcal{OC}$.

Proof. It is known that $L \in \mathcal{PNL}$; see [12]. Assume that $L \in \bigcap_{\text{fin}} \mathcal{OC}$, say

$$L = L_1 \cap \cdots \cap L_n$$

where the L_i are one-counter languages. Let K be the regular language $a^* b^*$.

Since $L \subseteq K$, we have $L = (L_1 \cap K) \cap \cdots \cap (L_n \cap K)$ and so, without loss of generality, we can assume that $L_i \subseteq K$ for all i (as the intersection of a one-counter language and a regular language is one-counter). Since the Parikh mapping $\Phi : \Sigma^* \to \mathbb{N}^2$ defined by $w \mapsto (|w|_a, |w|_b)$ is bijective on K, we have

$$L\Phi = L_1\Phi \cap \cdots \cap L_n\Phi.$$

By Parikh's theorem (see Theorem 2 in [18]) we know that any context-free language, and hence any one-counter language, has a semilinear Parikh image. Since the L_i are all one-counter, $L_i\Phi$ is semilinear for all i. Since any intersection of semilinear sets is semilinear, L would have a semilinear Parikh image. However, L does not have a semilinear Parikh image [12], a contradiction. □

There is a characterization of groups whose word problem is accepted by a BMM in [4], where it is shown that the word problem is accepted by such a machine with n counters if and only if the group G has a free abelian subgroup of rank n

of finite index in G. Whilst the class of groups is the same as that characterized here, the languages accepted by BMMs form a proper subclass of $\mathcal{PNL}$ (see [5]). The proof in [4] has some similarities with our approach; we are grateful to one of the referees for pointing out this connection.

We finish with a comment relating groups with a word problem in $\mathcal{PNL}$ to those with a word problem in $co\mathcal{CF}$. The latter is a very interesting class of groups (see [11,16] for example) but we do not yet have a classification as to which groups lie in this class. However, we can say the following:

Proposition 24. *If $G \in \mathcal{PNL}$ then $G \in co\mathcal{CF}$, but the converse is false.*

Proof. By Proposition 6 in [11], all virtually abelian groups are in $co\mathcal{CF}$, and so the inclusion follows from Proposition 20.

For the converse consider the free group on two generators; this is not virtually abelian, and so is not in $\mathcal{PNL}$, but it is in $co\mathcal{CF}$ (see [11] for example).   □

Acknowledgments. Some of the research for this paper was done whilst the authors were visiting the Nesin Mathematics Village in Turkey; the authors would like to thank the Village both for the financial support that enabled them to work there and for the wonderful research environment it provided that stimulated the results presented here. The authors would like to thank the referees for their helpful and constructive comments. The second author also would like to thank Hilary Craig for all her help and encouragement.

References

1. Anisimov, A.V.: Group languages. Cybernet. Syst. Anal. **7**, 594–601 (1971)
2. Autebert, J.M., Boasson, L., Sénizergues, G.: Groups and NTS languages. J. Comput. Syst. Sci. **35**, 243–267 (1987)
3. Dunwoody, M.J.: The accessibility of finitely presented groups. Invent. Math. **81**, 449–457 (1985)
4. Elder, M., Kambites, M., Ostheimer, G.: On groups and counter automata. Internat. J. Algebra Comput. **18**, 1345–1364 (2008)
5. Greibach, S.A.: Remarks on blind and partially blind one-way multicounter machines. Theor. Comput. Sci. **7**, 311–324 (1978)
6. Gromov, M.: Groups of polynomial growth and expanding maps. Publications Mathematiques de l'Institut des Hautes Etudes Scientifiques, vol. 53, pp. 53–78 (1981)
7. Hack, M.: Petri net languages. Computation Structures Group Memo 124, Project MAC, M.I.T. (1975)
8. Herbst, T.: On a subclass of context-free groups. RAIRO Theor. Infor. Appl. **25**, 255–272 (1991)
9. Herbst, T., Thomas, R.M.: Group presentations, formal languages and characterizations of one-counter groups. Theor. Comput. Sci. **112**, 187–213 (1993)
10. Holt, D.F., Owens, M.D., Thomas, R.M.: Groups and semigroups with a one-counter word problem. J. Aust. Math. Soc. **85**, 197–209 (2008)
11. Holt, D.F., Rees, S., Röver, C.E., Thomas, R.M.: Groups with context-free co-word problem. J. Lond. Math. Soc. **71**, 643–657 (2005)

12. Jantzen, M.: On the hierarchy of Petri net languages. RAIRO Theor. Inf. Appl. **13**, 19–30 (1979)
13. Jantzen, M.: Language theory of Petri nets. In: Brauer, W., Reisig, W., Rozenberg, G. (eds.) Petri Nets: Central Models and Their Properties. LNCS, vol. 254, pp. 397–412. Springer, Heidelberg (1987)
14. Johnson, D.L.: Presentations of groups. In: London mathematical society student texts, CUP, 2nd edn. Cambridge University Press, Cambridge (1997). http://books.google.co.uk/books?id=RYeIcopMH-IC
15. Lambert, J.: A structure to decide reachability in Petri nets. Theoret. Comput. Sci. **99**, 79–104 (1992)
16. Lehnert, J., Schweitzer, P.: The co-word problem for the Higman-Thompson group is context-free. Bull. Lond. Math. Soc. **39**, 235–241 (2007)
17. Muller, D., Schupp, P.: Groups, the theory of ends, and context-free languages. J. Comput. Syst. Sci. **26**, 295–310 (1983)
18. Parikh, R.J.: On context-free languages. J. ACM **13**, 570–581 (1966)
19. Peterson, J.L.: Computation sequence sets. J. Comput. Syst. Sci. **13**, 1–24 (1976)
20. Robinson, D.: A Course in the Theory of Groups, 2nd edn. Springer, New York (1995)

Regular Realizability Problems
and Context-Free Languages

A. Rubtsov[2,3]([✉]) and M. Vyalyi[1,2,3]

[1] Computing Centre of RAS, Moscow, Russia
[2] Moscow Institute of Physics and Technology, Moscow, Russia
[3] National Research University Higher School of Economics, Moscow, Russia
{rubtsov99,vyalyi}@gmail.com

Abstract. We investigate regular realizability (RR) problems, which are the problems of verifying whether the intersection of a regular language – the input of the problem – and a fixed language, called a filter, is non-empty. In this paper we focus on the case of context-free filters. The algorithmic complexity of the RR problem is a very coarse measure of the complexity of context-free languages. This characteristic respects the rational dominance relation. We show that a RR problem for a maximal filter under the rational dominance relation is **P**-complete. On the other hand, we present an example of a **P**-complete RR problem for a non-maximal filter. We show that RR problems for Greibach languages belong to the class **NL**. We also discuss RR problems with context-free filters that might have intermediate complexity. Possible candidates are the languages with polynomially-bounded rational indices. We show that RR problems for these filters lie in the class **NSPACE**$(\log^2 n)$.

1 Introduction

The context-free languages form one of the most important classes for formal language theory. There are many ways to characterize complexity of context-free languages. In this paper we propose a new approach to classification of context-free languages based on the algorithmic complexity of the corresponding regular realizability (RR) problems.

By 'regular realizability' we mean the problem of verifying whether the intersection of a regular language – the input of the problem – and a fixed language, called a filter, is non-empty. The filter F is a parameter of the problem. Depending on the representation of a regular language, we distinguish the deterministic RR problems $\mathrm{RR}(F)$ and the nondeterministic ones $\mathrm{NRR}(F)$, which correspond to the description of the regular language either by a deterministic or by a nondeterministic finite automaton.

A. Rubtsov—Supported in part by RFBR grant 14–01–00641.
M. Vyalyi—Supported in part RFBR grant 14–01–93107 and the scientific school grant NSh4652.2012.1.

J. Shallit and A. Okhotin (Eds.): DCFS 2015, LNCS 9118, pp. 256–267, 2015.
DOI: 10.1007/978-3-319-19225-3_22

The relation between algorithmic complexities of $RR(F)$ and $NRR(F)$ is still unknown. For our purpose – the characterization of the complexity of a context-free language – the nondeterministic version is more suitable. One of the reasons for this choice is a rational dominance relation $\leqslant_{rat}$ (defined in Sect. 2). We show below that the dominance relation on filters $F_1 \leqslant_{rat} F_2$ implies the log-space reduction $NRR(F_1) \leqslant_{log} NRR(F_2)$. So our classification is a very coarse version of the well-known classification of **CFL** by the rational dominance relation (see the book [2] for a detailed exposition of this topic).

Depending on a filter F, the algorithmic complexity of the regular realizability problem varies drastically. There are RR problems that are complete for complexity classes such as **L**, **NL**, **P**, **NP**, **PSPACE** [1,11]. In [12] a huge range of possible algorithmic complexities of the deterministic RR problems was presented. We prove below that for context-free nonempty filters the possible complexities are in the range between **NL**-complete problems and **P**-complete problems. Examples of **P**-complete RR problems are provided in Sect. 3. The filter consisting of all words provides an easy example of an **NL**-complete RR problem. In this case, the problem is exactly the reachability problem for digraphs. The upper bound by the class **P** follows from the reduction of an arbitrary NRR-problem specified by a context-free filter to the problem of verifying the emptiness of a language generated by a context-free grammar. We prove it in Sect. 3.

We will call a context-free language L *easy* if $NRR(L) \in$ **NL** and *hard* if $NRR(L)$ is **P**-complete. In Sect. 3 we present an example of a non-generator of the CFLs cone, which is hard in this sense. In Sect. 4 we provide examples of easy languages. They cover a rather wide class – the so-called Greibach languages introduced in [7].

The exact border between hard and easy languages is unknown. Moreover, there are candidates for an intermediate complexity of RR problems. They are languages with polynomially-bounded rational indices.

The rational index was introduced in [5]. Recall that *rational index* $\rho_L(n)$ of a language L is a function that returns the maximum length of the shortest word from the intersection of the language L and a language $L(\mathcal{A})$ recognized by an automaton $\mathcal{A}$ with n states, provided $L(\mathcal{A}) \cap L \neq \varnothing$:

$$\rho_L(n) = \max_{\mathcal{A}: |Q_\mathcal{A}|=n,\ L(\mathcal{A}) \cap L \neq \varnothing} \min\{|u| \mid u \in L(\mathcal{A}) \cap L\}. \tag{1}$$

The growth rate of the language's rational index is an another measure of the complexity of a language. This measure is also related to the rational dominance (see Sect. 5 for details).

In Sect. 5 we prove that the RR problem for a context-free filter having polynomially-bounded rational index is in the class **NSPACE**$(\log^2 n)$. Note also that there are many known CFLs having polynomially-bounded rational indices [10]. But the RR problems for these languages are in **NL**. It would be interesting to find more sophisticated examples of CFLs having polynomially-bounded rational indices.

2   Preliminaries

The main point of our paper is investigation of the complexity of the NRR-problem for filters from the class of context-free languages **CFL**.

Definition 1. The regular realizability problem NRR(F) is the problem of verifying non-emptiness of the intersection of the filter F with a regular language $L(\mathcal{A})$, where $\mathcal{A}$ is an NFA. Formally

$$\text{NRR}(F) = \{\mathcal{A} \mid \mathcal{A} \text{ is an NFA and } L(\mathcal{A}) \cap F \neq \varnothing\}.$$

It follows from the definition that the problem NRR(A^*) for the filter consisting of all words under alphabet A is the well-known **NL**-complete problem of digraph reachability. We will show below that NRR(L) $\in$ **P** for an arbitrary context-free filter L. So it is suitable to use deterministic log-space reductions in the analysis of algorithmic complexity of the RR problems specified by CFL filters. We denote the deterministic log-space reduction by $\leqslant_{\log}$.

Let us recall some basic notions and fix notation concerning the CFLs. For a detailed exposition see [2,3]. We will refer to the empty word as ε. Let A_n and $\bar{A}_n$ be the n-letter alphabets consisting of the letters $\{a_1, a_2, \ldots, a_n\}$ and $\{\bar{a}_1, \bar{a}_2, \ldots, \bar{a}_n\}$ respectively. A well-known example of a context-free language, the *Dyck language* D_n, is defined by the grammar

$$S \rightarrow SS \mid \varepsilon \mid a_1 S \bar{a}_1 \mid \cdots \mid a_n S \bar{a}_n.$$

Fix alphabets A and B. A language $L \subseteq A^*$ is *rationally dominated* by $L' \subseteq B^*$ if there exists a rational relation R such that $L = R(L')$, where $R(X) = \{u \in A^* \mid \exists v \in X \; (v, u) \in R\}$. We denote rational domination as $\leqslant_{\text{rat}}$. We say that languages L, L' are *rationally equivalent* if $L \leqslant_{\text{rat}} L'$ and $L' \leqslant_{\text{rat}} L$.

A rational relation is a graph of a multivalued mapping τ_R. We will call the mapping τ_R with a rational graph as a rational transduction. So $L \leqslant_{\text{rat}} L'$ means that $L = \tau_R(L')$. Such a transduction can be realized by a *rational transducer* (or finite-state transducer) T, which is a nondeterministic finite automaton with input and output tapes, where ε-moves are permitted. We say that u belongs to $T(v)$ if for the input v there exists a path of computation on which T writes the word u on the output tape and halts in the accepting state. Formally, a rational transducer is defined by the 6-tuple $T = (A, B, Q, q_0, \delta, F)$, where A is the input alphabet, B is the output alphabet, Q is the (finite) state set, q_0 is the initial state, $F \subseteq Q$ is the set of accepting states and $\delta \colon Q \times (A \cup \varepsilon) \times (B \cup \varepsilon) \times Q$ is the transition relation.

Let two rational transducers T_1 and T_2 correspond to rational relations R_1 and R_2, respectively. We say that a rational transducer $T = T_1 \circ T_2$ is the composition of T_1 and T_2 if the relation R corresponding to T such that $R = \{(u, v) \mid \exists y (u, y) \in R_1, (y, v) \in R_2\}$.

Define the composition of transducer T and automaton $\mathcal{A}$ in the same way: automaton $\mathcal{B} = T \circ \mathcal{A}$ recognizes the language $\{w \mid \exists y \in L(\mathcal{A}) \; (w, y) \in R\}$.

The following proposition is an algorithmic version of the Elgot-Mezei theorem (see, e.g., [2, Theorem 4.4]).

Proposition 1. *The composition of transducers and the composition of a transducer and an automaton are computable in deterministic log space.*

A *rational cone* is a class of languages closed under rational dominance. Let $\mathcal{T}(L)$ denote the least rational cone that includes language L and call it the *rational cone generated by* L. Such a cone is called *principal*. For example, the cone **Lin** of linear languages (see [2] for definition) is principal: $\mathbf{Lin} = \mathcal{T}(S)$, where the *symmetric language* S over the alphabet $X = \{x_1, x_2, \bar{x}_1, \bar{x}_2\}$ is defined by the grammar

$$S \to x_1 S \bar{x}_1 \mid x_2 S \bar{x}_2 \mid \varepsilon.$$

For a mapping $a \mapsto L_a$ the *substitution* σ is the morphism from A^* to the power set 2^{B^*} such that $\sigma(a) = L_a$. The image $\sigma(L)$ of a language $L \subseteq A^*$ is defined in the natural way. The *substitution closure* of a class of languages $\mathcal{L}$ is the least class containing all substitutions of languages from $\mathcal{L}$ to the languages from $\mathcal{L}$. We need two well-known examples of the substitution closure. The class **Qrt** of the *quasirational languages* is the substitution closure of the class **Lin**. The class of *Greibach languages* [7] is the substitution closure of the rational cone generated by the Dyck language D_1 and the symmetric language S.

It is important for our purposes that rational dominance implies a reduction for the corresponding RR problems.

Lemma 1. *If $F_1 \leqslant_{\mathrm{rat}} F_2$ then $\mathrm{NRR}(F_1) \leqslant_{\log} \mathrm{NRR}(F_2)$.*

Proof. Let T be a rational transducer such that $F_1 = T(F_2)$ and let $\mathcal{A}$ be an input of the $\mathrm{NRR}(F_1)$ problem. Construct the automaton $\mathcal{B} = T \circ \mathcal{A}$ and use it as an input of the $\mathrm{NRR}(F_2)$ problem. It gives the log-space reduction due to Proposition 1.

In particular, this lemma implies that if a problem $\mathrm{NRR}(F)$ is complete in a complexity class $\mathcal{C}$, then for any filter F' from the rational cone $\mathcal{T}(F)$ the problem $\mathrm{NRR}(F')$ is in the class $\mathcal{C}$.

We will use the following reformulation of the Chomsky-Schützenberger theorem.

Theorem (Chomsky, Schützenberger). $\mathbf{CFL} = \mathcal{T}(D_2)$.

In the next section, we prove that $\mathrm{NRR}(D_2)$ is **P**-complete under deterministic log-space reductions. Thus, it follows from the Chomsky-Schützenberger theorem and Lemma 1 that any problem $\mathrm{NRR}(F)$ for a CFL filter F lies in the class **P**.

3 Hard RR Problems with CFL Filters

In this section we present examples of hard context-free languages. The first example is the Dyck language D_2.

By use of Lemma 1 and the Chomsky-Schützenberger theorem, we conclude that any generator of the CFL cone is hard. But there are additional hard languages. We provide such an example, too.

We start with some technical lemmas. The intersection of a CFL and a regular language is a CFL. We need an algorithmic version of this fact.

Lemma 2. *Let $G = (N, \Sigma, P, S)$ be a fixed context-free grammar. Then there exists a deterministic log-space algorithm that takes a description of an NFA $\mathcal{A} = (Q_\mathcal{A}, \Sigma, \delta_\mathcal{A}, q_0, F_\mathcal{A})$ and constructs a grammar $G' = (N', \Sigma, P', S')$ generating the language $L(G) \cap L(\mathcal{A})$. The grammar size is polynomial in $|Q_\mathcal{A}|$.*

This fact is well-known. We provide the proof because the construction will be used in the proof of Theorem 5 below.

Proof (of Lemma 2). First, to make the construction clearer, we assume that automaton $\mathcal{A}$ has no ε-transitions. Let N' consist of the axiom S' and nonterminals $[qAp]$, where $A \in N$ and $q, p \in Q_\mathcal{A}$. Construct P' by adding for each rule $A \to X_1 X_2 \cdots X_n$ from P the set of rules

$$\{[qAp] \to [qX_1r_1][r_1X_2r_2] \cdots [r_{n-1}X_np] \mid q, p, r_1, r_2, \ldots, r_{n-1} \in Q_\mathcal{A}\}$$

to P'. Also add to P' rules $[q\sigma p] \to \sigma$ if $\delta_\mathcal{A}(q, \sigma) = p$ and $S' \to [q_0 S q_f]$ for each q_f from $F_\mathcal{A}$.

Now we prove that $L(G') = L(G) \cap L(\mathcal{A})$. Let G derive the word $w = w_1 w_2 \cdots w_n$. Then grammar G' derives all possible sentential forms

$$[q_0 w_1 r_1][r_1 w_2 r_2] \cdots [r_{n-1} w_n q_f],$$

where $q_f \in F_\mathcal{A}$ and $r_i \in Q_\mathcal{A}$. And $[q_0 w_1 r_1][r_1 w_2 r_2] \cdots [r_{n-1} w_n q_f] \Rightarrow^* w_1 w_2 \cdots w_n$ iff there is a successful run for the automaton $\mathcal{A}$ on w. If G' derives a word w then each symbol w_i of the word has been derived from some nonterminal $[q w_i p]$. Due to the construction of the grammar G' the word w has been derived from some sentential form $[q_0 w_1 r_1][r_1 w_2 r_2] \cdots [r_{n-1} w_n q_f]$, which encodes a successive run of $\mathcal{A}$ on w. Thus G' derives the word w only if G does as well.

The size of G' is polynomial in $Q_\mathcal{A}$. The size of N' is $|N| \cdot |Q_\mathcal{A}|^2 + 1$. Let k be the length of the longest rule in P. Then for each rule from P there are at most $|Q_\mathcal{A}|^{k+1}$ rules in P' and for rules in the form $[q\sigma p] \to \sigma$ or $S' \to [q_0 S q_f]$ there are at most $O(|Q_\mathcal{A}|^2)$ rules in P'.

Finally, the grammar G' is log-space constructible, because the rules of P' corresponding to the particular rule from P can be generated by inspecting all $(k + 1)$-tuples of states of $\mathcal{A}$ and $k = O(1)$. Adding ε-transitions just increases $k + 1$ to $2k$. For each rule $A \to X_1 \cdots X_n$ we add rules $[qAp] \to [qX_1q_1][q_2X_2q_3] \cdots [q_{2n-1}X_np]$, where $q_i = q_{i+1}$ or $q_i \xrightarrow{\varepsilon} q_{i+1}$ for all i. In the case of $[q\sigma p] \to \sigma$ rules we add all such rules that $q \xrightarrow{\varepsilon} q'$, $p' \xrightarrow{\varepsilon} p$ and $\delta(q', \sigma) = p'$. ∎

Note that if grammar G is in Chomsky normal form, then the number of nonterminals of the grammar G' is $O(|Q_\mathcal{A}|^2)$. Recall that for a grammar in the Chomsky normal form, the right-hand side of each rule consists of either two nonterminals, or one terminal. The empty word may be produced only by the axiom and the axiom does not appear in a right-hand side of any rule.

Also we need an algorithmic version of the Chomsky-Schützenberger theorem.

Lemma 3. *There exists a deterministic log-space algorithm that takes a description of a context-free grammar $G = (N, \Sigma, P, S)$ and produces a rational transducer T such that $T(D_2) = L(G)$.*

Now we are ready to prove hardness of the Dyck language D_2.

Theorem 1. *The problem* $\mathrm{NRR}(D_2)$ *is* **P**-*complete.*

Proof. To prove **P**-hardness we reduce the well-known **P**-complete problem of verifying whether a context-free grammar generates an empty language [6] to $\mathrm{NRR}(D_2)$. Based on a grammar G, construct a transducer T such that $T(D_2) = L(G)$ using Lemma 3. Let $\mathcal{A}$ be a nondeterministic automaton obtained from the transducer T by ignoring the output tape. Then $L(\mathcal{A}) \cap D_2$ is nonempty iff $L(G)$ is nonempty. The mapping $G \to \mathcal{A}$ is the required reduction.

To prove that $\mathrm{NRR}(D_2)$ lies in **P** we reduce this problem to the problem of non-emptiness of a language generated by a context-free grammar.

For an input $\mathcal{A}$ construct the grammar G such that $L(G) = L(\mathcal{A}) \cap D_2$ using Lemma 2.

Corollary 1. *Any generator of the* **CFL** *cone is a hard language.*

Now we present another example of a hard language. Boasson proved in [4] that there exists a principal rational cone of non-generators of the **CFL** cone containing the family **Qrt** of the quasirational languages.

Below we establish **P**-completeness of the nondeterministic RR problem for a generator of this cone. The construction follows the exposition in [3].

For brevity we denote the alphabet of the Dyck language D_1 by $A = \{a, \bar{a}\}^*$. Recall that the syntactic substitution of a language M into a language L is

$$L \uparrow M = \{m_1 x_1 m_2 x_2 \cdots m_r x_r \mid m_1, \ldots, m_r \in M, \, x_1 x_2 \cdots x_r \in L\} \cup (\{\varepsilon\} \cap L).$$

We also use the language $S_\# = S \uparrow \#^*$ which is the *syntactic substitution* of the language $\#^*$ in the symmetric language S.

Let $M = aS_\# \bar{a} \cup \varepsilon$. The language $M^{(\infty)}$ is defined recursively in the following way: $x \in M^{(\infty)}$ iff either $x \in M$ or

$$x = ay_1 a z_1 \bar{a} y_2 a z_2 \bar{a} \cdots y_{n-1} a z_{n-1} \bar{a} y_n \bar{a},$$

where $y_1, y_n \in X^*, y_i \in X^+$ for $2 \le i \le n-1$, $az_i\bar{a} \in M^{(\infty)}$ and $ay_1 y_2 \cdots y_n \bar{a} \in M$.

Let $\pi_X \colon (X \cup A)^* \to A^*$ be the morphism that erases symbols from the alphabet X. The language $M^{(+)}$ is defined to be $\pi_X^{-1}(A^* \setminus D_1)$.

Finally, we set $S_\#^\uparrow = M^{(\infty)} \cup M^{(+)}$.

Note that the languages S and $S_\#$ are rationally equivalent. So $S_\#$ is a generator of the cone **Lin** of the linear languages.

By combining this observation with Propositions 3.19 and 3.20 from [3], we get the following fact.

Theorem 2. $S_\#^\uparrow$ *is not a generator of the* **CFL** *cone, but the cone generated by* $S_\#^\uparrow$ *contains all quasirational languages.*

The language $S_\#^\uparrow$ is the union of two languages. In the proof of the **P**-completeness for the problem $\mathrm{NRR}(S_\#^\uparrow)$, we will use automata that do not accept words from the language $M^{(+)}$. For this purpose we need a notion of a marked automaton.

Definition 2. An NFA $\mathcal{A}$ over the alphabet $A_n \cup \bar{A}_n$ is *marked* if there exists a function $h: Q_{\mathcal{A}} \to \mathbb{Z}$ satisfying the relations

$$h(q') = h(q) + 1, \text{if there exists a transition } q \xrightarrow{a_j} q' \text{ in } \mathcal{A},$$

$$h(q') = h(q) - 1, \text{if there exists a transition } q \xrightarrow{\bar{a}_j} q' \text{ in } \mathcal{A},$$

$$h(q) = 0, \qquad \text{if } q \text{ is either the initial state or an accepting state of } \mathcal{A}. \quad (2)$$

In what follows we will identify for brevity the (directed) paths along the graph of an NFA and the corresponding words in the alphabet of the automaton. The vertices of the graph, i.e., the states of the automaton, are identified in this way with the *positions* of the word.

The *height* of a position is the difference between the number of the symbols a_i and the number of the symbols $\bar{a}_i$ preceding the position. In terms of the position heights, the words in D_1 are characterized by two conditions: the height of any position is nonnegative and the height of the final position is 0.

Proposition 2. *Let $\mathcal{A}$ be an NFA such that $D_2 \cap L(\mathcal{A}) \neq \varnothing$. Then there exists a word $w \in D_2 \cap L(\mathcal{A}) \neq \varnothing$ such that the height of any position in the word w is $O(|Q_{\mathcal{A}}|)^2$.*

Proof. The heights of positions are upperbounded by the height of the derivation tree in the grammar generating the language $D_2 \cap L(A) \neq \varnothing$.

It is easy to see that for any grammar generating a non-empty language there is a word such that the height of a derivation tree for the word is at most the number of nonterminals in the grammar.

To finish the proof, we use the grammar constructed by Lemma 2 from the grammar generating D_2 in the Chomsky normal form. This grammar has $O(|Q_{\mathcal{A}}|^2)$ nonterminals.

In the proof below we need a syntactic transformation of automata over the alphabet $A_2 \cup \bar{A}_2$.

Proposition 3. *There exists a transformation μ that takes a description of an automaton $\mathcal{A}$ over the alphabet $A_2 \cup \bar{A}_2$ and produces a description of a marked automaton $\mathcal{A}' = \mu(\mathcal{A})$ such that (i) $L(A) \cap D_2 \neq \varnothing$ iff $L(\mathcal{A}') \cap D_2 \neq \varnothing$ and (ii) for any $w \in L(\mathcal{A}')$ the height of any position is nonnegative and the height of the final position is 0. The transformation μ is computed in deterministic log space.*

Proof. Let m be an upper bound on the heights of the positions in a word $w \in L(A) \cap D_2$. By Proposition 2, m is $O(|Q_{\mathcal{A}}|^2)$. Note that m can be computed in deterministic log space.

The state set of the automaton $\mathcal{A}'$ is $Q_{\mathcal{A}} \times \{0, \ldots, m\} \cup \{r\}$, where r is the specific absorbing rejecting state.

If $q \xrightarrow{\alpha} q'$, where $\alpha \in \{a_1, a_2\}$, is a transition in the automaton $\mathcal{A}$ then there are transitions $(q, i) \xrightarrow{\alpha} (q', i+1)$ for all $0 \leq i < m$ and the transition $(q, m) \xrightarrow{\alpha} r$ in the automaton $\mathcal{A}'$.

If $q \xrightarrow{\alpha} q'$, where $\alpha \in \{\bar{a}_1, \bar{a}_2\}$, is a transition in the automaton $\mathcal{A}$ then there are transitions $(q, i) \xrightarrow{\alpha} (q', i-1)$ for all $0 < i \leq m$ and the transition $(q, 0) \xrightarrow{\alpha} r$ in the automaton $\mathcal{A}'$.

The initial state of the automaton $\mathcal{A}'$ is $(q_0, 0)$, where q_0 is the initial state of the automaton $\mathcal{A}$. The set of accepting states of the automaton $\mathcal{A}'$ is $F \times \{0\}$, where F is the set of accepting states of the automaton $\mathcal{A}$.

It is clear that the description of the automaton $\mathcal{A}'$ is constructed in deterministic log space.

Condition (ii) is forced by the construction of the automaton $\mathcal{A}'$. It remains to prove that condition (i) holds.

Note that if $L(\mathcal{A}) \cap D_2 = \varnothing$ then $L(\mathcal{A}') \cap D_2 = \varnothing$ too. In the other direction, if $L(\mathcal{A}) \cap D_2 \neq \varnothing$, then by Proposition 2 there exists a word $w \in L(\mathcal{A}) \cap D_2$ such that the height of any position in the word does not exceed m. So the word is accepted by the automaton $\mathcal{A}'$.

Theorem 3. $\mathrm{NRR}(S_\#^\uparrow)$ *is* **P***-complete under deterministic log space reductions.*

Proof. We reduce $\mathrm{NRR}(D_2)$ to $\mathrm{NRR}(S_\#^\uparrow)$.

Let $\mathcal{A}$ be an input of the problem $\mathrm{NRR}(D_2)$ and $\mathcal{A}' = \mu(\mathcal{A})$ be the marking transformation of the automaton $\mathcal{A}$.

We are going to construct the automaton $\mathcal{B}$ over the alphabet $A \cup X \cup \{\#\}$ such that $L(\mathcal{A}') \cap D_2 \neq \varnothing$ iff $L(\mathcal{B}) \cap S_\#^\uparrow \neq \varnothing$.

The morphism $\varphi \colon (A_2 \cup \bar{A}_2)^* \to (A \cup X \cup \{\#\})^*$ is defined as follows:

$$\begin{aligned} \varphi &: a_1 \mapsto a x_1, \\ \varphi &: \bar{a}_1 \mapsto \bar{x}_1 \bar{a} \#\#, \\ \varphi &: a_2 \mapsto a x_2, \\ \varphi &: \bar{a}_2 \mapsto \bar{x}_2 \bar{a} \#\#. \end{aligned} \qquad (3)$$

The automaton $\mathcal{B}$ accepts words of the form $a x_1 x_2 w \bar{x}_2 \bar{x}_1 \bar{a}$, where $w = \varphi(u)$. It simulates the behavior of the automaton $\mathcal{A}'$ on the word u and accepts iff $\mathcal{A}'$ accepts the word u.

It follows from the definitions that if $u \in D_2$ then $a x_1 x_2 \varphi(u) \bar{x}_2 \bar{x}_1 \bar{a} \in M^{(\infty)}$. So if $L(\mathcal{A}') \cap D_2 \neq \varnothing$ then $L(\mathcal{B}) \cap S_\#^\uparrow \neq \varnothing$.

Now we are going to prove the opposite implication. Let

$$w = a x_1 x_2 \varphi(u) \bar{x}_2 \bar{x}_1 \bar{a} \in S_\#^\uparrow \cap L(\mathcal{B}).$$

The automaton $\mathcal{A}'$ is marked and $\mathcal{B}$ simulates the behavior of $\mathcal{A}'$ on u. So the heights of positions in w are nonnegative and the height of the final position is 0. Thus $w \notin M^{(+)} = \pi_X^{-1}(A^* \setminus D_1)$. Take a pair of the corresponding parentheses $a, \bar{a}$ in the word w:

$$w = w_0 a x_i w_1 \bar{x}_j \bar{a} w_2.$$

If $i \neq j$ then $w \notin M^{(\infty)}$. So $i = j$ for all pairs of the corresponding parentheses. This implies $u \in D_2 \cap L(\mathcal{A}')$.

We just have proved the correctness of the reduction. It can be computed in log space due to the following observations. To produce the automaton $\mathcal{B}$ from the automaton $\mathcal{A}$ we need to extend the state set by a finite number of pre- and postprocessing states to operate with the prefix ax_1x_2 and with the suffix $\bar{x}_2\bar{x}_1\bar{a}$. Also we need to split all states in $Q_{\mathcal{A}'}$ in pairs to organize the simulation of $\mathcal{A}'$ while reading the pairs of symbols ax_i and $\bar{x}_i\bar{a}$. The transitions by the symbol $\#$ are trivial: $q \xrightarrow{\#} q$ for all q.

4   Easy RR Problems with CFL Filters

Now we present examples of easy languages. The simplest example is regular languages. Next we prove that the symmetric language and the language D_1 are easy. A simple observation shows that a substitution of easy languages into an easy language is easy. Thus we conclude that Greibach languages are easy.

Lemma 4. $\mathrm{NRR}(S) \in \mathbf{NL}$.

The proof of Lemma 4 is a slight modification of the arguments from [1] that prove a similar result for the language of palindromes.

Lemma 5. *Let L_c be a context-free language recognizable by a counter automaton. Then problem $\mathrm{NRR}(L_c)$ lies in $\mathbf{NL}$.*

In the proof we will use the following fact.

Lemma 6 ([13]) *Let M be a counter automaton with n states. Then the shortest word w from the language $L(M)$ has length at most n^3 and the counter of M on processing the word w doesn't exceed the value n^2.*

We now return to the proof of Lemma 5.

Proof. Let M be a counter automaton that accepts by reaching the final state such that M recognizes the language L_c. Let $\mathcal{A}$ be an automaton on the input of the regular realizability problem.

Construct the counter automaton $M_{\mathcal{A}}$ with the set of states $Q_M \times Q_{\mathcal{A}}$, the initial state $(q_0^M, q_0^{\mathcal{A}})$, with the set of accepting states $F_M \times F_{\mathcal{A}}$ and with the transition relation $\delta_{M_{\mathcal{A}}}$ such that $\delta_M(q, \sigma, z) \vdash (q', z')$, $\delta_{\mathcal{A}}(p, \sigma) = p'$ implies $\delta_{M_{\mathcal{A}}}((q,p), \sigma, z) \vdash ((q',p'), z')$. This is the standard composition construction.

The automaton $M_{\mathcal{A}}$ is a counter automaton with $|Q_M| \cdot |Q_{\mathcal{A}}| = c \times n$ states. Using Lemma 6 we obtain that the value of $M_{\mathcal{A}}$'s counter does not exceed $(cn)^2$ on the shortest word from $L(M_{\mathcal{A}})$. Then construct automaton $\mathcal{B}$ such that $L(\mathcal{B})$ contains all such words from $L(M_{\mathcal{A}})$ such that the counter of $M_{\mathcal{A}}$ does not exceed $(cn)^2$. The automaton $\mathcal{B}$ has $O(n^3)$ states and can be constructed in log space in the straightforward way similar to the proof of Proposition 3. Note that $L(M_{\mathcal{A}}) \neq \varnothing$ iff $L(\mathcal{B}) \neq \varnothing$. So the map $\mathcal{A} \to \mathcal{B}$ gives a reduction of the problem $\mathrm{NRR}(L_c)$ to the problem $\mathrm{NRR}(\Sigma^*)$, which is in $\mathbf{NL}$.

The language D_1 is recognized by a counter automaton in the obvious way.

Corollary 2. NRR$(D_1) \in$ **NL**.

Lemma 7. *If L, L_a for all $a \in A$, are easy languages then $\sigma(L)$ is also easy.*

Proof. Let $\mathcal{A}$ be an input for the problem NRR$(\sigma(L))$. Define the automaton $\mathcal{A}'$ over the alphabet A with the state set $Q_{\mathcal{A}'} = Q_{\mathcal{A}}$. There is a transition $q \xrightarrow{a} q'$ in the automaton $\mathcal{A}'$ iff there exists a word $w \in L_a$ such that $q \xrightarrow{w} q'$ in automaton $\mathcal{A}$.

It is clear from the definition that $L(\mathcal{A}) \cap \sigma(L) \neq \varnothing$ iff $L(\mathcal{A}') \cap L \neq \varnothing$. To apply an **NL**-algorithm for NRR(L) one needs the transition relation of $\mathcal{A}'$. The transition relation is not a part of the input now. But it can be computed by **NL**-algorithms for NRR(L_a). It is clear that the resulting algorithm is in **NL**.

Applying Lemma 7, Lemma 4 and Corollary 2, we deduce with the theorem.

Theorem 4. *Greibach languages are easy.*

5   The Case of Polynomially-Bounded Rational Index

We do not know whether there exists a CFL that is neither hard nor easy. In this section we indicate one possible class of candidates for an intermediate complexity: the languages with polynomially-bounded rational indices.

Rational index appears to be a very useful characteristic of a context-free language because rational index does not increase significantly under rational transductions.

Theorem (Boasson, Courcelle, Nivat, 1981, [5]). *If $L' \leqslant_{\mathrm{rat}} L$ then there exists a constant c such that $\rho_{L'}(n) \leq cn(\rho_L(cn) + 1)$.*

Thus the rational index can be used to separate languages w.r.t. the rational dominance relation. Note that the rational index of a generator of the **CFL** cone has rather good estimations.

Theorem (Pierre, 1992, [9]). *The rational index of any generator of the rational cone of* **CFL** *belongs to* $\exp(\Theta(n^2/\log n))$.

The examples of easy languages in Sect. 4 have polynomially-bounded rational indices. Moreover, context-free languages with rational index $\Theta(n^\gamma)$ for any positive algebraic number $\gamma > 1$ were presented in [10]. All of them are easy. The proof is rather technical and is skipped here. Thus it is quite natural to suggest that any language with polynomially-bounded rational index is easy.

Unfortunately we are able to give only a weaker bound on the algorithmic complexity in the case of polynomially-bounded rational index.

Theorem 5. *For a context-free filter F with polynomially-bounded rational index, the problem* NRR(F) *lies in* **NSPACE**$(\log^2 n)$.

We use a technique quite similar to the technique from [8]. First we need an auxiliary result.

Lemma ([8]). *For a grammar G in the Chomsky normal form and for an arbitrary string $w = xyz$ from $L(G)$ of length n there is a nonterminal A in the derivation tree, such that A derives y and $n/3 \le |y| \le 2n/3$.*

Let us return to the proof of the theorem.

Proof (of Theorem 5). Consider a grammar G' in the Chomsky normal form such that $L(G') = F$. Fix an automaton $\mathcal{A}$ with n states such that the minimal length of w from $L(\mathcal{A}) \cap F$ equals $\rho_F(n)$. The length of the word w is polynomial in n. Consider the grammar G such that $L(G) = L(\mathcal{A}) \cap F$ obtained from the grammar G' by the construction from Lemma 2.

The algorithm does not construct the grammar G itself, since such a construction expands the size of grammar G' up to n^3 times. Instead, the algorithm nondeterministically guesses the derivation tree of the word w in the grammar G, if it exists. Informally speaking, it restores the derivation tree starting from its 'central' branch.

The main part of the algorithm is a recursive procedure that checks correctness for a nonterminal $A = [qA'p]$ of the grammar G. We say that the nonterminal $A = [qA'p]$ is correct if A produces a word w in the grammar G.

If a nonterminal is $[q\sigma p]$, where σ is a terminal then the procedure should check that $q \xrightarrow{\sigma} p$ in the automaton $\mathcal{A}$.

In a general case the procedure of checking correctness nondeterministically guesses a nonterminal $A_1 = [\ell_1 A_1' r_1]$ such that $w = p_1 u_1 s_1$, and A_1 derives the word u_1 and $1/3|w| \le |u_1| \le 2/3|w|$. Then it is recursively applied to the nonterminal A_1. If successful the procedure sets $i := 1$ and repeats the following steps:

1. Nondeterministically guess the ancestor $A_{i+1} = [\ell_{i+1} A_{i+1} r_{i+1}]$ of A_i in the derivation tree. There are two possible cases:
   (i) either $A_{i+1} \to [q'C'\ell_{i+1}]A_i$ in the grammar G (set up $C := [q'C'\ell_{i+1}]$)
   (ii) or $A_{i+1} \to A_i[r_{i+1}C'p']$ (set up $C := [r_{i+1}C'p']$).
2. Recursively apply the procedure of checking correctness to the nonterminal C.
3. If successful set up $i := i + 1$.

Repetitions are finished and the procedure returns success if $A_j = A$. If any call of the procedure of checking correctness returns failure then the whole procedure returns failure.

In recursive calls the lengths of words to be checked diminish by a factor at most $2/3$. So the total number of recursive calls is $O(\log n)$, where n is the input length. Data to be stored during the process form a list of triples (an automaton state, a nonterminal of the grammar G', a automaton state). Each automaton state description requires $O(\log n)$ space and nonterminal description requires a constant size space since grammar G' is fixed. Thus the total space for the algorithm is $O(\log^2 n)$.

Acknowledgments. We are acknowledged to Abuzer Yakaryilmaz for pointing on the result of Lemma 5 and for reference to a lemma similar to Lemma 6.

References

1. Anderson, T., Loftus, J., Rampersad, N., Santean, N., Shallit, J.: Detecting palindromes, patterns and borders in regular languages. Inf. Comput. **207**, 1096–1118 (2009)
2. Berstel, J.: Transductions and Context-Free Languages. Teubner Verlag, Stuttgart/Leipzig/Wiesbaden (1979)
3. Berstel, J., Boasson, L.: Context-free languages. In: Leeuwen, J. (ed.) Handbook of Theoretical Computer Science, vol. B, pp. 59–102. Elsevier, Amsterdam (1990)
4. Boasson, L.: Non-générateurs algébriques et substitution. RAIRO Informatique théorique **19**, 125–136 (1985)
5. Boasson, L., Courcelle, B., Nivat, M.: The rational index, a complexity measure for languages. SIAM J. Comput. **10**(2), 284–296 (1981)
6. Greenlaw, R., Hoover, H.J., Ruzzo, L.: Limits to Parallel Computation: P-Completeness Theory. Oxford University Press, Oxford (1995)
7. Greibach, S.A.: An infinite hierarchy of context-free languages. J. ACM **16**, 91–106 (1969)
8. Lewis, P.M., Stearns, R.E., Hartmanis, J.: Memory bounds for recognition of context-free and context-sensitive languages. In: Switching Circuit Theory and Logical Design, pp. 191–202. IEEE, New York (1965)
9. Pierre, L.: Rational indexes of generators of the cone of context-free languages. Theor. Comput. Sci. **95**, 279–305 (1992)
10. Pierre, L., Farinone, J.M.: Rational index of context-free languages with rational index in $\Theta(n^\gamma)$ for algebraic numbers γ. Informatique théorique et applications **24**(3), 275–322 (1990)
11. Vyalyi, M.N.: On regular realizability problems. Probl. Inf. Transm. **47**(4), 342–352 (2011)
12. Vyalyi, M.N.: Universality of regular realizability problems. In: Bulatov, A.A., Shur, A.M. (eds.) CSR 2013. LNCS, vol. 7913, pp. 271–282. Springer, Heidelberg (2013)
13. Yakaryılmaz, A.: One-counter verifiers for decidable languages. In: Bulatov, A.A., Shur, A.M. (eds.) CSR 2013. LNCS, vol. 7913, pp. 366–377. Springer, Heidelberg (2013)

Generalization of the Double-Reversal Method of Finding a Canonical Residual Finite State Automaton

Hellis Tamm[✉]

Institute of Cybernetics, Tallinn University of Technology,
Akadeemia tee 21, 12618 Tallinn, Estonia
hellis@cs.ioc.ee

Abstract. Residual finite state automata (RFSA) are a subclass of nondeterministic finite automata with the property that every state of an RFSA defines a residual language (a left quotient) of the language accepted by the RFSA. Every regular language has a unique canonical RFSA which is a minimal RFSA accepting the language. We study the relationship of RFSAs with atoms of regular languages. We generalize the double-reversal method of finding a canonical RFSA, presented by Denis, Lemay, and Terlutte.

1 Introduction

Residual finite state automata (RFSAs), introduced by Denis, Lemay, and Terlutte [4,5], are a subclass of nondeterministic finite automata (NFA), such that every state of an RFSA defines a residual language, that is, a (left) quotient, of the language accepted by the RFSA. Every regular language has a unique canonical RFSA which is a minimal RFSA accepting the language. Denis et al. present two methods to find a canonical RFSA. The first method starts with a deterministic finite automaton (DFA) of a language, and applies two operators to it, called *saturation* and *reduction*, yielding a canonical RFSA. The other way they suggest to compute a canonical RFSA is similar to Brzozowski's double-reversal method of minimizing a DFA, which applies the reversal and determinization operations twice. Namely, Denis et al. define a modified subset construction operation which, when applied to an NFA, produces an RFSA, and such that when it is used in Brzozowski's double-reversal algorithm in place of determinization, results in a canonical RFSA.

We study the relationship of RFSAs with atoms of regular languages, introduced by Brzozowski and Tamm [1,2] as non-empty intersections of uncomplemented or complemented quotients of the language. In the same work, a related NFA called the átomaton was presented, whose states are the atoms of the language, and it was shown that the átomaton is isomorphic to the reverse automaton of the minimal DFA of the reverse language.

This work was supported by the ERDF funded CoE project EXCS and the Estonian Ministry of Education and Research institutional research grant IUT33-13.

© Springer International Publishing Switzerland 2015
J. Shallit and A. Okhotin (Eds.): DCFS 2015, LNCS 9118, pp. 268–279, 2015.
DOI: 10.1007/978-3-319-19225-3_23

We consider an NFA which we call the "maximized átomaton" of a language, that uses "maximized atoms", that is, certain unions of atoms, as its states. We show that the maximized átomaton is isomorphic to the reverse automaton of the saturated minimal DFA of the reverse language. In other words, a saturated minimal DFA of a language can be obtained by reversing the maximized átomaton of the reverse language. We note here that in the recent work by Myers, Adámek, Milius, and Urbat [6], a new canonical NFA was introduced, which they called the *distromaton*, and which appears to be the same NFA as the maximized átomaton.

We generalize the double-reversal method by Denis et al., characterizing the class of NFAs for which applying the modified subset construction operation results in a canonical RFSA. Namely, we show that the modified subset construction, when applied to some NFA, produces a canonical RFSA if and only if the left language of every state of that NFA is a union of left languages of the canonical RFSA. This generalization is similar to the result by Brzozowski and Tamm [1,2], who showed that determinization of an NFA results in a minimal DFA if and only if the right languages of the states of the reverse automaton of that NFA are unions of atoms of the reverse language.

In Sect. 2 we present definitions and some properties of automata, factorizations, quotients, and atoms of regular languages, and in Sect. 3, we recall basic properties of RFSAs. In Sect. 4, we define and study maximized atoms and the maximized átomaton of a language. In Sect. 5, we define canonical left languages of a regular language as the left languages of the states of the canonical RFSA, study their properties, and generalize the double-reversal method by Denis et al. of obtaining a canonical RFSA. Section 6 concludes the paper.

2 Automata, Factorizations, Quotients, and Atoms of Regular Languages

A *nondeterministic finite automaton (NFA)* is a quintuple $\mathcal{N} = (Q, \Sigma, \delta, I, F)$, where Q is a finite, non-empty set of *states*, Σ is a finite non-empty *alphabet*, $\delta : Q \times \Sigma \to 2^Q$ is the *transition function*, $I \subseteq Q$ is the set of *initial states*, and $F \subseteq Q$ is the set of *final states*. As usual, we extend the transition function to functions $\delta' : Q \times \Sigma^* \to 2^Q$, and $\delta'' : 2^Q \times \Sigma^* \to 2^Q$. We use δ for all three functions. An NFA $\mathcal{N}' = (Q', \Sigma', \delta', I', F')$ is a *subautomaton* of $\mathcal{N}$ if $Q' \subseteq Q$, $\Sigma' \subseteq \Sigma$, $I' \subseteq I$, $F' \subseteq F$, and $q \in \delta'(p, a)$ implies $q \in \delta(p, a)$ for every $p, q \in Q'$ and $a \in \Sigma'$.

The *language accepted* by an NFA $\mathcal{N}$ is $L(\mathcal{N}) = \{w \in \Sigma^* \mid \delta(I, w) \cap F \neq \emptyset\}$. The *right language* of a state q of $\mathcal{N}$ is $L_{q,F}(\mathcal{N}) = \{w \in \Sigma^* \mid \delta(q, w) \cap F \neq \emptyset\}$. A state is *empty* if its right language is empty. Two states of an NFA are *equivalent* if their right languages are equal. The *left language* of a state q of $\mathcal{N}$ is $L_{I,q} = \{w \in \Sigma^* \mid q \in \delta(I, w)\}$. A state is *unreachable* if its left language is empty.

A *deterministic finite automaton (DFA)* is a quintuple $\mathcal{D} = (Q, \Sigma, \delta, q_0, F)$, where Q, Σ, and F are as in an NFA, $\delta : Q \times \Sigma \to Q$ is the transition function, and q_0 is the initial state. A DFA is an NFA with obvious restrictions. A DFA is minimal if it has no unreachable states and no pair of equivalent states.

The following two operations on automata are most commonly used: the *determinization* operation D applied to an NFA $\mathcal{N}$, yielding a DFA $\mathcal{N}^D$, obtained by the well-known subset construction, where only subsets reachable from the initial subset of $\mathcal{N}^D$ are used and the empty subset, if present, is included, and the *reversal* operation R which, when applied to an NFA $\mathcal{N}$, yields an NFA $\mathcal{N}^R$, where the sets of the initial and the final states of $\mathcal{N}$ are interchanged and all transitions are reversed.

Let $L \subseteq \Sigma^*$ be a non-empty regular language. A *subfactorization* of a language L is a pair (X, Y) of languages from Σ^*, such that $XY \subseteq L$. A *factorization* of L is a subfactorization (X, Y), such that for every subfactorization (X', Y'), where $X \subseteq X'$ and $Y \subseteq Y'$, the equalities $X = X'$ and $Y = Y'$ hold.

The *left quotient*, or simply *quotient*, of a language L by a word $w \in \Sigma^*$ is the language $w^{-1}L = \{x \in \Sigma^* \mid wx \in L\}$. There is one *initial* quotient, $\varepsilon^{-1}L = L$. Left quotients are also known as *right residuals*, or simply *residuals*. It is well known that there is a one-to-one correspondence between the set of states $Q = \{q_0, \ldots, q_{n-1}\}$ of the minimal DFA $\mathcal{D} = (Q, \Sigma, \delta, q_0, F)$ accepting L and the set of quotients $\{K_0, \ldots, K_{n-1}\}$ of L, such that $L_{q_i,F}(\mathcal{D}) = K_i$ for $i = 0, \ldots, n - 1$.

An *atom* of a regular language L with quotients $K_0, \ldots, K_{n-1}$ is any non-empty language of the form $\widetilde{K_0} \cap \cdots \cap \widetilde{K_{n-1}}$, where $\widetilde{K_i}$ is either K_i or $\overline{K_i}$, and $\overline{K_i}$ is the complement of K_i with respect to Σ^*. Thus atoms of L are regular languages uniquely determined by L and they define a partition of Σ^*. They are pairwise disjoint and every quotient of L (including L itself) is a union of atoms. A regular language L with n quotients has at most 2^n atoms. An atom is *initial* if it has L (rather than $\overline{L}$) as a term; it is *final* if it contains ε. Since L is non-empty, it has at least one quotient containing ε. Hence it has exactly one final atom, the atom $\widehat{K_0} \cap \cdots \cap \widehat{K_{n-1}}$, where $\widehat{K_i} = K_i$ if $\varepsilon \in K_i$, and $\widehat{K_i} = \overline{K_i}$ otherwise. Let $A = \{A_0, \ldots, A_{m-1}\}$ be the set of atoms of L, let I_A be the set of initial atoms, and let A_{m-1} be the final atom.

We use a one-one correspondence $A_i \leftrightarrow \mathbf{A}_i$ between atoms A_i of a language L and the states $\mathbf{A}_i$ of the NFA $\mathcal{A}$ defined as follows:

Definition 1. *The* átomaton *of L is the NFA* $\mathcal{A} = (\mathbf{A}, \Sigma, \alpha, \mathbf{I}_A, \{\mathbf{A}_{m-1}\})$, *where* $\mathbf{A} = \{\mathbf{A}_i \mid A_i \in A\}$, $\mathbf{I}_A = \{\mathbf{A}_i \mid A_i \in I_A\}$, *and* $\mathbf{A}_j \in \alpha(\mathbf{A}_i, a)$ *if and only if* $A_j \subseteq a^{-1}A_i$, *for all* $\mathbf{A}_i, \mathbf{A}_j \in \mathbf{A}$ *and* $a \in \Sigma$.

It was shown in [1,2] that in the átomaton, the right language of any state $\mathbf{A}_i$ is the atom A_i.

The next theorem is a slightly modified version of the result by Brzozowski [3]:

Theorem 1. *If an NFA $\mathcal{N}$ has no empty states and $\mathcal{N}^R$ is deterministic, then $\mathcal{N}^D$ is minimal.*

By Theorem 1, for any NFA $\mathcal{N}$, $\mathcal{N}^{RDRD}$ is the minimal DFA equivalent to $\mathcal{N}$. This result is known as Brzozowski's double-reversal method for DFA minimization.

A new class of NFA's was defined in [1,2] as follows:

Definition 2. *An NFA $\mathcal{N} = (Q, \Sigma, \delta, I, F)$ is atomic if for every $q \in Q$, the right language $L_{q,F}(\mathcal{N})$ of q is a union of atoms of $L(\mathcal{N})$.*

In [1,2], a generalization of Theorem 1 was presented, providing a characterization of the class of NFAs for which applying determinization procedure produces a minimal DFA:

Theorem 2. *For any NFA $\mathcal{N}$, $\mathcal{N}^D$ is minimal if and only if $\mathcal{N}^R$ is atomic.*

3 Residual Finite State Automata

Residual finite state automata (RFSAs) were introduced by Denis, Lemay, and Terlutte in [4,5]. In this section, we state some basic properties of RFSAs. However, we note here that we usually prefer to use the term "quotient" over "residual".

An NFA $\mathcal{N} = (Q, \Sigma, \delta, I, F)$ is a *residual finite state automaton* (RFSA) if for every state $q \in Q$, $L_{q,F}(\mathcal{N})$ is a quotient of $L(\mathcal{N})$. Clearly, any DFA without unreachable states is an RFSA.

Let L be a regular language over Σ. Let $K = \{K_0, \ldots, K_{n-1}\}$ be the set of quotients of L. A quotient K_i of L is *prime* if it is non-empty and if it cannot be obtained as a union of other quotients of L.

The canonical RFSA of L is the NFA $\mathcal{C} = (K', \Sigma, \delta, I, F)$, where $K' \subseteq K$ is the set of prime quotients of L, Σ is an input alphabet, $I = \{K_i \in K' \mid K_i \subseteq L\}$, $F = \{K_i \in K' \mid \varepsilon \in K_i\}$, and $\delta(K_i, a) = \{K_j \in K' \mid K_j \subseteq a^{-1}K_i\}$ for every $K_i \in K'$ and $a \in \Sigma$.

Among all RFSAs of L, the canonical RFSA is minimal regarding to the number of states, with a maximal number of transitions. One way to build a canonical RFSA is to use the *saturation* and *reduction* operations defined in the following.

Let $\mathcal{N} = (Q, \Sigma, \delta, I, F)$ be an NFA. The *saturation* operation S, if applied to $\mathcal{N}$, produces the NFA $\mathcal{N}^S = (Q, \Sigma, \delta_S, I_S, F)$, with the transiton function defined as $\delta_S(q, a) = \{q' \in Q \mid aL_{q',F}(\mathcal{N}) \subseteq L_{q,F}(\mathcal{N})\}$ for all $q \in Q$ and $a \in \Sigma$, and with the set of initial states $I_S = \{q \in Q \mid L_{q,F}(\mathcal{N}) \subseteq L(\mathcal{N})\}$. An NFA $\mathcal{N}$ is saturated if $\mathcal{N}^S = \mathcal{N}$. Saturation may add transitions and initial states to an NFA, without changing its language. Also, if $\mathcal{N}$ is an RFSA, then $\mathcal{N}^S$ is an RFSA. Clearly, if $\mathcal{D}$ is a DFA, then $\mathcal{D}^S$ is an RFSA.

For any state q of $\mathcal{N}$, let $R(q)$ be the set $\{q' \in Q \setminus \{q\} \mid L_{q',F}(\mathcal{N}) \subseteq L_{q,F}(\mathcal{N})\}$. A state q is *erasable* if $L_{q,F}(\mathcal{N}) = \bigcup_{q' \in R(q)} L_{q',F}(\mathcal{N})$. If q is erasable, a *reduction* operator ϕ is defined as follows: $\phi(\mathcal{N}, q) = (Q', \Sigma, \delta', I', F')$ where $Q' = Q \setminus \{q\}$, $I' = I$ if $q \notin I$, and $I' = (I \setminus \{q\}) \cup R(q)$ otherwise, $F' = F \cap Q'$, $\delta'(q', a) = \delta(q', a)$ if $q \notin \delta(q', a)$, and $\delta'(q', a) = (\delta(q', a) \setminus \{q\}) \cup R(q)$ otherwise, for every $q' \in Q'$ and every $a \in \Sigma$. If q is not erasable, let $\phi(\mathcal{N}, q) = \mathcal{N}$.

If $\mathcal{N}$ is saturated and if q is an erasable state of $\mathcal{N}$, then $\phi(\mathcal{N}, q)$ is obtained by deleting q and its associated transitions from $\mathcal{N}$, without adding any transitions. An NFA $\mathcal{N}$ is *reduced* if there is no erasable state in $\mathcal{N}$. After applying the reduction operator ϕ to an NFA, its language remains the same. Also, if $\mathcal{N}$ is an RFSA, then $\phi(\mathcal{N}, q)$ is an RFSA.

The following proposition is from [4,5]:

Proposition 1. *If an NFA $\mathcal{N}$ is a reduced saturated RFSA of L, then $\mathcal{N}$ is the canonical RFSA of L.*

The canonical RFSA can be obtained from a DFA by using saturation and reduction operations.

Next we will discuss another method to compute the canonical RFSA, suggested by Denis et al. [4,5]. In Sect. 2, we presented the result that for any NFA $\mathcal{N}$, $\mathcal{N}^{RDRD}$ is a minimal DFA equivalent to $\mathcal{N}$. In [4,5], a similar double-reversal method is proposed to obtain a canonical RFSA from a given NFA. Namely, Denis et al. introduce a modified subset construction operation C to be applied to an NFA as follows:

Definition 3. *Let $\mathcal{N} = (Q, \Sigma, \delta, I, F)$ be an NFA. Let Q_D be the set of states of the determinized version $\mathcal{N}^D$ of $\mathcal{N}$. A state $s \in Q_D$ is coverable if there is a set $Q_s \subseteq Q_D \setminus \{s\}$ such that $s = \bigcup_{s' \in Q_s} s'$. The NFA $\mathcal{N}^C = (Q_C, \Sigma, \delta_C, I_C, F_C)$ is defined as follows: $Q_C = \{s \in Q_D \mid s \text{ is not coverable }\}$, $I_C = \{s \in Q_C \mid s \subseteq I\}$, $F_C = \{s \in Q_C \mid s \cap F \neq \emptyset\}$, and $\delta_C(s, a) = \{s' \in Q_C \mid s' \subseteq \delta(s, a)\}$ for any $s \in Q_C$ and $a \in \Sigma$.*

Applying the operation C to any NFA $\mathcal{N}$ produces an RFSA $\mathcal{N}^C$. This RFSA is not necessarily a canonical RFSA. However, Denis et al. [4,5] have the following result:

Theorem 3. *If an NFA $\mathcal{N}$ has no empty states and $\mathcal{N}^R$ is an RFSA, then $\mathcal{N}^C$ is the canonical RFSA.*

By Theorem 3, for any NFA $\mathcal{N}$, $\mathcal{N}^{RCRC}$ is the canonical RFSA equivalent to $\mathcal{N}$. So it seems that the operation C has a similar role for RFSAs as determinization D has for DFAs.

4   Maximized Atoms and Maximized Átomaton

One method to obtain a canonical RFSA of a language, is to apply the saturation and reduction operations to a DFA of the language. In this section we consider an NFA which we call the "maximized átomaton", and show that it is isomorphic to the reverse automaton of the saturated minimal DFA of the reverse language. Thus, a saturated minimal DFA of a language can be obtained by reversing the maximized átomaton of the reverse language. We note that the maximized átomaton is the same NFA as the recently introduced distromaton [6].

Let $K_0, \ldots, K_{n-1}$ be the set of quotients of a regular language L, and let $A_0, \ldots, A_{m-1}$ be the set of atoms of L. For every atom A_i, we define the corresponding *maximized atom* M_i, such that M_i is the union of all the atoms which occur in every quotient containing A_i:

Definition 4. *The* maximized atom M_i *of an atom* A_i *is the union of atoms* $M_i = \bigcup \{A_h \mid A_h \subseteq \bigcap_{A_i \subseteq K_k} K_k\}$.

Clearly, since atoms are pairwise disjoint, and every quotient is a union of atoms, $M_i = \bigcap_{A_i \subseteq K_k} K_k$.

Proposition 2. *Let* A_i *and* A_j *be some atoms of* L. *The following properties hold:*

1. $A_i \subseteq M_i$.
2. *If* $A_i \neq A_j$, *then* $M_i \neq M_j$.
3. $A_i \subseteq M_j$ *if and only if* $M_i \subseteq M_j$.
4. $A_j \subseteq a^{-1}M_i$ *if and only if* $M_j \subseteq a^{-1}M_i$.

Proof.

1. Clear from Definition 4.
2. Let $A_i \neq A_j$. Then by the definition of an atom, there is some quotient K_k, such that $A_i \subseteq K_k$ and $A_j \not\subseteq K_k$, or $A_j \subseteq K_k$ and $A_i \not\subseteq K_k$. If we suppose that $M_i = M_j$, then we get that for every quotient K_k, $A_i \subseteq K_k$ if and only if $A_j \subseteq K_k$, a contradiction. We conclude that $M_i \neq M_j$.
3. Suppose $A_i \subseteq M_j$. Then $A_i \subseteq \bigcap_{A_j \subseteq K_k} K_k$, and so $A_j \subseteq K_k$ implies $A_i \subseteq K_k$ for any quotient K_k. That is, $\{K_k \mid A_j \subseteq K_k\} \subseteq \{K_h \mid A_i \subseteq K_h\}$, and so we get $\bigcap_{A_i \subseteq K_h} K_h \subseteq \bigcap_{A_j \subseteq K_k} K_k$. Thus, $M_i \subseteq M_j$.
   Conversely, since by Part 1, $A_i \subseteq M_i$, we conclude that $M_i \subseteq M_j$ implies $A_i \subseteq M_j$.
4. First suppose $A_j \subseteq a^{-1}M_i$. We consider $a^{-1}M_i = a^{-1}\bigcap_{A_i \subseteq K_h} K_h = \bigcap_{A_i \subseteq K_h} a^{-1}K_h$. Since $A_j \subseteq a^{-1}M_i$, $A_j \subseteq \bigcap_{A_i \subseteq K_h} a^{-1}K_h$ holds. We get that if $A_i \subseteq K_h$, then $A_j \subseteq a^{-1}K_h$. Since for every quotient K_h, $a^{-1}K_h = K_k$ for some quotient K_k, we get that if $A_i \subseteq K_h$, then $A_j \subseteq K_k$, where $K_k = a^{-1}K_h$. We see by Definition 4 that if $A_j \subseteq K_k$, then $M_j \subseteq K_k$. Thus, if $A_i \subseteq K_h$, then $M_j \subseteq a^{-1}K_h$. We have that $M_j \subseteq \bigcap_{A_i \subseteq K_h} a^{-1}K_h$. This means, $M_j \subseteq a^{-1}M_i$.
   Conversely, if $M_j \subseteq a^{-1}M_i$, then $A_j \subseteq M_j \subseteq a^{-1}M_i$.   □

Let $\mathcal{A}$ be the átomaton of L. The following statement is from [2]:

Proposition 3. *The reverse* $\mathcal{A}^R$ *of* $\mathcal{A}$ *is the minimal DFA of* L^R.

Let $\mathcal{D}$ be the minimal DFA of L^R. By Proposition 3, the states of $\mathcal{A}^R$, and therefore also of $\mathcal{A}$, are the states of $\mathcal{D}$. Since the states of $\mathcal{A}$ correspond to atoms of L, and the states of $\mathcal{D}$ correspond to quotients of L^R, there is a one-one correspondence between atoms A_i of L and quotients Q_i of L^R.

Since by Proposition 2, Part 2, there is a one-one correspondence between the atoms A_i and the maximized atoms M_i, we also get a correspondence between maximized atoms M_i of L and quotients Q_i of L^R. We show the following properties:

Proposition 4. *Let* A_i *and* A_j *be some atoms of* L. *The following properties hold:*

1. *For any word* $u \in \Sigma^*$, $u \in A_i$ *if and only if* $(u^R)^{-1}L^R = Q_i$.
2. (Q_i^R, M_i) *is a factorization of* L.

3. $M_i \subseteq M_j$ if and only if $Q_j \subseteq Q_i$.
4. $M_j \subseteq a^{-1}M_i$ if and only if $Q_i \subseteq a^{-1}Q_j$, for $a \in \Sigma$.

Proof.

1. The property is a corollary of Proposition 3.
2. By Part 1, if $u \in A_i$, then $(u^R)^{-1}L^R = Q_i$. This implies that for every A_i, Q_i^R is the maximal set of words such that $Q_i^R A_i \subseteq L$. That is, $u \in Q_i^R$ if and only if $uA_i \subseteq L$. Since $uA_i \subseteq L$ is equivalent to $A_i \subseteq u^{-1}L$, we get that $u \in Q_i^R$ if and only if $A_i \subseteq u^{-1}L$. Clearly, M_i is maximal set of words such that $M_i \subseteq u^{-1}L$ for every quotient $u^{-1}L$, such that $A_i \subseteq u^{-1}L$. That is, M_i is maximal such that $M_i \subseteq u^{-1}L$ for every $u \in Q_i^R$. We get that M_i is the maximal set of words such that $Q_i^R M_i \subseteq L$. Thus (Q_i^R, M_i) is a factorization of L.
3. The property is implied by Part 2.
4. First, we consider $a^{-1}Q_j$. Since a quotient of any quotient of a language is some quotient of the language, let $a^{-1}Q_j = Q_k$ for some $k \in \{0, \ldots, m-1\}$. We consider the atoms A_j and A_k of L, according to the correspondence between quotients Q_j (Q_k) of L^R and atoms A_j (A_k) of L. By Proposition 3, $\mathcal{D}$ is isomorphic to $\mathcal{A}^R$, so we have $a^{-1}Q_j = Q_k$ if and only if $A_j \subseteq a^{-1}A_k$. We show that $M_j \subseteq a^{-1}M_i$ if and only if $A_k \subseteq M_i$. Let $M_j \subseteq a^{-1}M_i$ hold. Clearly, $A_j \subseteq M_j$ by Definition 4, so $aA_j \subseteq aM_j$ holds. Since $M_j \subseteq a^{-1}M_i$ is equivalent to $aM_j \subseteq M_i$, we get that $aA_j \subseteq M_i$. Since we had $A_j \subseteq a^{-1}A_k$, $aA_j \subseteq A_k$ holds, and because M_i is a union of atoms, it is implied that $A_k \subseteq M_i$. Conversely, if we suppose that $A_k \subseteq M_i$ holds, then $A_j \subseteq a^{-1}A_k \subseteq a^{-1}M_i$. By Proposition 2, Part 4, $A_j \subseteq a^{-1}M_i$ implies $M_j \subseteq a^{-1}M_i$.

   Now, by Proposition 2, Part 3, $A_k \subseteq M_i$ is equivalent to $M_k \subseteq M_i$. By Proposition 4, Part 3, $M_k \subseteq M_i$ holds if and only if $Q_i \subseteq Q_k$, that is, $Q_i \subseteq a^{-1}Q_j$ holds. $\qquad\square$

Let $A = \{A_0, \ldots, A_{m-1}\}$ be the set of atoms of L, with the set of initial atoms $I_A \subseteq A$, and the final atom A_{m-1}. Let $\mathcal{A}$ be the átomaton of L.

Let $M = \{M_0, \ldots, M_{m-1}\}$ be the set of the maximized atoms of L, let $I_M = \{M_i \mid A_i \in I_A\}$ be the set of maximized atoms corresponding to initial atoms, and let $F_M = \{M_i \mid A_{m-1} \subseteq M_i\}$ be the set of maximized atoms that contain the final atom A_{m-1}.

We define the maximized átomaton $\mathcal{M}$ of L, using a one-one correspondence $M_i \leftrightarrow \mathbf{M}_i$ between the sets M_i and the states $\mathbf{M}_i$ of the NFA $\mathcal{M}$ as follows:

Definition 5. *The* maximized átomaton *of L is the NFA defined by $\mathcal{M} = (\mathbf{M}, \Sigma, \mu, \mathbf{I}_M, \mathbf{F}_M)$, where $\mathbf{M} = \{\mathbf{M}_i \mid M_i \in M\}$, $\mathbf{I}_M = \{\mathbf{M}_i \mid M_i \in I_M\}$, $\mathbf{F}_M = \{\mathbf{M}_i \mid M_i \in F_M\}$, and $\mathbf{M}_j \in \mu(\mathbf{M}_i, a)$ if and only if $M_j \subseteq a^{-1}M_i$ for all $\mathbf{M}_i, \mathbf{M}_j \in \mathbf{M}$ and $a \in \Sigma$.*

We show that the maximized átomaton $\mathcal{M}$ of L is isomorphic to the reverse NFA of the saturated version of the minimal DFA $\mathcal{D}$ of L^R:

Proposition 5. *The* maximized átomaton $\mathcal{M}$ of L is isomorphic to $\mathcal{D}^{SR}$, where $\mathcal{D}$ is the minimal DFA of L^R.

Proof. Let $\mathcal{D} = (Q, \Sigma, \delta, q_{m-1}, F)$ be the minimal DFA of L^R, with its saturated version $\mathcal{D}^S = (Q, \Sigma, \delta_S, I_S, F)$. Let $Q = \{q_0, \ldots, q_{m-1}\}$, and let $Q_0, \ldots, Q_{m-1}$ be the quotients of L^R.

The initial states of $\mathcal{M}$ correspond to the initial atoms of L, or equivalently, to the initial states of the átomaton $\mathcal{A}$. Since by Proposition 3, $\mathcal{A}^R$ is isomorphic to $\mathcal{D}$, the initial states of $\mathcal{A}$ correspond to the final states of $\mathcal{D}$, which are the final states of $\mathcal{D}^S$, or equivalently, the initial states of $\mathcal{D}^{SR}$.

Any state $\mathbf{M}_i$ of $\mathcal{M}$ is final if and only if $A_{m-1} \subseteq M_i$. By Proposition 2, Part 3, $A_{m-1} \subseteq M_i$ is equivalent to $M_{m-1} \subseteq M_i$, and by Proposition 4, Part 3, $M_{m-1} \subseteq M_i$ is equivalent to $Q_i \subseteq Q_{m-1}$. Note that the quotient Q_{m-1} of L^R, corresponding to the final atom A_{m-1} of L, is the initial quotient of L^R, that is, $Q_{m-1} = L^R$. Now we have $Q_i \subseteq L^R$ which holds if and only if q_i is an initial state of $\mathcal{D}^S$, or equivalently, a final state of $\mathcal{D}^{SR}$.

We also have to show that $\mathbf{M}_j \in \mu(\mathbf{M}_i, a)$ if and only if $q_i \in \delta_S(q_j, a)$, for all $\mathbf{M}_i, \mathbf{M}_j \in \mathbf{M}$ and $a \in \Sigma$. Since $\mathbf{M}_j \in \mu(\mathbf{M}_i, a)$ if and only if $M_j \subseteq a^{-1}M_i$, and $q_i \in \delta_S(q_j, a)$ if and only if $aL_{q_i, F}(\mathcal{D}) \subseteq L_{q_j, F}(\mathcal{D})$ that is equivalent to $Q_i \subseteq a^{-1}Q_j$, we have to show that $M_j \subseteq a^{-1}M_i$ if and only if $Q_i \subseteq a^{-1}Q_j$. This equivalence holds by Proposition 4, Part 4.   □

Corollary 1. *The maximized átomaton $\mathcal{M}$ is isomorphic to $\mathcal{A}^{RSR}$.*

5   Obtaining a Canonical RFSA by Applying C

We know from Sect. 3 that applying the operation C to any NFA $\mathcal{N}$, produces an RFSA $\mathcal{N}^C$. This RFSA is not necessarily a canonical RFSA. However, Theorem 3 provides sufficient conditions for obtaining a canonical RFSA by applying C to an NFA. This is similar to the way how Theorem 1 provides sufficient conditions for obtaining a minimal DFA.

In Sect. 2, we recalled Theorem 2 from [1,2], a generalization of Theorem 1, characterizing the class of NFAs for which applying the determinization procedure produces a minimal DFA.

In this section we will present a theorem, similar to Theorem 2, that provides a characterization of NFAs for which applying C results in a canonical RFSA.

We interchange the languages L and L^R of Sect. 4. That is, we consider a language L, and assume that $A_0, \ldots, A_{m-1}$ are the atoms of the reverse language L^R, $M_0, \ldots, M_{m-1}$ are the maximized atoms of L^R, and $\mathcal{M}$ is the maximized átomaton of L^R. Also, let $\mathcal{D}$ be the minimal DFA of L, with the state set $Q = \{q_0, \ldots, q_{m-1}\}$, and let the quotients of L be $\{Q_0, \ldots, Q_{m-1}\}$. The NFA $\mathcal{D}^S$ is the saturated version of $\mathcal{D}$, and $\mathcal{D}^{SE}$ is the NFA that is obtained after removing all erasable states and their associated transitions from $\mathcal{D}^S$. Since by Proposition 1, $\mathcal{D}^{SE}$ is the canonical RFSA for L, let us denote it by $C = \mathcal{D}^{SE}$. Clearly, $\mathcal{D}^{SE}$ is a subautomaton of $\mathcal{D}^S$, and since by Proposition 5, $\mathcal{D}^S$ and $\mathcal{M}^R$ are isomorphic, C is isomorphic to a subautomaton of $\mathcal{M}^R$, where some states of $\mathcal{M}^R$ may have been removed, together with their transitions.

For every state q_i of C, let us denote the left language of q_i by L_i; the right language of q_i is Q_i. Since the canonical RFSA C is uniquely determined by L, so are L_i's, and we call these L_i's *canonical left languages* of L. Since the canonical RFSA is a state-minimal RFSA with maximal number of transitions, there is no RFSA with minimal number of states, such that the left language of any of its states is larger than the left language of the corresponding state of the canonical RFSA. That is, for every state-minimal RFSA, the left languages of its states are subsets of corresponding canonical left languages.

Since C is (isomorphic to) a subautomaton of $\mathcal{M}^R$, we can set a correspondence between L_i's and those M_i's which correspond to non-erasable states of $\mathcal{D}^S$. We can see by Proposition 4, Part 2, that (M_i^R, Q_i) is a factorization of L, therefore it is clear that for every L_i and its corresponding M_i, $L_i \subseteq M_i^R$ holds. Since there is a one-to-one correspondence between A_i's and M_i's, there is also a correspondence between L_i's and A_i's which correspond to M_i's associated with non-erasable states of $\mathcal{D}^S$.

Proposition 6. *The following properties hold for any canonical left languages L_i and L_j of L:*

1. $A_i^R \subseteq L_i \subseteq M_i^R$.
2. *If* $A_i^R \cap L_j \neq \emptyset$, *then* $L_i \subseteq L_j$.

Proof.

1. Suppose that there is some $u \in A_i^R$ such that $u \notin L_i$. By Proposition 4, Part 1, $u \in A_i^R$ implies that $u^{-1}L = Q_i$. Since $u \notin L_i$ and $u^{-1}L = Q_i$, there must be a set of quotients, $\{Q_j \mid j \in J_i\}$, $J_i \subset \{0, \ldots, m-1\}$, such that $Q_i = \bigcup_{j \in J_i} Q_j$ and $u \in L_j$, $j \in J_i$. We get that Q_i is not prime, a contradiction. Thus $A_i^R \subseteq L_i$, and since we saw above that $L_i \subseteq M_i^R$ holds, we have $A_i^R \subseteq L_i \subseteq M_i^R$.
2. Let $A_i^R \cap L_j \neq \emptyset$ for some L_i, L_j. That is, there is a word $u \in A_i^R$, such that $u \in L_j$. By Proposition 4, Part 1, $u \in A_i^R$ if and only if $u^{-1}L = Q_i$. Since $u \in L_j$, $uQ_j \subseteq L$ holds, implying that $Q_j \subseteq u^{-1}L$, that is, $Q_j \subseteq Q_i$.

   We show that $Q_j \subseteq Q_i$ implies $L_i \subseteq L_j$. First, if $\varepsilon \in L_i$, then q_i is an initial state of C, implying $Q_i \subseteq L$. Since $Q_j \subseteq Q_i$, we get $Q_j \subseteq L$, which implies $\varepsilon \in L_j$.

   Now, let $va \in L_i$, such that $v \in \Sigma^*$ and $a \in \Sigma$. Then there is some state q_h of C, such that $v \in L_h$ and $aQ_i \subseteq Q_h$. Since $Q_j \subseteq Q_i$, we get $aQ_j \subseteq Q_h$, implying $Q_j \subseteq a^{-1}Q_h$. By the definition of a canonical RFSA, there is a transition in C from its state q_h to q_j by a. Thus, $L_h a \subseteq L_j$, implying $va \in L_j$. We conclude that $L_i \subseteq L_j$. □

Now, let $\mathcal{N} = (Q, \Sigma, \delta, I, F)$ be an NFA. Let us consider the NFA $\mathcal{N}^C = (Q_C, \Sigma, \delta_C, I_C, F_C)$ obtained by applying the operation C to $\mathcal{N}$ as in Definition 3. Next, we will present a few propositions before we can prove our main theorem, Theorem 4, in the end of this section.

Proposition 7. *For any word $u \in \Sigma^*$ and any state s of $\mathcal{N}^C$, if $u \in L_{I_C, s}(\mathcal{N}^C)$, then $u \in L_{I, q}(\mathcal{N})$ for every $q \in s$.*

Proof. We prove the statement by induction on the length of u. Let s be a state of $\mathcal{N}^C$, and $u \in L_{I_C,s}(\mathcal{N}^C)$. If $|u| = 0$, then $u = \varepsilon$. That is, s is an initial state of $\mathcal{N}^C$. By Definition 3, s consists of some initial states of $\mathcal{N}$, that is, $\varepsilon \in L_{I,q}(\mathcal{N})$ for every $q \in s$.

Now, let $u = va$, where $v \in \Sigma^*$ and $a \in \Sigma$, and assume that the proposition holds for v. Let s be a state of $\mathcal{N}^C$ such that $va \in L_{I_C,s}(\mathcal{N}^C)$. Then there is some state t of $\mathcal{N}^C$ such that $v \in L_{I_C,t}(\mathcal{N}^C)$ and $s \in \delta_C(t, a)$. By the induction assumption, we have that $v \in L_{I,q}(\mathcal{N})$ for every $q \in t$. Also, by Definition 3, $s \subseteq \delta(t, a)$. Clearly, $va \in L_{I,q}(\mathcal{N})$ for every $q \in s$.   □

Proposition 8. *Let the left language of every state q of $\mathcal{N}$ be a union of L_i's. Assume that for every L_i there is a state s_i of $\mathcal{N}^C$, consisting of states q of $\mathcal{N}$ such that $L_i \subseteq L_{I,q}(\mathcal{N})$. Then $L_i \subseteq L_{I_C,s_i}(\mathcal{N}^C)$.*

Proof. We show that for $w \in \Sigma^*$, and for any L_i, if $w \in L_i$, then $w \in L_{I_C,s_i}(\mathcal{N}^C)$. The proof is by induction on the length of w. If $|w| = 0$, then $w = \varepsilon$. If $\varepsilon \in L_i$, then $\varepsilon \in L_{I,q}(\mathcal{N})$ for every $q \in s_i$. That is, every $q \in s_i$ is an initial state of $\mathcal{N}$, and by Definition 3, s_i is an initial state of $\mathcal{N}^C$. Thus, $\varepsilon \in L_{I_C,s_i}(\mathcal{N}^C)$.

Now, let $w = va$, where $v \in \Sigma^*$ and $a \in \Sigma$, and assume that our claim holds for v, that is, if $v \in L_h$ for any L_h, then $v \in L_{I_C,s_h}(\mathcal{N}^C)$. Let $va \in L_i$. Clearly, since L_i is the left language of some state of the canonical RFSA $\mathcal{C}$, there exists a state of $\mathcal{C}$ with the left language L_h, such that $v \in L_h$ and $L_h a \subseteq L_i$. We know that there is a state s_h of $\mathcal{N}^C$, consisting of all states p of $\mathcal{N}$ such that $L_h \subseteq L_{I,p}(\mathcal{N})$. Clearly, $v \in L_{I,p}(\mathcal{N})$ for every $p \in s_h$. By the induction assumption, $v \in L_{I_C,s_h}(\mathcal{N}^C)$.

We claim that for every $q \in s_i$ there exists some $p \in s_h$, such that $q \in \delta(p, a)$. Let $q \in s_i$. Then $L_i \subseteq L_{I,q}(\mathcal{N})$. Suppose that there is no state p of $\mathcal{N}$ such that $p \in s_h$ and $q \in \delta(p, a)$. Since we have that $L_h a \subseteq L_i$ holds, also the inclusion $A_h^R a \subseteq L_i$ holds by Proposition 6, Part 1. Then there must be some L_k such that $A_h^R \cap L_k \neq \emptyset$, and some state p of $\mathcal{N}$, such that $L_k \subseteq L_{I,p}(\mathcal{N})$ and $q \in \delta(p, a)$. By Proposition 6, Part 2, we get $L_h \subseteq L_k$. Consequently, $L_h \subseteq L_{I,p}(\mathcal{N})$ holds, implying $p \in s_h$. Thus, our claim holds, and therefore $s_i \subseteq \delta(s_h, a)$. By Definition 3, $\mathcal{N}^C$ has a transition from s_h to s_i by a. Thus, $va \in L_{I_C,s_i}(\mathcal{N}^C)$.   □

Proposition 9. *The RFSA $\mathcal{N}^C$ is a canonical RFSA if and only if for every state s_i of $\mathcal{N}^C$, $L_{I_C,s_i}(\mathcal{N}^C) = L_i$ for some L_i.*

Proof. If the RFSA $\mathcal{N}^C$ is canonical, then by definition, there is a one-one correspondence between its states s_i and canonical left languages L_i, such that $L_{I_C,s_i}(\mathcal{N}^C) = L_i$.

Conversely, let $\mathcal{N}^C$ be such that for each of its states s_i, $L_{I_C,s_i}(\mathcal{N}^C) = L_i$ holds for some L_i. Clearly, for any states s_i and s_j of $\mathcal{N}^C$, if $s_i \neq s_j$, then $L_{I_C,s_i}(\mathcal{N}^C) \neq L_{I_C,s_j}(\mathcal{N}^C)$, implying that there is exactly one state s_i of $\mathcal{N}^C$, such that $L_{I_C,s_i}(\mathcal{N}^C) = L_i$. Consequently, there is a one-one correspondence between the states of $\mathcal{N}^C$ and the states of the canonical RFSA. That means, $\mathcal{N}^C$ is a minimal RFSA. We also show that $\mathcal{N}^C$ is saturated. Indeed, if we suppose that $\mathcal{N}^C$ is not saturated, then we could add initial states and/or transitions to it, without modifying the language $\mathcal{N}^C$ accepts. First, if we make some non-initial state s_i of $\mathcal{N}^C$ initial, then an empty string is added to its left language, and so the left language of s_i is made larger than L_i, which is not possible as we discussed earlier in this section. Now consider adding a transition from s_i to s_j labelled by $a \in \Sigma$. Since this addition cannot change the left language of s_j, $L_{I_C,s_i}(\mathcal{N}^C)a \subseteq L_{I_C,s_j}(\mathcal{N}^C)$ must hold, implying $L_i a \subseteq L_j$. Since $A_i^R \subseteq L_i$, there has to be some state s_h of $\mathcal{N}^C$ with a transition from s_h to s_j by a, and $A_i^R \cap L_{I_C,s_h}(\mathcal{N}^C) \neq \emptyset$, that is, $A_i^R \cap L_h \neq \emptyset$. Then by Proposition 6, Part 2, $L_i \subseteq L_h$. We know by Proposition 7, that $L_i \subseteq L_{I,q}(\mathcal{N})$ for every $q \in s_i$, and $L_h \subseteq L_{I,q}(\mathcal{N})$ for every $q \in s_h$. And clearly, if $L_i \subseteq L_{I,q}(\mathcal{N})$, then $q \in s_i$, and if $L_h \subseteq L_{I,q}(\mathcal{N})$, then $q \in s_h$. Thus, $s_i = \{q \mid L_i \subseteq L_{I,q}(\mathcal{N})\}$ and $s_h = \{q \mid L_h \subseteq L_{I,q}(\mathcal{N})\}$. Since $L_i \subseteq L_h$, the inclusion $s_h \subseteq s_i$ holds. Now, because there is a transition from s_h to s_j by a, and $s_h \subseteq s_i$, there must also be a transition from s_i to s_j by a. We conclude that $\mathcal{N}^C$ is saturated. By Proposition 1, $\mathcal{N}^C$ is a canonical RFSA. □

Finally, we present the main theorem:

Theorem 4. *For any NFA $\mathcal{N}$ of L, $\mathcal{N}^C$ is a canonical RFSA if and only if the left language of every state of $\mathcal{N}$ is a union of some canonical left languages L_i of L.*

Proof. Let $\mathcal{N} = (Q, \Sigma, \delta, I, F)$ be an NFA and let $\mathcal{N}^C = (Q_C, \Sigma, \delta_C, I_C, F_C)$ be a canonical RFSA of L. We show that the left language of every state of $\mathcal{N}$ is a union of L_i's. Let q be any state of $\mathcal{N}$. Note that if the left language of q is empty, then it is an empty union of L_i's. Now consider the case where $L_{I,q}(\mathcal{N}) \neq \emptyset$. Let $u \in L_{I,q}(\mathcal{N})$. Then there is a state s_i of $\mathcal{N}^C$ such that $u \in L_{I_C,s_i}(\mathcal{N}^C)$ and $q \in s_i$. Since $\mathcal{N}^C$ is a canonical RFSA, $L_{I_C,s_i}(\mathcal{N}^C) = L_i$ for some L_i. We get that $u \in L_i$. Consider any other word $v \in L_i$. Clearly, $v \in L_{I_C,s_i}(\mathcal{N}^C)$, implying $v \in L_{I,q}(\mathcal{N})$ by Proposition 7. Thus, $L_i \subseteq L_{I,q}(\mathcal{N})$, and we conclude that the left language of q is a union of L_i's.

Conversely, assume that the left language of every state q of $\mathcal{N}$ is a union of L_i's. First, we claim that for every L_i there is some q, such that $L_i \subseteq L_{I,q}(\mathcal{N})$. To see this, suppose that for some L_i there is no such q. Let $u \in A_i^R$. Then by Proposition 4, Part 1, $u^{-1}L = Q_i$. Clearly, there is a state q of $\mathcal{N}$, such that $u \in L_{I,q}(\mathcal{N})$. Since by our assumption, $L_i \not\subseteq L_{I,q}(\mathcal{N})$, however, the left language of q is a union of canonical left languages, there is some L_j, such that $u \in L_j$. We get that $A_i^R \cap L_j \neq \emptyset$, which by Proposition 6, Part 2 implies $L_i \subseteq L_j$. Thus, $L_i \subseteq L_{I,q}(\mathcal{N})$.

Next, we will show that for every L_i, there is a state s_i of $\mathcal{N}^C$, such that $L_{I_C,s_i}(\mathcal{N}^C) = L_i$. So let us consider any L_i. Let s_i be the set of all states q of $\mathcal{N}$, such that $L_i \subseteq L_{I,q}(\mathcal{N})$. By the claim above, s_i is not empty. Let $q \in s_i$. Then by Proposition 6, Part 1, $A_i^R \subseteq L_{I,q}(\mathcal{N})$. Let $u \in A_i^R$. We note that if $u \in L_j$ for some $L_j \neq L_i$, then $L_i \subseteq L_j$ by Proposition 6, Part 2. Consequently, s_i is the set of all states q, such that $u \in L_{I,q}(\mathcal{N})$. We claim that s_i is a state of $\mathcal{N}^C$. Indeed, if we suppose that s_i is not a state of $\mathcal{N}^C$, then it is coverable, that is, a union of some states of $\mathcal{N}^C$. Since by Proposition 4, Part 1, $u^{-1}L = Q_i$, we then get that Q_i must be a union of some other quotients of L, which is a contradiction, because Q_i is the right language of a state of a canonical RFSA. Thus, for every L_i there is a corresponding state s_i of $\mathcal{N}^C$. Furthermore, since $L_i \subseteq L_{I,q}(\mathcal{N})$ for every $q \in s_i$, the inclusion $L_i \subseteq L_{I_C,s_i}(\mathcal{N}^C)$ holds by Proposition 8. Now, for every word $v \in L_i \setminus A_i^R$, let $s_v = \{q \mid v \in L_{I,q}(\mathcal{N})\} = \{q \mid v \in L_j, L_j \subseteq L_{I,q}(\mathcal{N})\}$. Then $s_v = \bigcup_{v \in L_j} s_j$. Clearly, s_v is coverable, and cannot be a state of $\mathcal{N}^C$.

We have obtained that $\mathcal{N}^C$ consists of states s_i, each corresponding to some L_i, therefore, $\mathcal{N}^C$ has the same number of states as does the canonical RFSA. Also, we obtained that for every state s_i of $\mathcal{N}^C$, the inclusion $L_i \subseteq L_{I_C,s_i}(\mathcal{N}^C)$ holds. Since a canonical RFSA has exactly one state for every L_i, with L_i being its left language, and because there is no RFSA with the same number of states and larger left languages, we get that $L_{I_C,s_i}(\mathcal{N}^C) = L_i$ for every s_i. We conclude by Proposition 9 that $\mathcal{N}^C$ is a canonical RFSA. □

6   Conclusions and Future Work

We started a study of the relationship of RFSAs with atoms of regular languages, defining the maximized atoms and the maximized átomaton of a language, and showing that the maximized átomaton is isomorphic to the reverse automaton of the saturated minimal DFA of the reverse language. It is easily implied that the canonical RFSA of a language is a subautomaton of the reverse automaton of the maximized átomaton of the reverse language. We believe that the maximized átomaton and its properties deserve to be studied further, also noting that the same NFA appeared recently in [6], where the study of its properties was begun.

We defined canonical left languages of a regular language as the left languages of the states of the canonical RFSA, and generalized the double-reversal method by Denis, Lemay, and Terlutte, of obtaining a canonical RFSA.

References

1. Brzozowski, J., Tamm, H.: Theory of átomata. In: Mauri, G., Leporati, A. (eds.) DLT 2011. LNCS, vol. 6795, pp. 105–116. Springer, Heidelberg (2011)
2. Brzozowski, J., Tamm, H.: Theory of átomata. Theor. Comput. Sci. **539**, 13–27 (2014)
3. Brzozowski, J.: Canonical regular expressions and minimal state graphs for definite events. In: Proceedings of the Symposium on Mathematical Theory of Automata. MRI Symposia Series, vol. 12, pp. 529–561. Polytechnic Press, Polytechnic Institute of Brooklyn, New York (1963)
4. Denis, F., Lemay, A., Terlutte, A.: Residual finite state automata. Fund. Inform. **51**, 339–368 (2002)
5. Denis, F., Lemay, A., Terlutte, A.: Residual finite state automata. In: Ferreira, A., Reichel, H. (eds.) STACS 2001. LNCS, vol. 2010, pp. 144–157. Springer, Heidelberg (2001)
6. Myers, R.S.R., Adámek, J., Milius, S., Urbat, H.: Canonical nondeterministic automata. In: Bonsangue, M.M. (ed.) CMCS 2014. LNCS, vol. 8446, pp. 189–210. Springer, Heidelberg (2014)

Quantum State Complexity of Formal Languages

Marcos Villagra and Tomoyuki Yamakami[✉]

Department of Information Science, University of Fukui,
3-9-1 Bunkyo, Fukui 910-8507, Japan
{mdvillagra,TomoyukiYamakami}@gmail.com

Abstract. In this extended abstract, our notion of state complexity concerns the minimal amount of descriptive information necessary for a finite automaton to determine whether given fixed-length strings belong to a target language. This serves as a descriptional complexity measure for languages with respect to input length. In particular, we study the minimal number of inner states of quantum finite automata, whose tape heads may move freely in all directions and which conduct a projective measurement at every step, to recognize given languages. Such a complexity measure is referred to as the quantum state complexity of languages. We demonstrate upper and lower bounds on the quantum state complexity of languages on various types of quantum finite automata. By inventing a notion of timed crossing sequence, we also establish a general lower-bound on quantum state complexity in terms of approximate matrix rank. As a consequence, we show that bounded-error 2qfa's running in expected subexponential time cannot, in general, simulate logarithmic-space deterministic Turing machines.

Keywords: Quantum finite automaton · Quantum state complexity · Approximate matrix rank · Minimal automaton · Advice · Timed crossing sequence · Permutation automaton

1 Background and Major Contributions

Finite (state) automata are unarguably one of the simplest machine models representing a realistic computation device, and the rich literature over the past 60 years has enthusiastically explored the power and limitation of such machines. Since the 1980s, in particular, a new wave of building non-conventional computing devices have arisen and, shortly, a concept of "quantum computer" emerged on the premise of quantum physics. As a simple model of such a device, Kondacs and Watrous [9] studied bounded-error *quantum finite automata* that make unitary evolution and also perform a projective (or von Neumann) measurement at every step. A striking feature that differentiates quantum finite automata from their classical counterparts is that all languages recognized by *1-way quantum finite automata* (abbreviated as 1qfa's) with bounded-error probability are still

M. Villagra is a research fellow of the Japan Society for the Promotion of Sciences (JSPS).

© Springer International Publishing Switzerland 2015
J. Shallit and A. Okhotin (Eds.): DCFS 2015, LNCS 9118, pp. 280–291, 2015.
DOI: 10.1007/978-3-319-19225-3_24

regular languages, while certain bounded-error *2-way quantum finite automata* (or 2qfa's) can recognize even non-regular languages [9]. In a classical setting, however, 1-way deterministic finite automata (or 1dfa's) are no less powerful than their 2-way counterparts. Moreover, the family of all languages recognized by bounded-error 1qfa's is not closed under union and intersection [2], although the family of regular languages is closed under those operations.

One of the focal points in the study of quantum finite automata has been to estimate the least number of *inner states* necessary for quantum finite automata to "accept" or "recognize" a target language. Such a number serves as a meaningful measure of "conciseness" or "succinctness" of the description of the automaton for the given language. Even equipped with no memory tape, inner states additionally provided to the finite automaton help store sufficient information to enhance their power of language recognition. For this reason, the minimal number of inner states has been used in general to classify the descriptional complexity of formal languages. Ambainis and Freivalds [1] exemplified the superiority of 1qfa's in the conciseness of their 1qfa's over 1dfa's by proving that a certain 1qfa can determine whether a given number is divisible by a fixed prime p using at most $O(\log p)$ inner states; on the contrary, any 1-way probabilistic finite automaton (or 1pfa) requires at least p inner states. Mereghetti et al. [12] explained how concise 1qfa's could be on unary languages in comparison with 1pfa's. Freivalds et al. [8], Yakaryilmaz and Say [18], and Zheng et al. [20] also studied the conciseness of quantum finite automata of different types.

By contrast, Ambainis, Nayak, Ta-Shma, and Vazirani [3] exhibited an architectural limitation of 1qfa's by presenting a special series $\{L_n\}_{n \geq 2}$ of finite languages, where $L_n = \{w0 \mid w \in \{0, 1\}^*, |w| < n\}$, each L_n of which is recognized by a certain $O(n)$-state 1dfa, but requires $2^{\Omega(n)}$ inner states on any bounded-error 1qfa that even performs superoperators. This result leads us to study the minimal number of inner states needed for quantum finite automata to recognize a given language L on inputs of each fixed length n (i.e., to determine whether or not input strings of length n belong to L) and we generally call this specific number the *state complexity*[1] *of* L *at* n on quantum finite automata and any finite automaton that achieves this state complexity is called a *minimal automaton*. To emphasize those underlying quantum finite automata, we prefer to use the new term of *quantum state complexity*. This notion enables us to discuss the complicated behaviors of finite automata that accept each segment of the given languages, in a more detailed fashion. Moreover, our quantum state complexity not only indicates a cost of recognizing languages but also helps describe a cost of performing language operations (Sect. 4), as well as a cost of simulating one type of machines on another type of machines (Sect. 5).

Despite the aforementioned research efforts on 1qfa's, there have been a few studies solely devoted to the quantum state complexity of languages, and thus a vast area of this emerging field is still uncultivated. This extended abstract

---

[1] Traditionally, "state complexity" refers to the minimal descriptional size of finite automata that recognize a language on *all* inputs and this notion has been proven to be useful to study the complexity of regular languages.

therefore intends to fill this void by initiating the development of a coherent and comprehensive theory of quantum state complexity, particularly, based on the "original" machine model of Kondacs and Watrous [9], because not only is this the most studied model in the literature, but also it is one of the simplest mathematical models whose structures and properties have been analyzed in great depth.

In the rest of this extended abstract, we wish to present three major contributions after exploring basic properties of quantum state complexity in Sects. 3–4. Hereafter, "n" expresses an arbitrary input size.

1. It was proven in [4] that measure-once 1-way quantum finite automata (or mo-1qfa's) of [13] are equivalent in recognition power to permutation automata. Not relying on a non-constructive argument of [4], we take a constructive approach toward proving a logarithmic lower bound on the state complexity of the language $L_{01} = \{0^m 1^n \mid m, n \geq 0\}$ on mo-1qfa's.
2. Quantum computation assisted by an external information source (called *advice*) was first discussed in [14]. Following the setting of [19], we prove that any language recognized by advised 2qfa's has at most $O(n)$ state complexity.
3. We introduce a new notion of *timed crossing sequence* and show a new lower bound on the state complexity of 2qfa's in terms of *approximate matrix rank*. As a consequence, we show that L (the deterministic log-space class) is not contained within a certain family of languages recognized by bounded-error 2qfa's running in expected subexponential time.

Omitted or abridged proofs in this extended abstract will appear shortly in its complete version.

2   Basic Notions and Notation

Let $\mathbb{Z}$ (resp., $\mathbb{R}$, $\mathbb{C}$) denote the set of all integers (resp., real numbers, complex numbers). Specifically, we write $\mathbb{N}$ for the set of all *natural numbers* (i.e., non-negative integers) and set $\mathbb{N}^+ = \mathbb{N} - \{0\}$. In addition, $\mathbb{A}$ denotes the set of all *algebraic complex numbers*. Given any number $\alpha \in \mathbb{C}$, α^* denotes its *conjugate*. The *real unit interval* between 0 and 1 is denoted by $[0, 1]$. Analogously, for any two numbers $m, n \in \mathbb{Z}$ with $m \leq n$, $[m, n]_{\mathbb{Z}}$ indicates the *integer interval* between m and n; that is, the set $\{m, m+1, m+2, \ldots, n\}$. The notation $|Q|$ expresses the *cardinality* of a finite set Q. Given an $m \times n$ complex matrix A, let A^T indicate the transposed matrix of A and $A^\dagger$ denotes the *adjoint* of A. In a finite-dimensional Hilbert space $\mathcal{H}$, $\|x\|$ denotes the ℓ_2-*norm* of vector x and $\|A\|$ expresses the *operator norm* of A, defined by $\|A\| = \max\{\|Ax\|/\|x\| : x \neq 0\}$.

An *alphabet* is a nonempty finite set. Given such an alphabet Σ, we write Σ^n (resp., $\Sigma^{\leq n}$) for the set of all strings over Σ of length exactly n (resp., of length at most n). In particular, $\Sigma^0 = \{\lambda\}$, where λ is the empty string. Set Σ^* to be $\bigcup_{n \in \mathbb{N}} \Sigma^n$. Given any language L over Σ and each $n \in \mathbb{N}$, L_n and $L_{\leq n}$ express the sets $L \cap \Sigma^n$ and $L \cap \Sigma^{\leq n}$, respectively. For brevity, we write $\overline{L}$ for $\Sigma^* - L$.

We assume the reader's basic knowledge regarding a model of *measure-many 2-way quantum finite automata* (abbreviated as 2qfa's) [9]. Such a 2qfa M has the form $(Q, \check{\Sigma}, q_0, \delta, A, R)$ where Q is a finite set of inner states, A and R are sets of accepting and rejecting inner states, q_0 is the initial inner state, δ is a (quantum) transition function mapping $Q \times \check{\Sigma} \times Q \times D$ to $\mathbb{C}$ (where the values of δ are called *amplitudes*), and $\check{\Sigma}$ stands for $\Sigma \cup \{\cent, \$\}$. Notice that M has a single read-only input tape on which any input string is written, surrounded by two designated endmarkers $\cent$ (left) and $\$$ (right). Our input tape is always assumed to be *circular* [9]. Let $D = \{-1, 0, +1\}$. A *configuration space* CONF_n of M on inputs of length n is a Hilbert space $\mathrm{span}\{|q\rangle|h\rangle \mid q \in Q, h \in [0, n+1]_{\mathbb{Z}}\}$. The transition (or time-evolution) matrices $\{U_\delta^{(x)}\}_{x \in \Sigma^*}$ acting on CONF_n are induced from δ by setting $U_\delta^{(x)}|q, h\rangle = \sum_{p \in Q, d \in D} \delta(q, x_h, p, d)|p, h+d \mod (n+2)\rangle$ with $x = x_1 x_2 \cdots x_n$, $x_0 = \cent$, and $x_{n+1} = \$$. Throughout this extended abstract, we implicitly demand that each $U_\delta^{(x)}$ should be *unitary*. Three projections Π_{acc}, Π_{rej}, and Π_{non} map quantum states onto the accepting, rejecting, and non-halting configuration spaces W_{acc}, W_{rej}, and W_{non}, respectively. A *computation* of M on input x proceeds by applying $U = U_\delta^{(x)}\Pi_{\mathrm{non}}$. Using a measurement operator Π_{acc} (resp., Π_{rej}), we define the acceptance (resp., rejection) probability $p_{M,\mathrm{acc},t}(x)$ (resp., $p_{M,\mathrm{rej},t}(x)$) of M on x at time t as $\|\Pi_{\mathrm{acc}}U^t|q_0, 0\rangle\|^2$ (resp., $\|\Pi_{\mathrm{rej}}U^t|q_0, 0\rangle\|^2$). Let ε be any function from $\mathbb{N}$ to $[0, 1/2]$. Given any language L over alphabet Σ, M is said to *recognize L with error probability at most $\varepsilon(n)$* if (i) for every $x \in L$, it accepts x with probability at least $1 - \varepsilon(|x|)$ (i.e., $\sum_{t=0}^\infty p_{M,\mathrm{acc},t}(x) \geq 1 - \varepsilon(|x|)$) and (ii) for every $x \in \overline{L}$, it rejects x with probability at least $1 - \varepsilon(|x|)$ (i.e., $\sum_{t=0}^\infty p_{M,\mathrm{rej},t}(x) \geq 1 - \varepsilon(|x|)$). When ε is a constant in $[0, 1/2)$, we intend to use the more familiar term of "bounded-error probability."

In contrast, we define a *measure-many 1-way[2] quantum finite automata* (or 1qfa's) M as $(Q, \check{\Sigma}, \{U_\sigma\}_{\sigma \in \check{\Sigma}}, s, A, R)$, where each unitary operator U_σ acting on $\mathrm{span}\{|q\rangle \mid q \in Q\}$ is applied whenever the tape head scans symbol σ, followed by a measurement as described above, and the machine must halt "absolutely" until or just after the right endmarker $\$$ is read.

Given a nonempty subset K of $\mathbb{C}$, we say that M has K-*amplitudes* if M uses only amplitudes in K. We then define $2\mathrm{BQFA}_K$ (resp., $1\mathrm{BQFA}_K$) as the collection of all languages that can be recognized by K-amplitude 2qfa's (resp., 1qfa's) with bounded-error probability. Similarly, $2\mathrm{BQFA}_K(t\text{-}time)$ is defined by K-amplitude 2qfa's that run in expected $t(n)$-time (i.e., the average runtime is upper-bounded by $t(n)$). We often drop the subscript K whenever $K = \mathbb{C}$.

3   Quantum State Complexity

As noted in Sect. 1, our "quantum state complexity" of a target language serves as a complexity measure for each input length n and indicates the minimal

---

[2] This model is sometimes referred to as "real time" because its tape head always moves to the right without staying still on any tape cell.

number of inner states used by quantum finite automata to recognize the language on length-n inputs with designated error probability using a given amplitude set.

To be more precise, first fix a language L over alphabet Σ, a subset K of $\mathbb{C}$, and a function ε from $\mathbb{N}$ to $[0, 1/2)$. Given any length $n \in \mathbb{N}$ and any 2qfa M, we say that M *recognizes*[3] L *at* n with error probability at most $\varepsilon(n)$ using amplitude set K if (1) M has K-amplitudes, (2) for all $x \in L_n$, M accepts x with probability at least $1 - \varepsilon(n)$, and (3) for all $x \in \Sigma^n - L_n$, M rejects x with probability at least $1 - \varepsilon(n)$. Note that no requirement is imposed on the outside of Σ^n. The *state complexity of* M is the total number of inner states used by M, not including the size of Σ.

For any function s on $\mathbb{N}$, L is of (or has) *state complexity* $s(n)$ on 2qfa's with error probability at most $\varepsilon(n)$ using amplitude set K if, for every $n \in \mathbb{N}$, $s(n)$ equals the smallest state complexity of any K-amplitude 2qfa M_n that recognizes L at n with error probability at most $\varepsilon(n)$. A 2qfa M that achieves this state complexity $s(n)$ is called a *minimal 2qfa* for L at n (or for L_n). In order to clarify the use of quantum finite automata, we intend to use a new term "quantum state complexity" in place of "state complexity."

In the past literature, 2qfa's were occasionally described by their "partial" transitions, because they can be easily expanded to "complete" transitions. To argue on the number of inner states, however, it is important to address that we should use only "completely specified" 2qfa's, which are described by their transition functions defined completely on the domain $Q \times \check{\Sigma} \times Q \times D$.

The special notation $2QSC_{K,\varepsilon}[L](n)$ stands for the number m for which L has quantum state complexity m at n with error probability at most ε using amplitude set K. As did before, we often drop the subscript K when $K = \mathbb{C}$. If we consider only 2qfa's whose expected running time is at most $t(n)$, then we specifically write $2QSC^t_{K,\varepsilon}[L](n)$ to emphasize "$t(n)$."

Lemma 1. *Let L be any language over Σ with $|\Sigma| \geq 2$, let ε be any function from $\mathbb{N}$ to $[0, 1/2)$, and let K be any nonempty amplitude set. For any length $n \in \mathbb{N}$, (1) $1 \leq 2QSC_{K,\varepsilon}[L](n) \leq |\Sigma|^{n+1} + 1$, (2) $2QSC_{K,\varepsilon}[\bar{L}](n) = 2QSC_{K,\varepsilon}[L](n)$, and (3) $2QSC_{\mathbb{C},\varepsilon}[L](n) \leq 2QSC_{\mathbb{R},\varepsilon}[L](n) \leq 2 \cdot 2QSC_{\mathbb{C},\varepsilon}[L](n)$.*

At this moment, it is helpful to introduce a slightly more lenient notion. A K-amplitude 2qfa M is said to *recognize L up to n* with error probability at most $\varepsilon(n)$ if, for all $x \in L_{\leq n}$ and for all $x \in \Sigma^{\leq n} - L_{\leq n}$, M accepts and rejects x with probability at least $1 - \varepsilon(n)$, respectively. To indicate the corresponding state complexity measure, we use another notation $2QSC_{K,\varepsilon}[L](\leq n)$. It is easy to see that $2QSC_{K,\varepsilon}[L](n) \leq 2QSC_{K,\varepsilon}[L](\leq n)$ for any $n \in \mathbb{N}$. Obviously, the following statement holds.

Lemma 2. *Given any language $L \in$ 2BQFA, there exist two constants $c \geq 1$ and $\varepsilon \in [0, 1/2)$ such that $2QSC_{\varepsilon}[L](\leq n) \leq c$ for any length $n \in \mathbb{N}$. A similar statement also holds for 2BQFA(t-time) and $2QSC^t_{\varepsilon}[L](\leq n)$.*

---

[3] In other words, M recognizes a so-called "promise problem" $(L_n, \Sigma^n - L_n)$ with error probability at most $\varepsilon(n)$.

Nishimura and Yamakami [15] noted that, following Watrous' argument [17], the family 2BQFA$_A$ is contained in PL, where PL is the family of all languages recognized by log-space probabilistic Turing machines with unbounded-error probability. With a use of this fact, it is possible to show that any bounded-error 2qfa with constant state complexity can be converted into another 2qfa incurring *no error* at a cost of subexponentially many inner states.

Lemma 3. *For any* $L \in 2BQFA_A$, $2QSC_{A,0}[L](\leq n) = 2^{O(\log^2 n)}$ *holds.*

Proof Sketch. Given any language L in 2BQFA$_A$, take a PL-machine M recognizing L with unbounded-error probability. Note that $\mathrm{PrSPACE}(s) \subseteq \mathrm{RevSPACE}(s^2)$ holds for any space bound $s(n) = \Omega(\log n)$ [16], where PrSPACE and RevSPACE respectively stand for unbounded-error probabilistic space (with all paths terminating absolutely) and reversible space. Hence, there exists a reversible Turing machine M' recognizing L using $O(\log^2 n)$ space. It is possible to simulate a use of such memory space by adding $2^{O(\log^2 n)}$ inner states to 2qfa's. We thus obtain an error-free 2qfa that recognizes L at a cost of $2^{O(\log^2 n)}$ inner states. Notice that the obtained error-free 2qfa runs in polynomial time.   □

Similarly to $2QSC_{K,\varepsilon}[L](n)$ and $2QSC_{K,\varepsilon}[L](\leq n)$, we define two more state complexity measures $1QSC_{K,\varepsilon}[L](n)$ and $1QSC_{K,\varepsilon}[L](\leq n)$ based on 1qfa's instead of 2qfa's. Notice that there is relatively rich literature on the state complexity of 1qfa's as noted in Sect. 1. Using some of the known results, we can show an exponential gap between $1QSC_\varepsilon[L](n)$ and $1QSC_\varepsilon[L](\leq n)$.

Lemma 4. *There is a regular language* L *such that, for any constant* $\varepsilon \in (0, 1/2)$, $\log_2(1QSC_\varepsilon[L](\leq n)) \geq \Omega(1QSC_\varepsilon[L](n))$ *holds for almost all* $n \in \mathbb{N}$.

Proof. Consider a regular language $L_0 = \{x0 \mid x \in \{0,1\}^*\}$ over an alphabet $\{0,1\}$. Ambainis et al. [3, Theorem 6.1] proved that any bounded-error 1qfa that allows superoperators requires $2^{\Omega(n)}$ inner states to recognize L_0 up to $n + 1$. This implies that $1QSC_\varepsilon[L_0](\leq n) = 2^{\Omega(n)}$ for any fixed constant $\varepsilon \in [0, 1/2)$. On the contrary, it is easy to see that $1QSC_\varepsilon[L_0](n) \leq n+2$, by setting $\delta_n(q_i, x_i, q_{i+1}) = \delta_n(q_n, 1, q_{acc}) = \delta_n(q_n, 0, q_{rej}) = 1$ for all $i \in [0, n-1]_{\mathbb{Z}}$, where $x_0 = ¢$ and $x = x_1 x_2 \cdots x_n$. The lemma thus follows.   □

We say that a regular language L *does not satisfy the partial order condition* (POC) [4] if the minimal 1dfa $(Q, \Sigma, \delta, q_0, F)$ for L satisfies the following: there are two distinct inner states $q_1, q_2 \in Q$ and three strings $x, y, z \in \Sigma^*$ for which (i) $\hat{\delta}(q_1, x) = \hat{\delta}(q_2, x) = q_2$ and $\hat{\delta}(q_2, y) = q_1$ and (ii) $\hat{\delta}(q_1, z) \notin F$ and $\hat{\delta}(q_2, z) \in F$ or vice versa, where $\hat{\delta}$ is an extended transition function of δ. As shown below, the POC gives a language having high state complexity on 1qfa's.

Lemma 5. *For any regular language* L, *if* L *does not satisfy the POC, then* $1QSC_\varepsilon[L](\leq n) = 2^{\Omega(n)}$ *holds for each fixed constant* $\varepsilon \in [0, 1/2)$. *Moreover, the converse does not hold.*

Proof Sketch. The proof of [4, Theorem 4.1] essentially shows the following claim. (*) Given a language L over alphabet Γ and a homomorphism $h : \Sigma \rightarrow \Gamma^*$, it holds that $1QSC_\varepsilon[h^{-1}(L)](\leq n) \leq (m+1)1QSC_\varepsilon[L](\leq n')$, where $m = \max_{\sigma \in \Gamma} |h(\sigma)|$ and $n' = \max_{x \in \Gamma^{\leq n}} |h(x)|$.

Next, we generalize [4, Theorem 4.3]. Take q_1, q_2 and x, y, z associated with the premise that L over Σ does not satisfy the POC. Assume that $\hat{\delta}(q_1, z) \notin F$ and $\hat{\delta}(q_2, z) \in F$. We also take the minimal string s for which $\hat{\delta}(q_0, s) = q_1$. Applying the above claim (*) twice, we can obtain $1QSC_\varepsilon[L](\leq an + b) = \Omega(1QSC_\varepsilon[L_0](\leq n))$, where $a = |xy|$ and $b = |sz|$. The aforementioned result of Ambainis et al. [3] implies that $1QSC_\varepsilon[L_0](\leq n) = 2^{\Omega(n)}$. Hence, by reassigning $an + b$ as "new" n, the lemma easily follows.

The converse of the lemma is not true because (i) there exists a language not in 1BQFA that does satisfy the POC [2] and (ii) for any $L \in$ 1BQFA, $1QSC_\varepsilon[L](\leq n) = O(1)$ for a certain constant $\varepsilon \in [0, 1/2)$.   □

4   Cost of Performing Language Operations

We begin our study by arguing on how costly (in terms of state complexity) it is to conduct "operations" among languages. An intersection as well as a union between two given languages is a fundamental operation in automata theory. The language family 1BQFA is closed under neither intersection nor union [2] while it is not yet known whether 2BQFA is closed under those operations.

Toward the above open problem, let us study the case of 2qfa's. Take two arbitrary languages L_1 and L_2 in $2BQFA_A$ and consider their witness 2qfa's M_1 and M_2. Simulate each machine on an appropriate log-space Turing machine as done in the proof of Lemma 3. Combine them into a single log-space machine that recognizes $L_1 \circ L_2$, where $\circ \in \{\cup, \cap\}$. Convert the resulting machine into a reversible Turing machine using $O(\log^2 n)$ space, and then into a 2qfa having state complexity of $2^{O(\log^2 n)}$. Therefore, we obtain $2QSC_{A,0}[L_1 \circ L_2](n) = 2^{O(\log^2 n)}$.

Concerning 1qfa's, by contrast, it is possible to obtain a much better upper bound on the quantum state complexity of union (as well as intersection) of languages.

Proposition 6. *Let L_1 and L_2 be two arbitrary languages on a common alphabet and let $\circ \in \{\cup, \cap\}$. Assume that $0 \leq \varepsilon(n) < \frac{3-\sqrt{5}}{2}$ for all $n \in \mathbb{N}$. If $1QSC_\varepsilon[L_1](n) = k_1(n)$ and $1QSC_\varepsilon[L_2](n) = k_2(n)$, then $1QSC_{\varepsilon'}[L_1 \circ L_2](n) \leq 8(n+3)k_1(n)k_2(n)$, where $\varepsilon'(n) = \frac{\varepsilon(n)(2-\varepsilon(n))}{1+\varepsilon(n)-\varepsilon(n)^2}$.*

5   Cost of Simulating 1QFAs on MO-1QFAs

Let us examine a specific type of 1qfa's, known as *measure-once 1qfa's* (or mo-1qfa's) [13], each of which conducts a projective measurement *only once* at the very end of its computation. In analogy to $1QSC_\varepsilon[L](n)$, we define $MO\text{-}1QSC_\varepsilon[L](n)$ based on those mo-1qfa's.

Lemma 7. *Let L be any language and let $\varepsilon : \mathbb{N} \to [0, 1/2)$ be any error bound. It holds that MO-1QSC$_\varepsilon[L](n) \leq (2n + 7)$1QSC$_\varepsilon[L](n)$ for all $n \in \mathbb{N}$.*

Let us demonstrate a lower bound of MO-1QSC$_\varepsilon[L](n)$. Consider a special language $L_{01} = \{0^m 1^n \mid m, n \in \mathbb{N}\}$ over a binary alphabet $\{0, 1\}$, which belongs to 1BQFA with error probability close to 0.32 [1]. Here, we estimate the quantum state complexity MO-1QSC$_\varepsilon[L_{01}](n)$ of this particular language L_{01}.

Proposition 8. $\frac{\log_2 n}{\log_2 (4/\varepsilon)} \leq$ MO-1QSC$_\varepsilon[L_{01}](n)$ *for any fixed constant $\varepsilon \in (0, 1/2)$ and for any sufficiently large number n.*

To prove this proposition, recall a result of Brodsky and Pippenger [4], who proved that bounded-error mo-1qfa's are equivalent in recognition power to *permutation automata* (also known as *group automata* [4]). Lemma 9 generalizes their result. For convenience, we write SC$_{\mathrm{per}}[L](n)$ for the (classical) state complexity of L at n for which an underlying computation model is limited to permutation automata.

Lemma 9. SC$_{\mathrm{per}}[L](n) \leq$ MO-1QSC$_\varepsilon[L](n)^2 \left(\frac{4}{\varepsilon}\right)^{\mathrm{MO\text{-}1QSC}_\varepsilon[L](n)/2}$ *holds for any language L, any fixed constant $\varepsilon \in (0, 1/2)$, and any sufficiently large n.*

Since the proof of [4] employs a non-constructive argument, it does not appear to serve as a helpful tool to prove Lemma 9. For the lemma, we instead make a constructive argument, which goes as follows. We first group together all superpositions of configurations generated by an mo-1qfa into a region given by a hyperspherical cap with half-chord $\sqrt{\varepsilon}/2$ on the hypersphere of unit radius. We then associate each of such regions with an inner state of a target permutation automaton. Hence, the state complexity is upper-bounded by the number of such regions that cover the entire hypersphere. The lemma follows from a result of [6] and a rough estimation of the area of hyperspherical caps.

To complete the proof of Proposition 8, it suffices to demonstrate a lower bound of SC$_{\mathrm{per}}[L_{01}](n)$ described below.

Lemma 10. SC$_{\mathrm{per}}[L_{01}](n) \geq \lfloor n/2 \rfloor + 1$ *holds for any length $n \geq 2$.*

6 2QFAs Equipped with Deterministic Advice

Our quantum state complexity measure naturally embodies a non-uniform nature, in the sense that a state-bounded language family $\mathcal{S}(s(n)) = \{L \mid \exists \varepsilon \in [0, 1/2) \; \forall n \in \mathbb{N} \; [2\mathrm{QSC}_{\mathrm{C},\varepsilon}[L](n) \leq s(n)]\}$ for a given $s : \mathbb{N} \to \mathbb{N}^+$ always contains *non-recursive languages* if $s(n) \geq 3$ for all $n \in \mathbb{N}$.

Such a non-uniform nature has been dealt in computational complexity theory with a notion of external information source, called *advice*. Advised quantum computation was initiated in [14] in the context of polynomial-time quantum Turing machines and in [19] for quantum finite automata. For classical finite automata, Damm and Holzer [5] and Freivalds [7] studied quite different types of advice.

Let Γ be any *advice alphabet* and let $\eta : \mathbb{N} \to \Gamma^*$ be any length-preserving *advice function* (i.e., $|\eta(n)| = n$ for all $n \in \mathbb{N}$). An *advised 2qfa* is a machine $M = (Q, \Sigma, \delta, q_0, A, R, \Gamma)$, which has a single tape split into two tracks, where an upper track contains the original input x and a lower track contains an advice string $\eta(|x|)$, except for two endmarkers. The machine M starts in inner state q_0 with the tape composed of the form $\mathcal{c} \left[\begin{smallmatrix} x \\ \eta(|x|) \end{smallmatrix} \right] \$$ and its tape head simultaneously scans both the ith symbol x_i of the input and the ith symbol $\eta(|x|)_i$ of the *advice string*, except for the endmarkers. We say that M recognizes a language L with a help of (or assisted by) advice with error probability at most $\varepsilon(n)$ if there exists a length-preserving advice function $\eta : \mathbb{N} \to \Gamma^*$ such that, for any $x \in L$, the machine M accepts $\left[\begin{smallmatrix} x \\ \eta(|x|) \end{smallmatrix} \right]$ with probability at least $1 - \varepsilon(n)$, and for any $x \notin L$, it rejects $\left[\begin{smallmatrix} x \\ \eta(|x|) \end{smallmatrix} \right]$ with probability at least $1 - \varepsilon(n)$. The notation 2BQFA/n is used for a family of all languages recognized by bounded-error advised 2qfa's.

Lemma 11. *Let L be any language in* 2BQFA/n *over alphabet Σ with $|\Sigma| \geq 2$. There exists a number $\varepsilon \in [0, 1/2)$ such that $2\mathrm{QSC}_\varepsilon[L](n) = O(n)$.*

This lemma can be contrasted with Lemma 2. The proof of the lemma relies on the fact that, given each index n, the advice string $\eta(n)$ is a fixed string and, when we encode the location of a tape head into a series of inner states, $\eta(n)$ can be also embedded into this series.

7   Use of Approximate Matrix Rank

For any given language, we wish to seek a reasonable lower bound of its quantum state complexity. For this purpose, we will estimate the quantum state complexity in terms of a notion of *approximate matrix rank*, which was first discussed in the context of communication complexity theory by Krause [10], who used it as a tool in analyzing randomized communication protocols. See, e.g., [11] for a survey of this field.

Fix a language L and consider its *characteristic matrix* M_L defined as follows: for any $x, y \in \Sigma^*$, the (x, y)-entry of M_L is 1 (resp., 0) if and only if $xy \in L$ (resp., $xy \in \overline{L}$). Given any number $n \in \mathbb{N}^+$, $M_L(n)$ expresses a submatrix of M_L whose entries are indexed by strings of length at most n. We say that a real matrix M ε-*approximates* $M_L(n)$ if $\|M - M_L(n)\|_\infty \leq \varepsilon$ (namely, for any entry (x, y) with $|xy| \leq n$, $|M(x, y) - M_L(x, y)| \leq \varepsilon$). For a given 2qfa $\mathcal{A}$, p_{xy} denotes the acceptance probability of $\mathcal{A}$ on input xy, and we define $P_n = (p_{xy})_{x,y}$ to be a matrix of all such acceptance probabilities, provided that xy is of length at most n. Note that $\mathcal{A}$ recognizes $L \cap \Sigma^{\leq n}$ with error probability at most $\varepsilon(n)$ if and only if P_n ε-approximates $M_L(n)$. Given any matrix M and any $\varepsilon > 0$, the ε-*approximate rank* of M, denoted by $\mathrm{rank}^\varepsilon(M)$, is defined to be $\min\{\mathrm{rank}(B) \mid B \ \varepsilon\text{-approximates } M\}$. The following key theorem establishes a relationship between quantum state complexity and approximate matrix rank.

Theorem 12. *Let t be any function on $\mathbb{N}$, let L be any language, and let ε and ε' be two arbitrary constants satisfying $0 < \varepsilon' < \varepsilon < 1/2$. For any input length $n \in \mathbb{N}$, letting $t'(n) = \lceil t(n)/(\varepsilon - \varepsilon') \rceil$, it then holds that*

$$2\text{QSC}^t_{\mathbb{R},\varepsilon'}[L](\leq n) \geq \frac{\sqrt{\text{rank}^\varepsilon(M_L(n))}}{\sqrt{t'(n)}(t'(n)+1)(n+1)}.$$

Using Theorem 12 together with Lemmas 1(3) and 2, we want to prove the following separation result.

Corollary 13. $L \nsubseteq 2\text{BQFA}(t\text{-}time)$, *where* $t(n) = 2^{n/6}/n^2$.

Proof Sketch. Let us consider the well-known *disjointness problem* $\text{DISJ} = \{x \natural y \mid x, y \in \{0,1\}^*, |x| = |y|, x \wedge y = 0^{|x|}\}$ defined over a ternary alphabet $\Sigma = \{0, 1, \natural\}$, where $\wedge$ denotes the *bit-wise AND operation*. Here, assume that $\text{DISJ} \in 2\text{BQFA}(t\text{-}time)$. Lemmas 1(3) and 2 imply that $2\text{QSC}^t_{\mathbb{R},\varepsilon'}[\text{DISJ}](\leq n) = O(1)$ for an appropriate constant $\varepsilon' \in (0, 1/2)$.

Choose an error bound ε from $(\varepsilon', 1/2)$. It was proven in [11] that $\text{rank}^\varepsilon(M_{\text{DISJ}}(n)) = \Omega(2^{n/2})$. Letting $t'(n) = \lceil t(n)/(\varepsilon - \varepsilon') \rceil$, we apply Theorem 12 and then obtain $2\text{QSC}^t_{\mathbb{R},\varepsilon'}[L](\leq n) \geq n/c$ for a certain constant $c \geq 1$. This is obviously a contradiction. Thus, $\text{DISJ} \notin 2\text{BQFA}(t\text{-}time)$ follows.

Since $\text{DISJ} \in L$, we immediately draw the desired conclusion. $\qquad\square$

Let us return to Theorem 12. This theorem is an immediate consequence of the statement given below. A *t-bounded 2qfa* refers to a 2qfa whose computation paths on input x are all "clipped" at step $t(|x|)$ (if not halted earlier) and any clipped computation path are treated as "non-halting" computation paths.

Lemma 14. *Let Σ be any alphabet and let L be any language over Σ. Assume that, for each $n \in \mathbb{N}$, an $\mathbb{R}$-amplitude t-bounded 2qfa M_n recognizes L up to n with error probability at most ε'. Let $P_{n,t(n)}$ be a matrix whose (x, y)-entry is the probability of M_n's reaching accepting inner states within time $t(n)$. There exist two matrices C and D satisfying the following conditions: (i) $P_{n,t(n)} = CD^T$ and (ii) C and D have rank at most $t(n)|Q_n|^2(n+2)^2(t(n)+1)^2$.*

Proof Sketch of Theorem 12. Choose a minimal $\mathbb{R}$-amplitude 2qfa M_n with a set Q_n of inner states satisfying that $|Q_n| = 2\text{QSC}^t_{\mathbb{R},\varepsilon'}[L](\leq n)$ and that M_n recognizes L up to n with error probability at most ε' in expected time $t(n)$. Let $c = 1/(\varepsilon - \varepsilon')$ and set $t'(n) = \lceil ct(n) \rceil$. Apply Lemma 14 and take $P_{n,t'(n)}$, C, and D in the lemma. Markov's inequality implies that M_n must be t'-bounded with error probability at most ε. Hence, it follows that $\|P_{n,t'(n)} - M_L(n)\|_\infty \leq \varepsilon$. Moreover, it follows that $\text{rank}(P_{n,t'(n)}) \leq \max\{\text{rank}(C), \text{rank}(D)\} \leq t'(n)|Q_n|^2(n+2)^2(t'(n)+1)^2$. Since $P_{n,t'(n)}$ ε-approximates $M_L(n)$, we conclude that $\text{rank}^\varepsilon(M_L(n)) \leq t'(n)2\text{QSC}_{\mathbb{R},\varepsilon'}[L](\leq n)^2(n+2)^2(t'(n)+1)^2$. From this, the theorem follows directly. $\qquad\square$

In the rest of this section, we are focussed on the proof of Lemma 14. For the proof, we first need to re-examine an old notion of *crossing sequence* in light of quantum computation. Here, we define our new crossing sequences so that they help us translate a computation of a given 2qfa into a 2-player communication game, in which the players exchange information on those new crossing sequences of the computation.

Fix any input length n and let x and y be two arbitrary strings satisfying $|xy| \leq n$. Let $U_\delta^{(xy)}$ be a unitary transition matrix of the given 2qfa M_n on input xy. Take an arbitrary number $t \in \mathbb{N}$ and define $S_n = Q \times [0,t]_\mathbb{Z}$. A *boundary* b refers to an imaginary line set between a tape cell indexed b and another cell indexed $b+1$. Recall that $\mathfrak{c}$ is at cell 0. A *timed crossing sequence (at boundary b) with respect to t* is a sequence $s = s_1 s_2 \cdots s_m$ for which (i) m (called the *length* and denoted $|s|$) is a number in $[0,t]_\mathbb{Z}$, (ii) each s_j ($j \in [1,m]_\mathbb{Z}$) is of the form (q_j, i_j) in S_n, and (ii) $i_j < i_{j+1}$ holds for every $j \in [1, m-1]_\mathbb{Z}$. We denote by $CR_{n,t}^{\mathrm{even}}$ (resp., $CR_{n,t}^{\mathrm{odd}}$) the set of all nonempty timed crossing sequences whose lengths are even (resp., odd) and we set $CR_{n,t} = CR_{n,t}^{\mathrm{even}} \cup CR_{n,t}^{\mathrm{odd}} \cup \{\lambda\}$, where λ is the *empty timed crossing sequence*.

Proof Sketch of Lemma 14. Let M_n be any $t(n)$-bounded 2qfa that recognizes L up to n. Given x and y, let $b \in [0, n+1]_\mathbb{Z}$ denote the boundary that splits x and y in M_n's input tape, namely, $|x| = b$. We set our configuration space $\mathcal{C}_n$ of M_n on inputs of length n to be $\mathrm{span}\{|q\rangle|d\rangle \mid q \in Q, d \in [0, n+1]_\mathbb{Z}\}$. We then partition $\mathcal{C}_n$ into two subspaces $\mathcal{A}_n = \mathrm{span}\{|q\rangle|d\rangle \mid q \in Q, d \in [0, b]_\mathbb{Z}\}$ and $\mathcal{B}_n = \mathrm{span}\{|q\rangle|d\rangle \mid q \in Q, d \in [b+1, n+1]_\mathbb{Z}\}$ satisfying $\mathcal{A}_n \oplus \mathcal{B}_n = \mathcal{C}_n$. Let $P_{\mathcal{A}_n}$ and $P_{\mathcal{B}_n}$ be respectively the projections onto $\mathcal{A}_n$ and $\mathcal{B}_n$ and define two linear operators $A_n^x = U_\delta^{(xy)} P_{\mathcal{A}_n} \Pi_{\mathrm{non}}$ and $B_n^y = U_\delta^{(xy)} P_{\mathcal{B}_n} \Pi_{\mathrm{non}}$.

Next, let $S_n = Q \times [0,t]_\mathbb{Z}$, where the second set relates to a "time period." Let us consider a timed crossing sequence $s = s_1 s_2 \cdots s_m$, where each s_j is of the form (q_j, i_j) in S_n. We view a computation of M_n on the input xy as a 2-player communication game between Alice and Bob who interact with each other in turn. Alice receives x and sees the tape region lying on the left side of the boundary b, while Bob receives y and sees the tape region lying on the right side of b. First, Alice computes $\alpha_{s_1}^x = \langle q_1, b+1|(A_n^x)^{i_1}|q_0, 0\rangle$ and sends $|q_1, b+1\rangle$ to Bob. In receiving it, Bob computes $\beta_{s_2}^y = \langle q_2, b|(B_n^y)^{i_2-i_1}|q_1, b+1\rangle$ and sends $|q_2, b\rangle$ to Alice. More generally, at step $2j$, Alice receives $|q_{2j-1}, i_{2j-1}\rangle$, computes $\alpha_{s_{2j}}^x = \langle q_{2j}, b+1|(A_n^x)^{i_{2j}-i_{2j-1}}|q_{2j-1}, b\rangle$, and sends $|q_{2j}, b+1\rangle$ to Bob. When Bob receives $|q_{2j}, i_{2j}\rangle$, he computes $\beta_{s_{2j+1}}^y = \langle q_{2j+1}, b|(B_n^y)^{i_{2j+1}-i_{2j}}|q_{2j}, b+1\rangle$ and sends $|q_{2j+1}, b\rangle$ back to Alice. This process continues until either $2j$ or $2j+1$ becomes m. Each nonempty timed crossing sequence $s \in CR_{n,t(n)}$ is associated with its amplitude $\gamma_s^{xy} = \alpha_s^x \beta_s^y$, where $\alpha_s^x = \prod_{j=1}^{m/2} \alpha_{s_{2j}}^x$ and $\beta_s^y = \prod_{j=1}^{m/2} \beta_{s_{2j+1}}^x$. Finally, we obtain the quantum state $\gamma_s^{xy}|q_m, b+1\rangle$ if m is odd; $\gamma_s^{xy}|q_m, b\rangle$ if m is even.

For simplicity, we assume that q_0 is a non-halting state. Our goal is to construct two matrices C and D that satisfy the desired two conditions. For our purpose, we make C a $|\Sigma^{n+1}| \times |U|$-matrix and D a $|\Sigma|^{n+1} \times |U|$-matrix, where U is composed of all triplets (k, s, s') for which $k \in [1, t(n)]_\mathbb{Z}$ and $s, s' \in CR_{n,t(n)}$. Note that $|U| = t(n)(t(n)+1)^2|Q|^2(n+2)^2$. We define p_{xy}^{even} and p_{xy}^{odd} to be two acceptance probabilities produced while (i) the tape head stays in the left tape region of the boundary b and while (ii) it stays in the right tape region of b, respectively. More precisely, let $p_{xy}^{\mathrm{even}} = \sum_{k=1}^{t(n)} \|W_{\mathrm{acc}} P_{\mathcal{A}_n}(A_n^x)^k|q_0, 0\rangle + \sum_{s \in CR_{n,t(n)}^{\mathrm{even}}: \ j_m \leq k} W_{\mathrm{acc}} P_{\mathcal{A}_n}(A_n^x)^{k-j_m}(\gamma_s^{xy}|q_m, b\rangle)\|^2$ and $p_{xy}^{\mathrm{odd}} = \sum_{k=1}^{t(n)} \|$

$\sum_{s \in CR^{odd}_{n,t(n)}: j_m \leq k} W_{acc} P_{\mathcal{B}_n}(B_n^y)^{k-j_m}(\gamma_s^{xy}|q_m, b+1\rangle)\|^2$, where $s = s_1 s_2 \cdots s_m$ and $s_m = (q_m, j_m)$. Note that each acceptance probability p_{xy} in $P_{n,t(n)}$ can be expressed as $p_{xy} = p_{xy}^{even} + p_{xy}^{odd}$. From this equality, we can construct the required matrices C and D from $\{p_{xy}^{even}\}_{x,y}$ and $\{p_{xy}^{odd}\}_{x,y}$. We omit further details due to the page limit. □

References

1. Ambainis, A., Freivalds, R.: 1-way quantum finite automata: strengths, weaknesses, and generalizations. In: Proceedings of FOCS 1998, pp. 332–342 (1998)
2. Ambainis, A., Ķikusts, A., Valdats, M.: On the class of languages recognizable by 1-way quantum finite automata. In: Ferreira, A., Reichel, H. (eds.) STACS 2001. LNCS, vol. 2010, pp. 75–86. Springer, Heidelberg (2001)
3. Ambainis, A., Nayak, A., Ta-Shma, A., Vazirani, U.: Dense quantum coding and quantum finite automata. J. ACM **49**, 496–511 (2002)
4. Brodsky, A., Pippenger, N.: Characterizations of 1-way quantum finite automata. SIAM J. Comput. **31**, 1456–1478 (2002)
5. Damm, C., Holzer, M.: Automata that take advice. In: Hájek, P., Wiedermann, J. (eds.) MFCS 1995. LNCS, vol. 969, pp. 149–158. Springer, Heidelberg (1995)
6. Dumer, I.: Covering spheres with spheres. Discret. Comput. Geom. **38**, 665–679 (2007)
7. Freivalds, R.: Amount of nonconstructivity in deterministic finite automata. Theor. Comput. Sci. **411**, 3436–3443 (2010)
8. Freivalds, R., Ozols, M., Mančinska, L.: Improved constructions of mixed state quantum automata. Theor. Comput. Sci. **410**, 1923–1931 (2009)
9. Kondacs, A., Watrous, J.: On the power of quantum finite state automata. In: Proceedings of FOCS 1997, pp. 66–75 (1997)
10. Krause, M.: Geometric arguments yield better bounds for threshold circuits and distributed computing. Theor. Comput. Sci. **156**, 99–117 (1996)
11. Lee, T., Shraibman, A.: Lower bounds in communication complexity. Found. Trends Theor. Comput. Sci. **3**, 263–398 (2009)
12. Mereghetti, C., Palano, B., Pighizzini, G.: Note on the succinctness of determinsitic, nondeterminsitic, probabilistic and quantum finite automata. RAIRO–Theor. Inf. and Applic. **35**, 477–490 (2001)
13. Moore, C., Crutchfield, J.: Quantum automata and quantum languages. Theor. Comput. Sci. **237**, 275–306 (2000)
14. Nishimura, H., Yamakami, T.: Polynomial time quantum computation with advice. Inf. Process. Lett. **90**, 195–204 (2004)
15. Nishimura, H., Yamakami, T.: An application of quantum finite automata to interactive proof systems. J. Comput. Syst. Sci. **75**, 255–269 (2009)
16. Watrous, J.: Space-bounded quantum complexity. J. Comput. Syst. Sci. **59**, 281–326 (1999)
17. Watrous, J.: On the complexity of simulating space-bounded quantum computations. Comp. Complex. **12**, 48–84 (2003)
18. Yakaryilmaz, A., Say, A.C.C.: Succinctness of two-way probabilistic and quantum finite automata. Disc. Math. Theor. Comput. Sci. **12**, 19–40 (2010)
19. Yamakami, T.: One-way reversible and quantum finite automata with advice. Inf. Comput. **239**, 122–148 (2014)
20. Zheng, S., Gruska, J., Qiu, D.: On the state complexity of semi-quantum finite automata. RAIRO–Theor. Inf. and Applic. **48**, 187–207 (2014)

Author Index

Printed in the United States
By Bookmasters